Henry Minton

A Practical Homeopathic Treatise on the Diseases of Women and Children

Henry Minton

A Practical Homeopathic Treatise on the Diseases of Women and Children

ISBN/EAN: 9783337371425

Printed in Europe, USA, Canada, Australia, Japan

Cover: Foto ©berggeist007 / pixelio.de

More available books at **www.hansebooks.com**

A PRACTICAL

HOMŒOPATHIC TREATISE

ON THE

DISEASES OF WOMEN AND CHILDREN.

INTENDED FOR

INTELLIGENT HEADS OF FAMILIES

AND

STUDENTS IN MEDICINE.

BY

HENRY MINTON, M. D.

Similia similibus curantur;
Remedium singulum cuique morbo;
Pars minima, sano homino tentato.

NEW YORK:
BLELOCK AND COMPANY.
No. 19 BEEKMAN STREET.
1866.

To the Memory of

DR. ROBERT ROSMAN,

THE

PIONEER OF HOMŒOPATHIA

IN THE

CITY OF BROOKLYN,

THIS WORK IS RESPECTFULLY

DEDICATED:

IN PROFOUND RESPECT FOR HIS HIGH PROFESSIONAL ATTAINMENTS; IN HOMAGE TO HIS
LONG AND SUCCESSFUL DEVOTION TO THE ADVANCEMENT OF HOMŒOPATHIA, AND
IN REMEMBRANCE OF MANY ACTS OF KINDNESS SHOWN TO
THE AUTHOR IN THE EARLY PERIOD OF HIS
PROFESSIONAL CAREER.

PREFACE.

THERE are several Homœopathic works upon general domestic practice now before the public. Their practical utility, and the estimation in which they are held by the public, are amply demonstrated by the manner in which they have been received. There is scarcely a family in the community, entertaining views favorable to the cause, but possesses one or more of them.

They are accepted and welcomed wherever advice is needed, and medical aid cannot be obtained. Their use is not, however, exclusively confined to those who are the friends and open advocates of Homœopathia. A great many families, always employing an allopathic physician as their regular medical adviser, keep one of these manuals, with a few of the more common remedies, and themselves prescribe for all the minor ailments, especially those to which the children are liable.

The first practical information, often carrying conviction with it, of the art of Homœopathia is, not unfrequently, obtained from these manuals. They thus become the harbingers of its great and abiding truths; the silent, but efficient missionaries, working their way through the wilderness of error and prejudice, carrying the glad tidings of great joy to the suffering and afflicted, preparing the way for the speedy entrance of the regular practitioner of Homœopathia. They are often far in the advance, where, indeed, Homœopathia would never reach, were it obliged to wait until it were openly and publicly accepted in the person of a regular Homœopathist.

With their aid, the people mitigate suffering, and conquer disease, and in the experimentation — for, at first, it is nothing else — themselves become conquered, and ever after stand forth as open advocates of a cause which they never could have embraced, had it not been for these silent workers, which had thus undermined their prejudices, and eventually, by direct, practical demonstration, rooted out the errors which precept and example had long before established.

It is not at all to be wondered at, that those, who, from early teaching, have been led to believe that large doses of nauseous drugs are necessary for the eradication of slight ailments, should hesitate to believe, upon simple hearsay, that minute, nay, even infinitesimal doses are sufficient, when properly administered, successfully to combat diseases of the most violent and fatal tendencies.   It is not to be presumed that people imbued with doctrines so entirely opposite to those which we believe and teach, will at once, upon the recommendation of an acquaintance, abandon their old belief, and become converts to the new, even though the old were not entirely satisfactory.   Such proselytes would be a detriment rather than an advantage to any profession.   To convince an intelligent man that you have a way superior to his, it is necessary to bring forward a fair amount of evidence in support of the assertion.   He is not going to believe it simply because you say so.   He may not doubt your word or the sincerity of your belief; but he doubts your judgment.   He may think he is just as competent to form an intelligent opinion as you are; and so, perhaps, he is.   But he may be looking through an obscure medium ; prejudice may have drawn a film over his eyes, or he and the object of which he judges may each be standing in opposite valleys, with an intervening hill between.   Now, bring him upon the same level with yourself, — remove his prejudices, and clear his vision ; and, quite likely, you will both arrive at the same conclusion.   No man has a right to pronounce a statement false, especially when made by another equally as intelligent and honest as himself, simply because it, or the arguments advanced to support it, appear to *him* to be at variance with reason and common sense.   He is in duty bound, in justice to himself and those who depend upon his judgment and decision, to give every subject that is honestly advanced, that touches upon his particular sphere of life, a thorough and candid investigation ; not to argue the case with another as firmly opposed to it as himself, but to put it to the actual, practical test, which alone can decide its truth or falsity, and which, if it be advanced with truth and sobriety, its friends and advocates will always court.   How many of those physicians in this city, who are our most bitter opponents, have ever given Homœopathia the least practical investigation ?   Few, indeed.

Tell a person, unfamiliar with the laws of chemistry, that water can be made to burn, or one ignorant of the fact, that every drop of water he drinks contains minute, often hideous-looking animals ; and your veracity or sanity will at once be doubted. You may use your best arguments and illustrations, and still fail to convince him ; but actually burn the water before his eyes, and, as the flame arises, it will have a wonderful effect upon his perceptive faculties ; or, place a drop under the microscope, and the little animalculæ will soon wriggle a firm belief of their actual existence into his imagination.

So it is with Homœopathia ; you may talk all day, and in the end fail to make one believe that a few globules, looking and tasting, for all the world, like sugar, and of which you could eat a pound, perhaps, without harming you in the least, are sufficient, when administered according to known laws, to remove the most serious diseases ; you simply excite the ridicule of doubters.

There was a time when all learned men disbelieved and doubted, what I have no doubt nine-tenths of the fair, intelligent part of this community would, to-day, were the fact for the first time presented to them, "that an ounce-weight, and a ton-weight, will fall down a pit with the same speed and in an equal time." About three hundred years ago, Galileo, an Italian, taught views contrary to the popular belief. His teachings, however, met with little success, until the University of Pisa challenged him to the proof. That was just what he wanted, and the leaning tower of Pisa was just the place for the experiment. Two balls were procured, one exactly double the weight of the other. Both were taken to the top, and all the dignitaries of the place assembled, to see the then obscure and despised young Galileo proven a false teacher. The two balls were dropped at the same instant. Old theory and all the world said, that the large ball, being twice as heavy as the other, would come down in half the time ; any other result was, in their estimation, "contrary to all reason." All eyes watched, and lo ! all eyes beheld them strike the earth at the same time. The experiment, though oft repeated, always produced the same result. The little ball was sufficient to destroy a theory two hundred years old. Had Galileo's teachings never been tested, they would, to this day, have been called vagaries. Argument would never have convinced a doubting world.

Hahnemann, when he presented his discoveries to a suffering people, was also confronted with the objection, that it was "contrary to all reason and common sense." He hurled his little pills against a theory as time-honored as that which opposed Galileo, and with the same result.

Those who honestly came out to observe the effect were convinced; many refused to look, as did some in Galileo's day, and cried, "Humbug." It would seem absurd for two men to spend their time in arguing the point whether an apple is sweet or sour, when each of them can soon decide the case by tasting for himself. Strange as it may sound, there are those, who, doubting the truth that there has been a better way discovered of treating diseases than that handed down to them through several generations, should, in this enlightened age, flatly refuse to put the theory to a practical test; nevertheless, such is the fact.

Perhaps there never was a more bitter opponent to the vagaries, as he then thought them, of Hahnemann than the writer of these pages, when he was first matriculated at the College of Physicians and Surgeons, in the City of New York. Before the term had expired, however, his faith in the collected wisdom of past ages was somewhat shaken, as the comparison of the two modes of treatment, that pursued at the cliniques of the College, and that at his preceptor's office — who was a Homœopathist — was too unmistakably in favor of the latter. Demonstration here taught in weeks, what argument the most logical, backed by statistics incontrovertible, had failed to do in years.

Experience has long since demonstrated that public opinion is shaped mainly by the inaudible and invisible teachings of passing events, rather than by any appeals, however urgent, or by any arguments, however forcible.

The work of reformation in medicine is being carried on to a great extent by manuals like this. Not a few, otherwise opposed to Homœopathia, think it an excellent practice for children; some even go so far as to admit females into the list of those to whom it is specially applicable; but, that it is sufficient for all cases met with in general practice, they will not admit.

The favorable impression they possess, has not been forced upon them by argument, but, rather more likely, by *seeing* a neighbor's child instantly relieved of colic by two or three globules of chamomilla or colocynth; or, a female friend, suffering from some annoying complaint, restored to permanent health, by doses tasteless and infinitesimally small.

They argue that the organization, and, especially, that of the nervous system, of delicate females and children, is so sensitive, that small doses are sufficient, and, therefore, they think it an excellent system for such persons. It is a noticeable fact, however, that, when a family begins to give chamomilla to the baby, and have learned the names of some of the more common remedies for colds, coughs, and children's complaints, and have actually seen diseases cured with them, it soon becomes the owner of a book and case of medicines, and afterwards treats all those slight ailments which are constantly occurring in every family, according to the teachings of Hahnemann.

All this they can do without exciting the taunts and ridicule of friends, or hurting the feelings of their family physician. But this mode of medication cannot be continued long, without convincing those who have thus far given it a trial, that it can be used with safety, as well, in complicated and serious cases. By their prompt, pleasant and satisfactory action, — the convenience and ease with which they are administered, — these remedies gradually establish their own claims to superiority in the estimation of many a one whom arguments would have never convinced, and whom nothing but practical demonstration, and that too, without pomp or ceremony, could.

For a while, books upon domestic medicine were looked upon with distrust by many in the profession. It was feared that the honor and dignity of the cause would be injured, were the common people allowed free access to those mines of knowledge, which for ages had been so scrupulously and religiously guarded. This feeling, however, it is but fair to state, was principally entertained by those who had been educated under old-school regime, in a practice where the confidence of its patrons was in direct ratio to their ignorance of an art they were required to believe. At present, as far as I am aware, the only antipathy to them exists among the drones of the profession; those who are too indolent to keep in advance of an intelligent public, when this public is not debarred from acquiring that knowledge upon medical subjects, which has, heretofore, to too great an extent, been the exclusive privilege of a chosen few. In general, the profession seems to accept them as necessary allies in the vast work of instructing the people, and carrying forward the great truths, readily seeing the important part which they have taken in disseminating the beneficent doctrines of our art, and facilitating their application by enlightening the public upon the principles which we profess.

For our own part, we are of the decided opinion that the more intelligent the people become as to the great principles of medicine, the higher its standard among the profession will be elevated. As the teacher must keep in advance of his pupils and patrons, so an intelligent public demands and will have intelligent physicians.

Perhaps there never has been a time, in the history of medicine, since Hahnemann first gave his discovery to the world, when the public mind was so generally detaching itself from its old moorings, and yielding more and more to the current of Homœopathia.

The old practice is becoming distasteful to the people; it is unsatisfactory; it will not bear comparison with the new. The fact is, Allopathy has been, for some time, like an old ship at sea, without chart or compass, running from one hypothetical fog into another, until now she is stranded, high and dry, upon Galen's theoretical sand-bar of contraries, her every timber creaking out the echo of her dying groans, while the sea-gulls — druggists — catch the mournful requiem, and Hygeia weeps for joy.

The present appears to be a most auspicious time to erect sign-boards, that those may read, who will not listen, — to place at the disposal of those who wish to test the correctness of our doctrine, the necessary instructions for conducting the experiment.

In the spirit of the foregoing remarks, implied if not always expressed, the following manual is presented to the public. The author is not unmindful of the responsibilities he has assumed, in thus entering himself upon the list of public instructors; however, an experience of over ten years in active practice awards him some grounds of assurance, in giving his opinion and advice upon that specific branch of a science, to which he has devoted the larger part of his attention.

The author saw, what every other practising physician must also have seen, the imperative necessity of a work upon this particular branch of medicine; he has endeavored to supply the want to the best of his abilities. Perhaps, in other and abler hands, the matter might have been presented in a more concise and agreeable manner, but other hands neglected or refused the task; therefore the responsibility has fallen where it has.

The author does not wish to apologize, or "crave the indulgence of a generous public?" the work must stand upon its own merits, or fall.

Many of these pages were written without a thought of publication, and, therefore, no pains were taken to note authorities. It would not, therefore, be at all strange if the author had used facts, reasonings, and even long quotations from other works, without acknowledgment. This causes him much annoyance, but is now quite beyond his power to remedy.

Before closing this preface, the author would remark, that it must not be presumed that this work is intended to supersede the physician ; on the contrary, we do not believe that this book, or any that has been, or ever will be written, can supply the place of the intelligent physician at the bedside of the patient.

Should these pages attract the attention of any who have already passed the *gradum tyronis,* it is requested that all such will bear in mind for whom these are intended, as they do not profess, by any means, to present a complete and formal treatise upon the particular branch of medicine to which they are devoted, but are simply offered, according to the title-page, " as a text-book for intelligent heads of families and students in medicine."

With the hope that the work will prove acceptable, and fulfil its design in answering all the requirements of the lay practitioner, the author bids it " God speed."

                                        HENRY MINTON.
No. 138 REMSEN ST.,   *March,* 1866.
    BROOKLYN, N. Y.

INTRODUCTION.

As this manual is intended for the people at large, — for those who are, and for those who are not, familiar with the doctrines and practice of Homœopathia, — for those confessing humble literary attainments, as well as those more favored in mental culture, — we have endeavored to avoid all ambiguous and technical terms, and to use only such language as would, in our opinion, be within the comprehension of all. We are well aware, that, in giving the *definition* and *causes* of diseases, we have, in many instances, used terms not familiar in ordinary language. This, it has been impossible to avoid. In the *treatment* of diseases, however, which is the *practical*, and, therefore, most important and useful part of the work, we have been able to come nearer our desires. It has constantly been the endeavor of the writer to state in the clearest and simplest possible form, the precise ideas wished to be conveyed. To do this, grace of style has, in many instances, been sacrificed. This immolation of language, however, is of but little account, provided the object aimed at is attained, namely, commanding the attention and the assent of the mass of the people, or of the common people, and bringing the whole subject within the easy grasp of ordinary intellectual capacity. If we have succeeded in doing this, — and we have been so explicit in the practical part of the work, that it seems impossible for one who attentively reads, to be left in any doubt as to the proper course to pursue, — we have no fear whatever that remedies will be misapplied, or that it will ever be necessary to grope in darkness, and prescribe by chance. For the above reasons, we have not seen fit to append a glossary of medical terms, deeming it quite evident that those who are unable to understand the text, are most assuredly not sufficiently intelligent to have charge of the sick.

Some knowledge of Anatomy, Physiology, Pathology, and Hygiene may unquestionably be regarded as useful information for every female in the community. All these, though incidentally, are embraced in the body of the present work. In the consideration of a number of diseases, it has been deemed expedient, for the purpose of conveying a better understanding of the subject, to give a concise description of the anatomy of the parts involved. This, in many instances, will be found of immense value, especially to medical students, and all those who wish to obtain a definite, correct, and practical knowledge of diseases. As all these divisions are distinctly marked, they occasion no confusion, but, on the contrary, render the whole article at once intelligible, and its different parts of easy access.

In order to facilitate their study, every article in the book has been so divided that any of its parts may be consulted, without the necessity of reading the whole; so that a person, wishing to ascertain the *nature, treatment,* or *causes* of a given disease, can at once turn directly to the desired head.

For reasons already given, the most common names of the various diseases have been given, as well as those adopted for the same, in nosological works.

DIRECTIONS FOR PRESCRIBING.

THE fundamental principle governing all prescriptions in Homœopathic practice, is that " diseases are cured most quickly, safely and effectually, by medicines which are capable of producing symptoms *similar* to those existing in the patient, and which characterize his disease."

This law of cure is expressed by the maxim, "*similia similibus curantur;*" and in the adaptation of our remedies to the sick, this therapeutic law of similarity must be scrupulously observed, or our success in practice will not be commensurate with our reasonable expectations.

In prescribing for a given case, great care should be taken that the first prescription be as near Homœopathic * as possible, to the disease; not that there is any danger of doing harm, should a wrong remedy be selected. The medicines are given in such minute doses, that they affect only such portions of the system as are morbidly susceptible to their action, raised to this condition of idiosyncracy by the disease; and, therefore, if they are not Homœopathic to the disease for which they are prescribed, the whole force of their action will be expended upon healthy parts, which are always capable of resisting the action of doses much larger than any we ever think of giving. Hence, if a remedy does no good, it will do no harm, which is considerably more than can be said of other modes of practice. Still, if the right remedy is not selected at the outset of the disease, valuable time will be wasted.

In all cases, where the cause of the difficulty can be readily detected, the course of treatment to be pursued, becomes at once very much simplified. Unfortunately, however, it is not always possible to say exactly

---

* By this term is meant that the drug, selected for the cure of a disease, should possess the power of exciting, in the healthy subject, a series of phenomena similar to the symptoms of the disease for which it is administered.

14

what is the cause: frequently there are many causes, and besides, after the cause has been removed, or passes away, the effect does not always immediately disappear.

Hahnemann has directed us to form a "correct image of the disease," by taking cognizance of every symptom of which the patient complains, and noting down every detail of the case. Then, from the remedies given, select the one that covers the largest number, and the most prominent of the symptoms present.

The remedy that corresponds to the *totality* of the symptoms, and especially to all the most prominent characteristic and leading ones, should be administered in preference to any other, no matter whether such remedy be found under the head of the disease of which the patient is complaining, or elsewhere.

If, in making the selection of a remedy, the choice should fall between two or more remedies which seem equally indicated, the preference should be given to the one which corresponds most nearly to the symptoms which appeared last.

It is often a difficult point, to decide as to which a preference shall be given of two remedies, that seemingly are equally indicated in a given case, still neither of them corresponding to all the symptoms, or even to *all* the prominent ones, but each covering important points, which the other fails to. Many physicians, in such cases, compromise the matter by giving the two in alternation. This I believe to be usually a bad practice, though, in some few instances, I have recommended it in these pages. There is always one remedy that is preferable to another, and that remedy should be diligently sought for. By the layman, it may not easily be found, but, by the regular practitioner, it generally should. Where more than one remedy is required to complete a cure, as is frequently the case, they should be given in rotation or succession, not in alternation.

I have heard of physicians prescribing half a dozen different remedies at the same time ; — one for fever, one for headache, one indeed for every symptom of which the patient complained. Such ignorance is inexcusable. The physician who is too indolent to hunt up the appropriate remedy, or too ignorant to recognize it, when found, should not be tolerated for a moment in charge of a patient.

ON THE ADMINISTRATION OF MEDICINES, AND THE REPETITION OF DOSES.

HOMŒOPATHIC medicines are prepared in the form of tinctures, triturations, dilutions, and globules. The latter we consider the most convenient for general domestic use, and therefore recommend them in this work. Some prefer the dilutions, thinking they act with greater certainty, or are less liable to deteriorate, or lose their active principle. We are confident, however, from long experience and observation, that the globules are just as reliable as any other form in which the medicines can be prepared. We have found them to act promptly and satisfactorily seven years after having been once thoroughly and carefully medicated. Many of our best physicians prefer them, and use nothing else; and indeed our higher potencies are used in no other form. If the medicines are put up by a reliable pharmaceutist, it matters but little, except for convenience' sake, whether they are in the form of dilutions or globules. Tinctures and triturations are but the primary, or preparatory stages, and by pure homœopathists are seldom, if ever, prescribed.

The medicines should be kept in a chest made for the purpose, in a dry place, free from all odors. When uncorking a phial, care should be taken to replace the identical cork, or a new one, should the first get broken. Medicine should never be put into a phial that has been once used, unless indeed it be the same medicine that the phial formerly contained. Old corks should not be used. Every phial should be plainly labelled, and the potency marked upon it.

The medicines may be administered by placing three or four globules, dry, upon the tongue, and let them dissolve. For small children, a less number will be sufficient; for an infant, one globule will be quite sufficient. Where the medicine is to be given for a length of time, at short intervals, it is best they should be dissolved in water. For this purpose, take a clean tumbler containing about twelve spoonfuls of water; put twelve globules of the selected remedy into the water; after they are dissolved, stir the solution with a clean spoon for several minutes, until the medicine is thoroughly incorporated with the water. Of this solution, one spoonful may be given as a dose for an adult; for an infant, half the quantity. If the tincture is used, put one or two drops in a tumbler about one third full of water, and administer as above. Should the trituration be used, the size of the dose will be about as much as could be placed on a three-cent piece, or taken up by the point of a knife.

Too much care cannot be taken that the vessel in which the medicines are prepared are *perfectly clean*. A tumbler is always to be preferred to a cup. Cups which are cracked, or have particles of the glazing worn off, are objectionable. The spoon, if convenient, should be a silver one, and it should never be allowed to remain in the medicine. If more than one remedy is used at the same time, each should have a separate spoon. The same vessel and spoon should never be used for different medicines, until they have been thoroughly washed and rinsed out.

In most instances, specific directions have been given in the body of the work for the repetition of doses. It is difficult to say much upon the subject here, as the frequency with which a remedy should be repeated, must necessarily be, in a measure, governed by the severity and nature of the disease. In mild cases, one dose will often be sufficient to remove the trouble. In acute cases, the remedy may be repeated at intervals of one or two hours. In those severe and dangerous diseases which rapidly run their course to a fatal termination, unless arrested in their progress, it may be necessary to repeat the dose as often as every half-hour, until amelioration takes place. In chronic cases, a well-chosen remedy need not be repeated oftener than once in one, two, or three days.

A general rule may be here laid down, which, if followed, will be productive of good results. It is this: no matter what the disease, its nature or severity, if, after one dose has been given, the patient begins to feel better, no matter how slight the improvement may at first be, discontinue the medicine at once, and wait the result. In many instances, one dose will be sufficient to complete the cure. In all cases, however, as soon as the improvement ceases, repeat the dose, or, in case the symptoms have assumed a new form, select and administer another remedy.

If, after the administration of a remedy, a slight aggravation of the symptoms should speedily become manifest, it may be taken as a good sign. The medicine should be discontinued, and the result, which will soon follow, will be a permanent improvement in the condition of the patient, which should not be interfered with, as long as it continues, by a repetition of the remedy. As soon, however, as improvement ceases, and it becomes necessary to repeat the medicine, give it in smaller doses. Should the symptoms have altered, select another remedy.

A *medicinal aggravation* may be known by the symptoms becoming *suddenly* worse after the exhibition of a remedy. The aggravation of the disease is *gradual* and *progressive*, manifesting such symptoms as belong to the advanced stage of the malady.

Should a medicinal aggravation be severe, it may be necessary to counteract it by administering an antidote. See " LIST OF REMEDIES."

Sometimes, during the course of a disease, the action of remedies is interrupted by some extraneous circumstance, as errors in diet, taking cold, and so forth. Whenever this happens, the interruption should be removed by appropriate treatment, after which the previous remedies should be resumed.

A most pernicious and mischievous habit, too often observed in the practice of laymen and beginners in Homœopathia, is the frequent changing from one remedy to another. A medicine is hastily and inconsiderately chosen, of which two or three doses are given in quick succession. If the patient does not begin immediately to improve, it is quickly abandoned for another, which perhaps bears no more resemblance to the case than the first, and this, in turn, without having had an opportunity to exert a salutary influence, were it perfectly capable of so doing, is as hastily disposed of. And thus they fly from one remedy to another, obtaining but little, if any, beneficial results from any. This practice cannot be too strongly condemned.

It may be set down, as an *invariable* rule, that, after a remedy has been judiciously selected, and administered in accordance with directions already given, it *should not be changed as long as benefit results from its employment, or until a reasonable length of time has been allowed for its action.* Still, after a reasonable length of time has been allowed a remedy, and it produces no good results, it should be at once exchanged for another that has been carefully selected.

ON POTENCIES.

No definite rule can be given as to the potency to be used in all cases. Difference in age, sex, temperament, constitution, and habits, renders variations, both in the potency used, and the frequency with which the doses should be repeated, absolutely essential. Females and children are, generally speaking, more susceptible to medicinal influence than others, and, therefore, the higher potencies are better adapted to their diseases.

The potencies, used by physicians in this city, range all the way up from the first to the two-hundredth, and higher; none being willing to confine himself to any one particular point. Our own convictions incline us to the higher, though we have cured a great many cases with the third, that we could not with the two-hundredth, but we have cured vastly more with the two-hundredth that we could not touch with the third.

It is recommended that laymen who use this volume should make use of the *thirtieth* potency. This will be found to correspond with the vast majority of the cases treated of in the work. Still, we would prefer that each one should consult the family physician upon the subject, and then act in accordance with his directions.

OBSERVATIONS ON DIET DURING TREATMENT.

To obtain the most satisfactory results from remedies, much depends on the adoption of a proper course of diet by the patient during treatment. It is of the utmost importance, that the food partaken of should be *light*, of *easy digestion*, and *nutritious*, and in quantities just sufficient to satisfy hunger.

The idea is an erroneous one, still very generally entertained, that a patient should eat a certain amount of food daily, whether he feels an inclination to or not. Food that is not relished, or at least when taken with absolute repugnance, — "forced down," as the saying is, — adds but little, if any, to the patient's strength.

Our advice always is, to wait until the patient calls for nourishment, or, at least, has some little desire for it; then its *judicious* administration will afford valuable assistance in his restoration from disease.

The idea is a prevalent one, that the reason Homœopathic physicians place such rigid restrictions upon their patients' diet is, that they imagine most articles of food contain sufficient medicinal properties to antidote, or turn aside the health-renewing action of their remedies. In regard to a few dishes which modern cookery has invented, this objection may be reasonably advanced, but its universal application, as generally supposed, is entirely a mistake. At least, speaking personally, and for ourselves alone, it is by no means the principal consideration that governs us in our dietetic instructions to patients. Had we no more faith in the reliability of Homœopathic medicines, than for a moment to suppose that their action could be neutralized by the simple seasoning of ordinary food, or by the mineral salts held in solution by the water we drink, or that certain remedies, *arsenicum,* for instance, could not be relied on, when given to persons living upon the coast, for fear a breeze from seaward might contain *natrum muriaticum,* sufficient to act as an antidote, we would abandon the practice, and seek some other employment.

We place restrictions upon diet, because we believe the system is incapacitated by disease, from carrying on, with the same amount of energy, the process of converting food into the necessary material for supplying the ordinary waste of the body, as in health. We wish to tax the system as little as possible, because we believe its capabilities are restricted. The process of repair is, in a measure, arrested by every disease, and the first intimation we have of a loss of this power is manifested by a loss of appetite.

It should ever be remembered, that we " eat to live," and, therefore, no more food should be taken than can be properly digested ; for all food taken into the system and not digested is like coals thrown upon the fire that are not ignited by the flame ; they clog up the furnace, and extinguish what little fire was there burning.

The necessity, therefore, of regulating a patient's diet, so that he shall have nothing but what is of easy digestion, is of prime importance. It is not only important that the food should be of easy digestion, but no more of it should be taken than can be easily disposed of.

Deeming this subject an important one, we have treated it at large in another part of this work. The reader is referred to that portion of the article on " DYSPEPSIA " which speaks of the " DIET AND REGIMEN."

ARTICLES OF DIET THAT MAY BE ALLOWED WHILE UNDER TREATMENT.

Beverages. Pure cold water : this should have the preference over all other drinks. There is no disease, where its free use in moderate quantities is not allowable. Should it be preferred, the water may be sweetened with sugar, currant jelly, raspberry or strawberry syrup, or quince and apple jelly ; barley-water, rice-water, toast-water; cocoa, milk, and all other beverages of a non-medicinal nature.

Gruels, made of oat-meal, wheat-flour, farina, rice-flour, corn-meal, corn starch, pearl barley, tapioca.

Soup or Broth, made from beef, mutton, or chicken, to which may be added rice or barley, or any other farinaceous article ; also vegetable soups.

Meats. Beef, mutton, chickens, pigeons, turkeys, all kinds of tongue and venison, game of all kinds in its season.

Fish. Cod, rock-fish, perch, flounders, haddock, pike, perch, trout, mackerel, and herrings.

Salt fish should be well soaked in cold water before it is used.

Oysters roasted in the shell, made into soup, or raw, are not only nutritious, but of easy digestion.

Vegetables. Potatoes, beets, green peas, all kinds of beans, when young and tender, carrots, turnips, tomatoes, spinach, cauliflower, cabbage, and, in some cases, asparagus, mushrooms, dried peas and beans. All kinds of vegetables ought to be well cooked ; potatoes are best when roasted.

Puddings, made of arrow-root, rice, sago, tapioca, Indian meal, corn-starch, farina, oat-meal, barley flour, and so forth. Puddings should not be made too rich ; eggs, milk, and sugar, ought to be used sparingly.

Bread and Cakes. All kinds of light bread not recently baked, biscuits, simple cakes.

Eggs, lightly dressed, either boiled, poached, or made up into custards.

Fruits. Baked, stewed, or preserved apples or pears, raspberries, whortleberries, blackberries, strawberries, peaches, oranges, plums, apricots, watermelons, muskmelons, &c. Also, some kinds of dried fruit, as dates, prunes, figs, or, in fact, any fruit not of too acid a quality.

No fruit whatever, except, perhaps, peaches and blackberries, should be used in case of bowel-complaints.

Milk. Milk, either raw or boiled, may be used, provided it agrees ; the same may, also, be said of fresh buttermilk.

The above list is given to convey an approximate idea of what is wholesome, and will not disagree, under ordinary circumstances, with an invalid, provided it is taken in proper quantities, and at regular intervals. Still, all such regulations are subject to considerable modifications, for, as it is frequently said, " What is one's meat is another's poison," so individual peculiarities must be studied and consulted. Whatever is known or found upon trial to disagree, should be scrupulously avoided.

Regularity in the hours of meals should be observed. Too long fasting, as well as too frequent eating, is to be deprecated.

ARTICLES FORBIDDEN,

UNLESS SPECIALLY ALLOWED BY THE ATTENDING PHYSICIAN.

Beverages. All kinds of liquors, coffee, green tea, lemonade, and all acidulated drinks. — See " DIET DURING NURSING."

Meats. Pork, veal, sausages, kidney, geese, ducks, mince-pies, and every kind of salted or fat meat.

Soups. All high-seasoned soups, such as turtle, mock-turtle, &c.

Fish. Crabs, lobsters, clams, and all kinds of fish not mentioned in " ARTICLES ALLOWED."

Vegetables. Cucumbers, onions, radishes, celery, parsnips, garlic, all kinds of pepper, pickles, and salads of every description.

· *Pastry* of every description, whether boiled, baked, or fried.

Spices and *artificial sauces* of every kind. All condiments, as catsup, vinegar, and mustard.

Perfumery of every description, as well as medicated or scented tooth-powders.

Rancid cheese and *butter.*

All kinds of nuts, and fruits not mentioned among the articles allowed.

All patent medicines, and every description of domestic medicine, no matter how simple, as well as all external applications, medicated plasters or poultices, and all irritants.

TABLE OF MEDICINES:

THEIR SYNONYMES AND ANTIDOTES.

1. ACONITUM or ACONITE — Aconitum-napellus. Monk's Hood (plant). *Antidotes* —Wine, Vinegar, Camphor, Nux-vom.

2. ALUMINA — Oxide of Aluminum (mineral). *Antidotes* — Bryonia, Chamomilla, Ipecac.

3. ANTIMONIUM-CRUD. —Antimony (mineral). *Antidotes* — Hepar-sulph., Mercurius, Pulsatilla.

4. APIS-MELLIFICA — Poison of the honey-bee. *Antidotes* — Arsenicum, Cantharis.

5. ARNICA — Arnica-montana, Leopard's Vane (plant). *Antidotes* — Camphor, Ignatia, Ipecac.

6. ARSENICUM — Arsenicum-album (mineral). *Antidotes* for its poisonous effects — Rust of Iron. For its dynamic effects — China, Hepar-sulph., Ipecac. Nux-vom., Veratrum.

7. AURUM — Aurum-metallicum, Gold (metal). *Antidotes* — Belladonna, China, Cuprum, Mercurius.

8. BELLADONNA — Deadly Night-shade (plant). *Antidotes* — Coffea, Hyoscyamus, Hepar-sulph., Pulsatilla.

9. BORAX — Borax-veneta, Sub-borate of Soda (mineral). *Antidotes* — Chamomilla, Coffea.

10. BROMINE — (Chemical Element). *Antidotes* — Coffea, Opium, Camphor, Ammonia.

11. BRYONIA — Bryonia-alba, White Bryonia (plant). *Antidotes* — Aconite, Chamomilla, Ignatia, Nux-vom.

12. CALCAREA-CARB.— Carbonate of Lime. *Antidotes* — Camphor, Nitric acid, Sulphur.

13. CAMPHORA — Camphor. *Antidotes* — Opium, Nitris spiritus dulcis.

14. CANNABIS — Cannabis-sativa, Hemp (plant). *Antidote* — Camphora.

15. CANTHARIS — Spanish Fly (animal). *Antidote* — Camphora.

16. CAPSICUM — Cayenne Pepper (vegetable). *Antidote* — Camphor.

17. CARBO-VEGETABILIS — Wood-Charcoal. *Antidotes* — Arsenicum, Camphor, Lachesis.

18. CAUSTICUM — Caustic of the Alkalies. *Antidotes* — Coffea, Nux-vom., Colocynth.

19. CHAMOMILLA — Common Chamomile. *Antidotes* — Aconite, Cocculus, Coffea, Ignatia, Nux-vom., Pulsatilla.

23

18. CHINA — Cinchona, Peruvian Bark (vegetable). *Antidotes* — Arnica, Arsenicum, Belladonna, Calcarea-carb., Ipecac., Carbo-veg., Sulphur.

19. CICUTA — Cicuta-virosa, Water-Hemlock (plant). *Antidotes* — Arnica, Tabacum.

20. CINA —Wormseed (vegetable). *Antidotes* —Ipecac., Veratrum.

21. CINNAMONUM — Cinnamon (vegetable). *Antidotes* —

22. COCCIONELLA — Cochineal (animal).

23. COCCULUS — Indian Cockle (vegetable). *Antidotes* — Camphor, Nux-vom.

24. COFFEA — Coffea-cruda, Raw Coffee (vegetable). *Antidotes* — Aconite, Chamomilla, Nux-vom.

24. COLCHICUM — Meadow Saffron (plant.) *Antidote* — Nux-vom. Cocculus, Pulsatilla.

24. COLOCYNTHIS — Colocynth, Bitter Cucumber (vegetable). *Antidote* — Camphor, Causticum, Coffea, Chamomilla.

25. CANIUM — Hemlock (vegetable). *Antidotes* —Coffea, Spiritus nitris dulcis.

26. CORALLIA — Red Coral (mineral). *Antidote* — Calc-carb.

27. CROCUS — Crocus-sat., Saffron (vegetable). *Antidote* —

28. CUPRUM — Cuprum-metallicum, Copper (mineral). *Antidotes* — Belladonna, China, Ipecac., Mercurius, Nux-vom.

29. DIGITALIS — Digitalis-purpurea, Fox-glove (plant). *Antidotes* —Nux-vom., Opium.

30. DROSERA — Round-leaved Sun-dew (vegetable). *Antidote* — Camphor.

31. DULCAMARA — Bitter-sweet, Woody Nightshade (plant). *Antidotes* — Camphor, Ipecac., Mercurius.

32. EUPHRASIA — Eye-bright (plant). *Antidote* — Pulsatilla.

33. FERRUM — Ferrum-metallicum, Iron (metal). *Antidotes* — Arnica, Arsenicum, Ipecac., Mercurius, Belladonna, Pulsatilla.

34. GLONOINE — A Chemical Preparation from Glycerine, Nitrate of Oxide of Glycil.

35. GRAPHITES — Plumbago, Pure Black-Lead (metal). *Antidotes* — Arsenicum, Nux-vom., Wine.

36. HELLEBORUS — Helleborus-niger, Black Hellebore (plant). *Antidotes* — Camphor, China.

37. HEPAR-SULPH. — Sulphuret of Lime. *Antidotes* —Vinegar, Belladonna.

38. HYOSCYAMUS — Hyoscyamus-niger, Henbane (plant). *Antidotes* — Belladonna, Camphor, China.

39. HAMAMELIS — Hamamelis-virginica, Witch Hazel (plant).

40. IGNATIA— St. Ignatius' Bean (plant). *Antidotes* — Pulsatilla, Chamomilla, Camphor, Cocculus, Vinegar, Arnica.

41. IPECAC. — Ipecacuanha (vegetable). *Antidotes* — Arsenicum, Arnica, China.

42. KALI-BICHROMICUM — Bichromate of Potash (mineral). *Antidote* — Ammonium-carb.

43. KALI-CARBONICUM — Carbonate of Potash (mineral). *Antidotes* — Camphor, Coffea.

44. KALMIA — Kalmia-latifolia, Mountain Laurel, Broad-leaved Laurel (plant).

45. KREOSOTUM — Creosote, an oxy-hydro-carburet (vegetable). *Antidotes* — Aconite, Nux-vom., China, Arsenicum.

46. LACHESIS — Trigonocephalus Lachesis, the chemical extract of the virus of the South American Snake (animal). *Antidotes* — Arsenicum, Belladonna, Nux-vom., Rhus-tox.

47. LEDUM — Ledum-palustre, Marsh-tea (vegetable). *Antidote* — Camphor.

48. LYCOPODIUM — Club Moss, Wolf's Claw (vegetable). *Antidotes* — Camphor, Pulsatilla.

49. MERCURIUS — Mercurius-solubilis Hahnemanni, Hahnemann's Preparation of Mercury (mineral). *Antidotes* — Arnica, Belladonna, Camphora, Hepar-sulph., Iodine, Lachesis, Sulphur. [There are several preparations of Mercury; those generally used are the Mercurius-sol., Mercurius-vivus (Quicksilver), Mercurius-cor., Mercurius-iodatus.]

50. MEPHITIS PUTORIUS — Skunk (animal). *Antidote* — Camphor.

51. MEZEREUM — (plant). *Antidotes* — Camphor, Mercurius.

52. MURIATIC-ACID — Acidum Muriaticum (mineral). *Antidotes* — Camphor, Bryonia.

53. NATRUM — Natrum-muriaticum, Muriate of Soda, Kitchen-Salt (mineral). *Antidotes* — Arsenicum, Camphor, Nitris-spiritus.

54. NITRIC-ACID — Nitri Acidum (mineral). *Antidotes* — Calcarea, Camphor, Conium, Hepar-sulph., Sulphur.

55. NUX-MOSCHATA — Myristica Moschata, Nutmeg, from the East Indies, (vegetable). *Antidote* — Camphor.

56. NUX-VOMICA — Strychnos, Nux-vomica (vegetable). *Antidotes* — Aconite, Camphor, Coffea, Pulsatilla.

57. OPIUM — White Poppy (vegetable). *Antidotes* — Camphor, Coffea, Calcarea, Hepar-sulph., Sulphur.

58. PETROLEUM — Mineral Oil. *Antidotes* — Aconite, Cocculus, Nux-vom.

59. PHOSPHORUS — (mineral). *Antidotes* — Camphor, Coffea, Nux-vom.

60. PLATINA — (Mineral). *Antidotes* — Belladonna, Pulsatilla.

61. PLUMBUM — Lead (mineral). *Antidotes* — Belladonna, Opium.

62. PODOPHYLLUM — May Apple (vegetable). *Antidote* — Nux-vom.

63. PULSATILLA — Meadow Anemone (plant). *Antidotes* — Chamomilla Coffea, Ignatia, Nux-vomica.

64. RHEUM — Rhubarb (vegetable). *Antidotes* — Camphor, Coffea, Ignatia, Nux-vom.

65. RHUS — Rhus-toxicodendron, Sumach, Poison Oak (vegetable). *Antidotes* — Belladonna, Bryonia, Camphor, Coffea, Sulphur.

66. SABINA — Savin (plant). *Antidote* — Camphor.

67. SAMBUCUS — Sambucus Nigra, Elder Flowers (vegetable). *Antidotes* — Arsenicum, Camphor.

68. SANGUINARIA — Sanguinaria Canadensis, Common Blood-root (vegetable).

69. SECALE — Secale-cornutum, Ergot of Rye (vegetable). *Antidotes* — Camphor, Opium.

70. SEPIA — Inky Juice of the Cuttle-Fish (animal). *Antidotes* — Aconite, Vinegar.

71. SILICEA — Silicious Earth (mineral). *Antidotes* — Camphor, Hepar-sulph.

72. SPIGELIA — Spigelia Anthelmintica, Indian Pink (vegetable). *Antidotes* — Camphora, Aurum.

73. SPONGIA — Spongia-tosta, Burnt Sponge. *Antidote* — Camphora.

74. STANNUM — Pure Tin (metal). *Antidotes* — Coffea, Pulsatilla.

75. STAPHYSAGRIA — Stavisacre (vegetable). *Antidote* — Camphora.

76. STRAMONIUM — Thorn Apple (vegetable). *Antidotes* — Belladonna, Nux-vom.

77. SULPHUR — (mineral). *Antidotes* — Aconite, Camphora, Mercury, Nux-vom., Pulsatilla.

78. SULPHURIC-ACID — Oil of Vitriol (mineral). *Antidote* — Pulsatilla.

79. TARTAR EMETIC — Tartarus Emeticus, Stibium, Tartarized Antimony (mineral). *Antidotes* — Cocculus, Ipecac., Pulsatilla.

80. TEUCRIUM — Teucrum Marum Verum, Germander (plant). *Antidotes* — Camphora, Ignatia.

81. THUJA — Thuja-occidentalis, Arbor Vitæ, Tree of Life (vegetable). *Antidotes* — Camphora, Pulsatilla.

82. VERATRUM — Veratrum-album, White Hellebore (vegetable). *Antidotes* — Ipecac., Arsenic, Camphora, Coffea, Aconite, China.

83. VIOLA TRI-COLOR — Jacea, Heart's-ease, Violet (plant). *Antidote* — Camphora.

TINCTURES FOR EXTERNAL USE.

84. CANTHARIS TINCTURE.

85. ARNICA TINCTURE.

86. CALENDULA TINCTURE.

87. RUTA TINCTURE.

88. URTICA-URENS TINCTURE.

89. STAPHYSAGRIA TINCTURE.

90. HYPERICUM PERFOLIATUM TINCTURE.

DIRECTIONS.

The use of these tinctures is specially referred to in the book whenever needed. They are intended for external use only.

HOMŒOPATHIC TREATISE

.DISEASES OF WOMEN AND CHILDREN.

CHAPTER I.

MENSTRUATION AND ITS DERANGEMENTS.

GENERAL REMARKS. — Perhaps no function of the female economy is so imperfectly understood by those most interested — the
females themselves — as that of menstruation.   Indeed, this is not
to be wondered at, when we remember that all the investigations
and writings upon the subject have been confined to that branch
of literature, which seldom comes within their range of observation.   Besides, what has been written upon the subject, till within
a comparatively short time, has been so speculative and hypothetical, and so mystified with verbiage, which writers have seen fit to
use in endeavoring to explain what they themselves but imperfectly
understood, that none but an earnest seeker after truth would
ever have the patience to wade through it.   It would take a scholar
indeed to extract the wheat from the chaff.

Deeming it next to impossible to make any one understand
the *disorders* of menstruation without first imparting some little
information in regard to the normal condition of the function, I
shall endeavor, in this article, to lay before the reader, in as short
and comprehensive a manner as possible, the true theory of menstruation, its causes, and its use, as settled by scientific investigations within the last thirty years.

Perhaps no question has excited so much controversy and
speculation, or such earnest desire and constant inquiry after
truth, as what could be the cause of the regular menstruation in
woman.   The ancients had many superstitious notions in regard

to it; and to this day, many of their rules and observances are
still regarded by not a few among the lower classes of society.
The wonderful periodicity and regularity of the menses, their
recurrence once in each lunar month, led to a general conviction,
that, like the tides of the ocean, their return was governed by the
moon.   It is somewhat strange that this opinion should have
gained such universal credence, when, at any day in the month,
millions of persons could testify to its fallacy.   For, here, in this
city, there is not an hour in the whole year, but that women are,
one or another, in every stage of the term.   Did the influence of·
the moon have anything to do with it, this would not be so, but
all women would be unwell at the same time.   .

Other hypotheses have been advanced, and have found favor,
more, perhaps, from the ability of their authors than from any
merit which they possess.   But so obscure, indeed, lay that unknown force, the real cause of the menstrual hemorrhage, in spite
of all the researches, conjectures, and observations of the most
learned men in the profession, that near twenty-five hundred years
passed before a satisfactory conclusion was arrived at, and the
essential truth was discovered.

Perhaps no one has done more to satisfy the inquiring mind that
the true theory of menstruation has been arrived at, than Dr.
Meigs, of Philadelphia.   He has exhausted the whole subject in his
writings and lectures before his class.   To him I am largely
indebted for the material of this article.

"*Omne vivum ex ovo*" (the germ of all life is the egg or cell),
is the universal, sole law or principle of reproduction.   This has
been controverted by some, and we have had set forth several
theories of generation; but we might as well suppose half a doz-
en laws of therapeutics, of elective attraction, or of gravitation, as
to suppose a number of reproductive laws or principles.   There
is but one law.   Men of eminence and ability, who have made
this subject their study, have brought forward an overwhelming
amount of evidence in support of our assertion.

The violet by the wayside, the sturdy oak upon the mountain-
top, the sparrow that chirps in the hedge, the lion that roams
through the forest, and man, — God's last and noblest work, —
are all produced by the one, same, inherent, generic law.

Every grain of wheat, every kernel, every seed, and every egg,
whether it be of the humming-bird or of the ostrich, contains a
germ within it which, when brought within known and proper

conditions, is capable of reproducing organized bodies, each after its own kind. When we plant wheat, we expect to grow wheat from it; and when we plant an acorn, we expect an oak. So far it is all plain enough; but the great mystery of generation, especially in all the zoölogical series, was the production of germs. Whence came they? It is now ascertained that each animal, as well as each plant, that comes into existence, is provided with an organ for the evolution of germs or cells. This organ in the human female — and it is of her only that we deem it necessary to speak — is the ovary. The ovaries are two in number, small, oval bodies, about an inch in length, rather more than half an inch in breadth, and a third of an inch in thickness. In general terms, this measurement would answer; but the size of the ovaries differs in different women: some are larger, and some are smaller. Each ovary is attached to an angle of the womb, by means of a ligament, which is about one inch in length. Their situation, in relation to adjacent parts, varies, at different ages, and also according to the state of the uterus. Being attached to within about an inch of the upper portion of the womb, they are low down in the pelvis, or up in the abdomen, according as that organ is distended by pregnancy, or otherwise.

Now, the physiological function and sole duty of the ovary is, to mature and deposit its ova, or eggs, which contain the germs, once in every twenty-eight days; and this it does, in the majority of healthy females, with great regularity.

The microscopic egglet, thus produced, measuring no more than the two-hundred-and-fortieth of an inch in diameter, is, in every essential, an egg. In fact, the egg of a canary bird is no more a perfect egg, than is this ovum of the human female.

During the maturition and depositing or discharging of the ovum into the canal or tube, which conducts it into the womb, the whole vascular apparatus of the generative organs is turgid or congested, looking almost as if inflamed. This local congestion rises to such a height that it, as it were, overflows and produces the outward signs of menstruation, — a discharge of bloody fluid from the genitalia. This fluid exudes from the vessels on the inner surface of the womb, and completely relieves the engorgement. As the flow commences, the heat and aching in the region of the ovaries, and the weight and dragging sensation in the pelvis, which almost all women experience, gradually diminish, and finally all disappear.

You will therefore observe, that the whole act of menstruation consists simply in the depositing or discharging of an ovum or egg, which, when not impregnated, is washed away by the menstrual fluid, — the menstrual fluid being nothing more nor less than an exudation of blood from the inner surface of the womb.

The periodicity of menstruation has excited a good deal of wonder and controversy. And, really, I cannot see why it should have done so; for if we look about us, and consider for a moment, we shall observe that, throughout the animal and the vegetable kingdom, germ-production takes place, not continuously, but at stated times. Why should it be otherwise with our race? It is the same.

We all know when fruit and vegetables are ripe; when the farmer mows and reaps his harvest; when the tulips bud and blossom; when birds and animals produce their young. Every lady knows at what season of the year to go out to gather the seeds from her flowers. The farmer does not dig his potatoes in spring, cut his corn nor reap his wheat in June, — and why? Because he knows they are not ripe. The germ-production is not completed. All this should teach us that the paroxysmal, periodical maturing of germs is a law of nature. And if every vegetable and every other animal obeys it, why should not woman, in her monthly term.

It may also be remarked, that, by the same wonderful law of which we have been speaking, all forms of existence are kept within prescribed and proper limits. The blade of grass of to-day is an exact copy of its primitive pattern. Nature makes no improvements. Her laws are perfect from the first; every animal that inhabits this globe, all plants and trees, are exact copies of those which preceded them ten thousand years ago. The same force which governs one governs all, and keeps them in their pristine purity, as free from admixture as when they first issued from the hands of the Creator.

It is only when civilized man interferes with nature that we find deviations from her laws. We find mongrels and crosses only among domestic plants and animals. It is the work of man.

Adopting this theory, that the essential feature of menstruation consists in the periodical maturition and spontaneous deposit of an ovum, — the menstrual flow being but the outward visible sign of such an act, — it is possible that a woman may menstruate regularly and truly without having the least show. This may sound strange to one who has read or thought but little upon the subject. But there is not the least doubt that many women are perfectly regular who give not the least outward sign of it. This is proved by the fact that many a female has become pregnant who has never menstruated at all. There are not a few cases upon record, where a young woman reaching the age of puberty, gets married, becomes pregnant, and has a child, without ever having seen the first outward sign or show. Again, it is not at all uncommon for a woman to have two, three, or more children, without, in the meantime, ever having had a sign of regular menstruation. We have every reason to suppose, that germ-production takes place regularly, though perhaps feebly, through the whole nursing period, although the woman does not perceive that she menstruates, — the function of supplying milk being sufficient to turn aside or draw off the nervous and sanguine determination from the organs of generation.

The foregoing remarks may be sufficient to show, that because a woman does not have her menses just at the time it is expected, she does not necessarily need " doctoring;" because though she does not *seem* regular, she really may be.

Great alarm is often created in the minds of parents, because their child fails to menstruate as early in life as they expected she should. And, not unfrequently, all sorts of quack nostrums are resorted to, to induce or force the discharge. If the young lady be bright, cheerful and active, if her health appears good, and no derangement is observable, though she does not have her menses, medical interference is unnecessary and unjustifiable. In such a condition of affairs, the work should be left to nature.

THE FIRST ACCESSION OF THE MENSTRUA. — Menstruation, menses, catamenia, courses, monthly periods, and being "unwell," are some of the conventional terms by which this periodical deposit of an ovum, and its accompanying discharge of blood and mucus, are designated.

In this country, the sexual function is not assumed until the fourteenth or fifteenth year; in warm climates, it appears somewhat earlier, and in cold ones later. Perhaps situations and conditions have as much to do with the early or late appearance of the catamenia as has the climate. It has been observed that those who are brought up luxuriously, and whose moral and physical training has been such as to exaggerate the susceptibilities of their nervous system, are unwell at an earlier period than those who are brought up roughly, and are accustomed to coarse food and laborious employment.

The appearance of the menses previous to the fourteenth year is to be regretted, because it indicates a premature development of certain parts; and, on the other hand, a late or retarded first appearance is always to be regarded as an evidence of weakness, or disorder. An undeveloped state of the uterine organs, indicated by a procrastinated, or non-eruption of the menstrua, always, in the minds of physicians, excites apprehensions for the welfare and security of the person in whom it is observed. In such cases, we often find the general character and appearance of the body blighted, the mind dull and weak, with the chest and lungs insufficiently developed,—all of which render the patient an easy prey to disease.

The first accession of the menses is generally preceded by the following symptoms: headache, heaviness, languor, pains in the back, loins, and down the thighs, and an indisposition to exertion. "There is a peculiar dark tint of the countenance, particularly under the eyes, and occasionally uneasiness, or a sense of constriction in the throat. The cutaneous perspiration has often a faint or sickly odor, and the smell of the breath is peculiar. The breasts are enlarged and tender. The appetite is fastidious and capricious, and digestion impaired. These symptoms continue one, two, or three days, and subside as the menses appear." The period continues during three, five, or seven days, according to the peculiarities of the constitution. The quantity of the menstrual discharge varies in different women. Some have to make but one change during the whole period. This, however, is uncommon, as they generally average from ten to fifteen.

These monthly periods return with great regularity, from the age of fourteen or thereabout, to the forty-fifth year, except during pregnancy. At about the forty-fifth year, the final cessation of menstruation takes place. This period is one which often excites the fears of females; not unfrequently the symptoms, which present themselves at this time, such as sickness at the stomach, capricious appetite, swelling and pain in the breast, etc., are mistaken for those of pregnancy. Menstruation rarely ceases at once; more commonly the change is gradual, and, for a time, is attended with irregularities. Sometimes the discharge returns every two or three weeks, then ceases for two months, or perhaps more, returning again, perhaps as regular as ever, for a few months, when it finally disappears altogether.

During the menstrual period, especially in young persons, great care should be taken to ward off all influences, whether mental or physical, which may have the least possible tendency either to interrupt or to increase the discharge. Because, upon the healthy and regular action of the discharge depends so much of the beauty, perfection, and security of the female. During this period there is an increased susceptibility and excitability of the system, and consequently a greater liability to derangements and to diseases of various kinds.

Serious and even dangerous results often follow a sudden suppression of the menses. Among the causes which produce trouble during this period, we may mention, — sudden frights, fits of anger, great anxiety, and all powerful mental emotions. Excessive exertions of every kind, long walks or long rides, especially over rough roads, dancing, frequent running up and down stairs, have a tendency not only to increase the discharge, but to produce falling of the womb. The discharge is not unfrequently morbidly increased, or entirely arrested by taking purgatives, emetics, stimulants, and the various patent medicines recommended for female weaknesses. Cold and warm bathing, hip and foot baths should be discontinued during the period. Care should also be taken not to expose the feet to cold or wet. Females subject to leucorrhœa and who are taking vaginal injections, should discontinue them shortly before and during this period.

During the menstrual period in a healthy person, there is little required besides carefully avoiding the injurious mental and physical influences above-mentioned. If, however, the female be delicate, and suffering from any of the various derangements of menstruation, medical treatment should at once be secured.

DELAYED AND OBSTRUCTED MENSTRUATION.

(FIRST MENSES.)

DEFINITION. — When the menses do not appear at the period of life, at which they may be naturally expected, we call it *delayed* or *retained menstruation*. This delay may be owing to a disordered condition of the general system, or to functional inactivity, or weakness of the uterine organs themselves. By *obstructed* menstruation is meant an actual impediment to the flow. In such cases there is periodically an evident effort on the part of nature to produce the change, the patient having all the premonitory signs, the pains, aches, etc., the ovaries and womb perform their part, but still the flow does not appear; there is an obstruction, either at the mouth of the womb or somewhere within the vagina. This obstruction may be congenital, — that is, having existed from birth; or it may be the result of some former disease, — inflammation, perhaps, of the parts.

CAUSES. — Delayed and obstructed menstruation may arise from so many causes, and present such a complication of symptoms, that it requires the skill of an experienced physician to treat them successfully. I shall, therefore, simply enumerate some of the causes and symptoms of the disease, only tendering the advice that all these cases should be placed under the care of an intelligent homœopathic physician.

It has already been stated that, as a general thing, regular menstruation, in this country at least, does not take place until about the fourteenth or fifteenth year. Whenever it is delayed beyond the sixteenth year we have good reason to have serious apprehensions for the security and future health of the individual in whom it is observed.

Among the causes of delayed menstruation we have enumerated an imperfect or late development of the uterine organs, functional inactivity, weakness or disorder of the uterus or ovaries, or an entire absence of these two organs. Perhaps, in the majority of cases, the direct or immediate cause lies in some peculiar condition of the ovaries. It is most frequently met with in those who lead indolent and sedentary lives, who indulge in luxurious and gross diet, and who have been accustomed to hot rooms, soft beds, and much sleep.

As a cause, we may also mention the vicious system of modern fashionable education, over-taxation of the mental faculties; want of exercise in the open air; badly lighted, cold, and damp dwellings; sudden and extreme atmospheric changes.

The disease is not unfrequently a mere symptomatic or sympathetic condition, dependent upon some disease existing in a distant part or organ of the body.

The causes of *obstructed* menstruation are, an imperfect hymen, total or partial obliteration of the vagina, or closure of the mouth of the womb. Either of these causes, you will readily observe, could only be detected by a physician. Hence the necessity of having counsel in these cases.

SYMPTOMS. — In *delayed* menstruation the symptoms of uterine congestion, such as an aching pain in the region of the ovaries, a weight in the lower part of the abdomen, aching in the tops of the hips, etc., are almost or entirely wanting. Or, occasionally, or even monthly, an apparent attempt at menstruation occurs, characterized by the usual symptoms preceding the discharge. After two or three days, these symptoms all subside without any menstrual evacuation, or merely with leucorrhœal discharge.

These cases always present a great many symptoms of constitutional derangement. There is a general lassitude, with great aversion to exercise, either mental or physical. The bowels become irregular, the countenance pale, the skin discolored, the general appearance bloated and flabby, the pulse small and frequent, the breathing short and hurried. There is headache and buzzing in the ears. Digestion becomes deranged; there is a taste for strange things, — to eat chalk, slate-pencils, pickles, vinegar, etc. Nausea and vomiting are frequent accompaniments. Various hysterical symptoms manifest themselves through the course of the disease.

If relief is not obtained, the general health gradually declines, and the patient passes into chlorosis, tubercular consumption, or some one of the various nervous ailments, hysteria, epilepsy, etc.

The symptoms of *obstructed* menstruation are somewhat different from those which attend delayed menstruation. In these cases, we have monthly exacerbations of all the symptoms preceding menstruation. All the outward signs of puberty are present. The patient complains of pain in the back, loins, and hips; of distention of the lower part of the abdomen, and of a sense of weight and bearing down. Still there is no show. These symptoms are increased every month; the fulness in the abdomen is augmented at each monthly period, and the general health becomes very much affected. In these cases menstruation takes place regularly, but the flow is prevented by some obstruction. The menstrual blood is confined within either the womb or vagina; of course, each monthly period increases the distention of the parts, and, if rupture of the parietes does not take place, as sometimes occurs, the general health gives way, and the patient dies from the general disorder thus induced.

TREATMENT. — It is not necessary to put a girl under a course of medical treatment, simply because a regular monthly evacuation does not take place as early in life as it is usually expected, because a female may be regular in her menstruation and still have no show at all. — " See *Menstruation.*" If the girl be bright,

2

cheerful, and active; if her general health appear good, and no
derangement is observable, leave the case to nature. When,
however, the symptoms above enumerated begin to show themselves, and it is evident that nature, without assistance, is unable
to effect the change, judicious medical treatment should at once
be instituted. But, of all things beware of what are called *forcing* medicines. Many a female has ruined her health by resorting
to quack medicines of this description.

My remarks upon treatment shall be brief, and they shall be
confined to the first stage of the disease. When the remedies
here recommended do not afford relief, apply to a homœopathic
practitioner without delay. Where the difficulty apparently arises
from close application to study, or close confinement in school,
or from sedentary habits, proper attention to out-door exercise,
and a little restriction of diet, will generally produce a healthy
state of the function. The girl should be removed from school;
studies should be banished; a cheerful disposition should be cul-
tivated; out-door exercise, either walking or riding, is a capital
remedy; wear warm stockings and thick shoes; change the
clothing to suit the variations of temperature, but always have it
warm; avoid all stimulating food and drinks; eschew tea and
coffee; avoid exposure to night air; keep regular hours; retire
early, and rise early in the morning.

The remedies most applicable to this disorder, are, *Pulsatilla,
Bryonia, Lycopodium, Phosphorus, Arsenicum, Sulphur.*

Bryonia. — Where there is congestion to the head, with frequent nose-bleeding, palpitation of the heart and constipation.
Lycopodium is suitable for similar symptoms.

Pulsatilla. — This remedy is especially adapted to females of a
mild, easy disposition, inclined rather to weep, than to be angry
or fretful. Paleness of the face, with flushes of heat; loss of appetite with desire for acids; sour taste in the mouth after eating;
nausea and vomiting. After *Pulsatilla* give *Sulphur.*

Phosphorus. — Where there is a predisposition to lung disease;
delicate constitution; weak chest; cough; pain in the chest; expectoration of blood.

Arsenicum. — In those cases attended with dropsical swellings
either of the feet or about the eyes; pale complexion; great
weakness.

ADMINISTRATION OF REMEDIES. — Give either of the above remedies, according to the indications, a dose every morning for one

week; if the symptoms abate, wait one week without medicine, and afterwards give one dose of *Sulphur* every evening, for a week. If not better, select another remedy and apply in the same manner.

Dose, six globules, or dissolve twelve globules in six spoonfuls of water, and take one spoonful of the solution at a time. — See "*Chlorosis.*"

MENORRHAGIA — PROFUSE MENSTRUATION.
TOO FREQUENT AND OF TOO LONG DURATION.

DEFINITION. — By the term *menorrhagia* is to be understood an immoderate flow of the menses. The quantity of blood eliminated from the uterus at each menstrual period cannot be definitely ascertained. In this country, perhaps, the amount varies from four to eight ounces. However, every woman in good health, has a rate of her own; some discharge one ounce, and some discharge eight, or ten, or more, at each period. Therefore, what would be excessive for one woman, might be quite moderate for another. Less than four, I think, may be considered scanty, and more than eight or ten, excessive. The quantity discharged at each period may be approximately estimated by the number of napkins used. Half an ounce, or about one table spoonful would render a napkin quite uncomfortable. Then eight napkins would contain about four ounces; and twenty, ten ounces.

Excessive menstruation may occur in a variety of ways; the evacuation may return too frequently, or too copiously, and continue for too long a period.

CAUSES. — Some females are hereditarily predisposed to uterine hemorrhages. There is a weak condition, or rather a relaxed or flabby state of the texture of the uterus. The same condition, however, may be induced by frequent child-bearing, by abortions, by prolonged or too frequent suckling, by rich living, indolence, hot rooms, and soft beds.

Among the exciting causes of profuse menstruation, we may mention over-exertion, lifting heavy weights, running up and down stairs, dancing, local injuries, falls, the use of drastic medicines, irritating purgatives, exposure to cold, and mental excitement and moral emotions. Young girls, as well as grown and married women, are subject to menorrhagia. In them, the disorder is generally dependent upon a weak, enervated condition of the uterine organs. This is especially the case with school-girls, whose mental application is severe, and whose opportunity for physical out-door exercise is limited.

SYMPTOMS. — The symptoms are similar to those of ordinary menstruation, except the flow is greatly increased, and consists of a purer blood than that of the regular catamenial evacuation. Sometimes the discharge, though natural enough otherwise, continues for too long a period, lasting two, or even three weeks. Again, although natural, at least in quantity, it may return too frequently, as every two or three weeks. As a general thing, the patient is also afflicted with leucorrhœa.

This excessive drain upon the system, as a matter of course, cannot continue a great while without producing debility, to a greater or less extent. If the disease goes on unchecked, the patient even begins to complain of languor; she is always tired; there is an indisposition to all exertion. Headache, throbbing in the temples, weakness across the loins and hips are common symptoms. As the exhaustion and languor increase, the countenance becomes pale and sallow, the pain about the loins and hips increases; pain in the left side, repeated and severe headache, derangement of the stomach and bowels follow. The general appearance of the patient is such that her friends remark that she is going into a decline.

In some cases the flow commences with sudden and violent gushes, after which, it stops for some hours, and then recurs, and thus alternates during the whole menstrual period. Sometimes the hemorrhage continues so excessive that the patient is completely prostrated, and sinks, fainting, from exhaustion.

TREATMENT. — Whatever other treatment shall be instituted, it is absolutely necessary, during the flow, that the patient maintain a recumbent position. She should go to bed in a cool room, and cover herself lightly with bedclothes. If the flowing be *excessive*, cloths wrung out of vinegar and water may be applied to the lower bowels.

Crocus. — This is a most important remedy, especially in cases where the menses have returned too soon, and the discharge, which is copious, consists of dark-colored clots.

Chamomilla. — When, in connection with the symptoms calling for Crocus, there are also griping pains through the abdomen, severe colic, or pains like those of labor, extending from the back forwards towards the abdomen, great thirst, and paleness of the face.

Ipecacuanha — When the blood consists of bright-red blood, and when the menses return too early, may be followed by *Sabina*.

Belladonna. — When the menses return too soon, are accompanied with bearing-down pains, and pressing outward. Also severe headache with flushed face and cold extremities.

Nux-vomica. — When the flow commences with sudden, violent gushes, or stops for a short period, and returns again; is too copious; continues too long, and returns before the usual time; when the periods are attended with spasms in the abdomen, nausea, sickness at the stomach, and fainting; also when there is a sensation of heaviness, or bloatedness, with pain and soreness as from a bruise. It may follow *Chamomilla*.

Platinum. — When the discharge consists of thick, dark-colored blood, and is attended with bearing-down pains.

Cinchona. — This remedy is called for in those cases where the catamenial discharge has been so excessive as to produce great weakness. Should be given after other remedies have controlled the discharge.

Calcarea. — This is a valuable remedy in obstinate cases, — those which have resisted other remedies, continued a long time, and caused general constitutional sufferings. It is especially adapted to persons of relaxed muscular fibre, to weak, cachectic or scrofulous subjects.

Secale. — Profuse menses, with violent cramp; tingling in the legs, cramps in the feet.

Sulphur. — When other remedies fail to act, though apparently well indicated.

ADMINISTRATION OF REMEDIES. — Of the selected remedy dissolve twelve globules in twelve spoonfuls of water, and of this solution take one spoonful at a dose, and repeat every hour, or two, three, or four hours, according to the severity of the attack.

DIET. — The diet should be light; all hot drinks should be avoided.

TARDY AND SCANTY MENSTRUATION.

DEFINITION. — After menstruation has once become established, it may show itself less frequently than at the regular periods. That is, instead of returning once in twenty-eight days, it may be delayed to five, six, or more weeks. And, besides, the discharge, when it does appear, is not free, and does not afford that relief, which regular, healthy menstruation does. This is especially the case in young females. Scanty menstruation is not always accompanied by a tardy eruption; on the contrary, the periods may return as often as every two or three weeks, and the discharge scarcely amount to anything.

Elderly females — those who are about experiencing a " change of life" — are almost always subject to intermissions and delays in their menstrual periods. In persons thus situated, when the discharge does make its appearance, it is apt to be very severe, amounting sometimes to almost flowing.

TREATMENT. — For irregular menstruation, be it either too late or too early, too scanty or too profuse, *Pulsatilla* is most generally the appropriate remedy. It is especially adapted to females of a mild and easy disposition, and particularly if the following symptoms are present: pain low down in the abdomen and through the small of the back; nausea and vomiting; shivering; pale face; alternate crying and laughing; sadness and melancholy; giddiness; fulness about the head; semilateral headache; roaring in the ears; drawing pains, extending to the face; tendency to diarrhœa and leucorrhœa.

All the symptoms are worse in the afternoon and at evening; the pains frequently change from place to place, and are relieved by the patient's being out in the open air.

Natrum muriaticum. — Menses too late and too scanty with rush of blood to the head.

Sepia. — Delayed menses, with violent colic; fits of fainting; leucorrhœa and pressure in the abdomen; uterine cramps; face bloated, and marked with yellow spots, especially across the cheeks and nose.

Belladonna. — If there is a rush of blood to the head, nose-bleed, redness of the eyes, intolerance to light, and giddiness, especially after stooping, the menses delayed, but when they do occur, too profuse.

Bryonia. — If, instead of menstruation, there is bleeding at the nose, with congestion of the head and chest.

Causticum. — Menses delayed, but profuse; pains in the sides; yellow complexion.

Cocculus. — Scanty discharge of dark clots; excessive nervousness; contracting, pinching pains in the pelvis.

Graphites. — Menses too late, too little, and too pale. During the menses, spasms and violent cutting pains in the abdomen; labor-like pains; nausea and violent headache.

Lachesis. — Especially applicable during the " critical period." Short and scanty menses ; before the discharge, vertigo and headache ; colic and leucorrhœa.

Phosphorus. — For females of a delicate constitution ; those predisposed to lung difficulties, and when, in place of regular menstruation, there is expectoration of blood, or menses too soon and too profuse. During the menses, there is weariness, shivering, headache, pain in the back.

Sulphur. — When other remedies have been insufficient, or when there is great heat in the head, giddiness ; palpitation of the heart, shortness of breath ; abdominal cramps, pains in the side and bowels ; loss of appetite, sickness at the stomach, emaciation, and depression of spirits.

Compare, also, " *Suppression of the Menses,*" " *Delayed and Obstructed First-Menses,*" " *Chlorosis.*"

ADMINISTRATION OF REMEDIES. — Dissolve, of the remedy selected, six globules in twelve spoonfuls of water ; take of this solution one spoonful every six hours, until amelioration or change.

DYSMENORRHŒA.

PAINFUL MENSTRUATION — MENSTRUAL COLIC.

DEFINITION AND CAUSES. — The term dysmenorrhœa, which is a Greek word, signifies a difficult monthly flow. The menstrual discharge is preceded by severe pains through the loins and in the lower part of the abdomen. These pains are sometimes so intense, as to be almost insupportable, compelling the patient to go to bed, forcing from her tears and groans, as she writhes under the agonizing pain.

The pain is in the nerves of the womb, and perhaps may be fairly attributed to the compression which they receive from the congested state of the organ which exists during the period of menstruation. The pains may also be neuralgic or rheumatic in their character.

The most common exciting causes are exposures to cold, sudden fright, or shocks, or violent mental emotions. Indeed, any cause which would excite inflammation, or produce a suppression of the menses, would be sufficient to cause dysmenorrhœa.

The *neuralgic* variety of dysmenorrhœa occurs chiefly in unmarried females, and in the married who have not borne children. It is almost confined to women of irritable, hysterical, and nervous temperament. The *inflammatory* variety of the disease most frequently attacks those of a full habit and of a sanguine temperament, married or unmarried, whether they have borne children or not.

SYMPTOMS. — The majority of females are notified of the approach of their terms by a sense of weight and dragging in the region of the pelvis, — symptoms which gradually disappear, as soon as the flow fairly commences. But still not a few never experience the least trouble or inconvenience from these monthly visitants. This is not the case, however, with the woman who has an irritable, a rheumatic, or a neuralgic womb.

The pain of dysmenorrhœa, according to my observations, presents all the characteristics of neuralgia. For a short time previous to the flow, there is a sense of general uneasiness, a deep-seated feeling of coldness, restlessness, sometimes chills, flushes of heat and headache. The pain commences, usually, in the region of the sacrum; — that is, the lower portion of the backbone, — and extends round and through the lower part of the bowels, and down the thighs. The amount of suffering varies, but it is sometimes very severe ; the forcing or bearing-down sensation, which is often present, is not unfrequently so great, that it seems as though the whole contents of the pelvis would be forced out. In some instances, the torture is so extreme, that the patient cannot lie still, but constantly keeps rolling about, lies or presses upon the abdomen, endeavoring to get ease from her sufferings. Nausea, sickness at the stomach, and severe retching are not uncommon.

These sufferings may continue only for a few hours, or they may continue for two or three days, before the discharge commences. As a general thing, the discharge appears slowly, and is at first scanty, or it may appear in slight gushes. The quantity varies in different persons, or even in the same persons at different periods. It is often scanty, rarely too much. In color, it is frequently quite natural ; sometimes it is light and mixed with small clots.　·

Between these monthly periods, the health of the patient is little affected, although headaches are not unfrequent ; but these are generally transient. Leucorrhœa is sometimes present, during the interval, and, especially if the attacks are of an inflammatory nature, is often persistent and difficult to control.

TREATMENT. — If it be possible, the patient should lie down, or at least she should keep very quiet. The indications in the treatment are twofold: first, to reduce the pain during the attack, and secondly, to prevent its return by appropriate remedies during the interval.

Belladonna.—Is called for, when the pains in the abdomen are as if the parts were clutched, with severe pain in the back; also, when the pains are in the lower portion of the abdomen, and are of a bearing-down nature, as if the parts were about to fall out. The pains come on before the appearance of the menses, with strong determination of blood to the head, accompanied with frightful visions, confusion of sight, headache, red and puffed face, and violent thirst.

Coffea. — When there is great nervous excitement, wringing of the hands, grinding the teeth, screaming, great restlessness, twisting of the whole body; distressing colic, fulness and pressure in the abdomen, as if it would burst; coldness of the body, numbness and stiffness. May be given in alternation with *Pulsatilla* or may be followed by *Cocculus.*

Chamomilla.—For pains resembling those of labor, commencing in the small of the back, and extending around to the front and lower part of the abdomen; menses premature and too profuse, attended with violent abdominal cramps, great sensitiveness of the abdomen, discharges of dark, coagulated clots.

Pulsatilla.—For abdominal spasms with discharge of dark clots, or of pale blood; weight in the bowels, as from a stone; pressing pain in the abdomen and the small of the back, pains in the sides, nausea and vomiting; frequent inclination to pass water, and ineffectual efforts to evacuate the bowels.

Nux-vomica.—For cramps of the womb; writhing pains in the abdomen, or pains in the back, as if dislocated; lameness of the sides; forcing pains, nausea, and sickness at the stomach; frequent urging to urinate.

Cocculus.—For colic pains; abdominal cramps; flatulency; nausea and faintness; cramps in the chest; premature menses with abdominal spasms.

Secale cornutum.—Tearing and cutting colic, with profuse and long-continued flow; pale face; cold limbs; cold sweat.

Veratrum.—Menstrual colic, with nausea and vomiting; hysteric distress; nervous headache; coldness of the nose, feet, and hands; fainting fits; headache before the flow commences.

Platina.—Leucorrhœa before and after the menses. The

3

menses are too frequent, too profuse, last too long, and are accompanied by cramp, colic, and forcing pain.

Sulphur. — Abdominal cramps and colic during the menses. The menses are preceded by headache.

ADMINISTRATION OF REMEDIES. — The remedies above enumerated are to be given at short intervals during the continuance of the pain. The remedy selected may be dissolved in water, twelve globules in twelve spoonfuls, and one spoonful given every half hour, until the pain abates ; then the intervals can be lengthened.

A preventive course of treatment should be adopted between the monthly returns of these severe attacks. Every effort should be made to strengthen the patient, and to diminish the general and local irritability. The diet should be plain and nourishing ; stimulants of all kinds, condiments, high-seasoned dishes should be strictly avoided ; coffee, under no consideration whatever, should be used.

The patient should take plenty of out-door exercise. The habitual daily use of cold or tepid baths, and vaginal injections are often of great benefit. Cold bathing and injections should, however, be discontinued a few days previous to the expected menstrual eruption, and should not be resumed, until three or four days after its cessation. It is sometimes advisable to make use of warm hip and foot-baths, for two or three nights in succession, preceding the eruption of the menses.

The remedies to be made use of, during the intervals, are *Sulphur, Pulsatilla,* or *Nux-vomica.* One dose of six globules, every third or fourth evening.

SUPPRESSION OF THE MENSES.

DEFINITION. — By suppression of the menses is understood a disappearance of the same, after having been regularly established for a longer or shorter period, independent of pregnancy, or old age.

Suppression of the menses may take place suddenly or gradually ; that is, it may be acute or chronic.

CAUSES. — Among the causes of acute suppression, may be mentioned cold caught during the flow, by exposure to damp night-air, by wet feet, etc., by fear, bodily shocks, or by sudden, violent mental emotions, either just previous to or during the menstrual discharge. Anxiety, depressing passions, insufficient clothing, and exposure to want will also produce it. Fevers, inflammations or almost any acute disease, occurring just before the period, will have the same effect.

Chronic suppression of the menses is commonly a consequence of the acute, or it may arise from the gradual supervention of delicate health. It may also proceed from disease of the ovaria or womb. This form of the disorder comes on gradually. The quantity of the discharge becomes diminished, the periods irregular and uncertain, until at length the function ceases altogether.

SYMPTOMS. — The amount of disturbance, consequent upon the sudden arrest of the menses, varies with the habit of body and temperament of the patient. In robust females, — those of full habits, — most frequently a degree of fever results, with headache, hot skin, thirst, quick pulse, etc. And, not unfrequently, severe attacks of acute disorders, such as inflammation of the lungs, brain, bowels, or womb are produced. Females of spare habit and nervous temperament are liable to be seized with various hysterical affections; palpitations, loss of voice, nervous cough, headache, pains in the side, back, and abdomen, or attacks resembling inflammation, or neuralgia of different parts, but which are purely nervous.

The symptoms following chronic suppression are gradual impairment of the general health, disorder of the digestive organs, loss of appetite, constipation, pains in the back and side, and headache.

TREATMENT. — When the suppression of the menses is caused by the presence of some other disease in the system, the cure of such disease must first be effected by appropriate treatment, before we can look for a return of the catamenia.

If, however, the suppression is sudden, — the result of cold, or fright, or other morbific cause, — our first endeavor should be to recall the discharge, if possible. For this purpose, the following remedies are recommended ; *Aconitum, Bryonia, Belladonna, Pulsatilla, Opium, Veratrum, Sepia, Sulphur.* It is sometimes advisable for the patient to take, just before retiring, a warm foot or sitz-bath.

Should the attempt to reëstablish the discharge fail, we must content ourselves with mitigating the severer symptoms, until the approach of the next menstrual period.

Aconitum. — This remedy may be given, when a sudden suppression of the menses is occasioned by fright, and, especially, if there is congestion of blood to the head and chest, with redness of the face, pains in the head, giddiness, nausea, and faintness. It may be given in alternation with *Bryonia.* Should these remedies fail, or afford but partial relief, they may be succeeded by *Opium.*

Belladonna. — For females of a full habit, when there is congestion of the head or chest with throbbing headache, red and bloated face, and burning thirst.

Bryonia. — When the suppression is followed by headache and giddiness, which is aggravated by stooping and motion; also when there is pain in the pit of the stomach, sour eructations, rising of food, constipation, and bleeding at the nose.

Pulsatilla. — This is the chief remedy in this disorder, and, in the majority of cases, will be the only one required to make a prompt and perfect cure. It is especially called for when the suppression has been occasioned by exposure to dampness or cold, or by getting the feet wet, and the following symptoms are present: severe headache, which is chiefly confined to one side, with shooting pains, which extend to the ears, face, and teeth; pale complexion; vertigo, with buzzing in the ears; palpitation of the heart; feeling of suffocation, especially upon the least exertion; coldness of the hands and feet; flushes of heat; nausea and vomiting; pressure in the lower abdomen; frequent desire to urinate; swelling of the feet; leucorrhœa; tendency to sadness.

Sepia. — This is another important remedy, especially when there is leucorrhœa. Also, when the following symptoms are present: nervous headache, with alternate shuddering and heat; colic and pain in the sides; bearing-down pains; pale complexion, or yellow spots in the face; sad mood; weeping; hysteric complaints; lameness and soreness of the limbs.

Veratrum. — For nervous headache, and hysteric ailments; frequent nausea and vomiting, with coldness of the hands and feet; great weakness, and fainting turns.

Sulphur. — Headache, chiefly in the back part of the head; feeling of fulness and weight in the head; pain over the eyes; confusion of the head, with throbbing and buzzing in the head; pale face; red spots on the cheeks; bluish circles around the eyes; sour stomach; sour eructations; pressure, weight, and fulness in the stomach; constipation, with ineffectual urging to stool, or else mucous diarrhœa, with slimy evacuations; tendency to piles; leucorrhœa, with itching in the parts; numbness of the limbs; tendency to take cold; pains in the loins; weariness after talking; difficulty of breathing; irritability of temper, or sad and weeping disposition.

Chronic cases, — those of long standing, — where the patient has run down, other remedies will be called for; such as *China, Arsenicum, Natrum-muriaticum, Graphites,* but it will be much safer to place the patient under the care of an intelligent homœopathic physician.

See, also, " *Tardy Menstruation,*" and " *Chlorosis.*"

ADMINISTRATION OF REMEDIES. — Dissolve twelve globules in twelve teaspoonfuls of water; of this solution give one spoonful every hour or every two hours, according to the urgency of the symptoms, until the discharge returns, or the symptoms are removed. If the discharge does not return in a day or two, and yet the pain, headache, and other symptoms subside, discontinue the medicine, and resume it again a few days before the next period.

CHANGE OF LIFE — CESSATION OF THE MENSES.

DEFINITION OF SYMPTOMS. — By the phrase, " change of life," or the " critical period," as it is also called, is understood the final cessation of menstruation. Public opinion holds that this change takes place at the forty-fifth year, and all women look for the change at that time, but not a few are disappointed; for all women do not change at that time; some women definitely cease to menstruate at thirty, and some even earlier, while in others, again, the change may be postponed to the fiftieth year, or even later. However, as a general thing, we can anticipate the change at *about* the forty-fifth year.

Women of delicate constitutions, and those who have been in the habit of living well, enjoying the good things of life, as they went along, and whose habits have been sedentary, who have been confined to the house, and especially to warm rooms, experience the change earlier than those of a more robust organization, or those who have led a temperate, active life, avoiding all dissipation.

This period, which is rightly considered a critical one for every female, may pass without a single untoward symptom, the monthly evacuation gradually ceasing, without being attended by any unpleasant consequence, and leaving the patient enjoying better health than she ever experienced before. On the contrary, it may be fraught with peril, through which she can be safely conducted only by a skilful and experienced physician. It is, therefore, highly important that all the unpleasant sensations, which may be experienced, during this time, should receive a careful consideration, and not be hushed up with the unsatisfactory reply, that such complaints are owing to the " change of life," and likely to vanish whenever that change shall become complete.

If proper attention is not paid to the various complaints which may and frequently do manifest themselves during 'this period, the seeds of endless miseries and even early death will be allowed to germinate, and cut short a life that, by proper foresight and care, might have been conducted to a ripe old age.

As the change approaches, the menses gradually becomes irregular both in regard to the time of their recurrence and the quantity discharged. They may return too soon or be delayed beyond the usual time. The quantity discharged is at times much less than common. Sometimes the discharge returns every two weeks, then ceases for several weeks, or even months, and afterward recurs for a few periods as regularly as ever, and then altogether ceases.

Perhaps, in the majority of women, while this change, which lasts usually from a year to a year and a half, is in progress, there is more or less disturbance of the general health. And not unfrequently it is difficult, sometimes quite impossible, to say exactly what is the matter with the patient, except that she is generally out of health. A host of symptoms present themselves ; the patient complains of headache, vertigo, biliousness, indigestion, flatulency, acidity of the stomach, diarrhœa, costiveness, irregularity in the urinary discharge, piles, pruritis, — violent itching of the private parts, — cramps and colics in the abdomen, palpitation of the heart, hysterics, nervousness, pains in the back and loins, swelling of the abdomen, swelling of the extremities, paleness and general debility. To unravel all these, and to decide what is the best plan of treatment to pursue, requires the skill of an intelligent physician. I would, therefore, recommend you to take the early and constant advice of your family physician during this time.

The remedies called for during this change, or those which we most frequently have occasion to use, are *Pulsatilla, Lachesis, Bryonia, Cocculus, Ignatia,* and *Sulphur.*

As a general thing, treatment may commence with *Pulsatilla* and *Lachesis.* One dose, — six globules, — of *Pulsatilla* may be given for four days, then omit all medicine four days ; then give Lachesis in the same manner. If the symptoms abate, wait as long as the improvement continues ; if they do not abate, repeat the remedy as before, or select a new one.

It is very important in these cases to pay strict attention to the dress, diet, and exercise. The diet should be light and easily digested. Everything of a stimulating nature, unless otherwise ordered by the attending physician, should be studiously avoided. Daily exercise in the open air, either by walking or riding, will be found highly beneficial. The clothing should be warm and comfortable, and changed to suit the weather.

CHAPTER II.

DISPLACEMENTS OF THE WOMB.

GENERAL REMARKS. — It is my desire, in this article, to impart a more definite idea of what uterine displacements are, than that possessed by the generality of females. Indeed, very few females know anything about the womb, — its size, shape, position, or even location within the body. Every one has heard, it is true, of falling of the womb, but where it fell from, or what made it fall, few have the remotest idea, — not as much, indeed, as they have in regard to the fall of man. Why Eve ate the apple, when tempted by the serpent, is necessarily a mystery, and always must remain so; but why a womb can or does turn topsy-turvy, or sink down into the pelvis, need not be a mystery; because a very trifling knowledge of anatomy and physiology which can, with but little labor, be acquired by any one, will explain it all, and make it perfectly clear and plain to even the most obtuse understanding. . Many physicians studiously avoid explaining to their patients the nature or extent of their maladies; they prefer blind faith and obedience to their behests, rather than an intelligent and willing compliance, when a small amount of reasoning would not only convince the patient what was the nature of the indisposition which had reduced her from perfect health to a bundle of aches and pains, but would cause her gracefully to yield to the proposed course of treatment, — which at times is anything but agreeable, — necessary to reëstablish a healthful condition of the parts diseased. For my own part, I prefer treating my patients as rational, reasoning beings; and, when I can make them understand, by appeals to their common sense, by illustrations, or, when possible, by actual demonstration, that I perfectly comprehend their disorder, and can assure them that the disease is curable, I think I have gained a great deal; I have won their confidence, and removed from the mind that pernicious impression, which, in most diseases, does more toward retarding a cure than all else put together, that they are to live and to suffer year after year, by a disease which banishes all their charms, their splendor, and their gayety, and which, eventually, is destined to bring them to an untimely grave. As I have before frequently remarked, I do not expect, nor do I desire, to make anatomists, or physiologists, or practising physicians of all who may do me the favor to read these pages; but I do wish to impart such a knowledge in regard to diseases, and especially that class of diseases which affects the uterine organs, that the patient may in a measure so understand her malady, as not to expect a spontaneous cure, and, in the fond anticipation of soon receiving so desirable a boon, allow her fastidious delicacy to prevent her from applying to a physician for relief, until her disease becomes so firmly established that its removal will be difficult and tedious, or even, perhaps, impossible.

The difficulty of treating uterine disease is greatly enhanced, by the unwillingness on the part of the patient to apply for relief early, and by the reluctance, which is always present and natural enough, to a thorough investigation of the disorder.

Now, every woman who complains of pelvic pain has not a prolapsion, although she may think she has; and, if she has complained of these pains for a long while, it is quite likely that her physician has himself got in the habit of speaking of her complaint as a prolapsed uterus. I contend that it is absolutely impossible for a physician to say, positively, whether a woman has or has not a dislocated womb, simply from her enumeration of her aches, pains, and *all-gone* feelings. No sane man would think of rendering an opinion in regard to a supposed luxation of a joint, however much the patient might complain of feelings indicative of such a condition, without first making a complete and thorough examination of the part.

Physicians are in many instances not permitted to institute the only inquiries which can possibly reveal the precise nature and indications of the case. Even in those few cases where the patient has consented to an investigation, which alone can form the basis of a rational diagnosis, there often remains more or less obscurity, because the physician, in too many instances, feels a delicacy in making free use of all, or in employing every means of research in his explorations, and in repeating those explorations sufficiently often.

4

There are plenty of physicians, who, knowing the host of objections sure to be met with, almost never think of proposing an examination to their patients, but go on treating them in a haphazard manner, never effecting a cure, and scarcely ever expecting one ! Now, a physician who prescribes for his patients without knowing what is the matter (and how can he know, if he is not permitted to make the necessary inquiries?) is like the hunter, who shoots up into a tree, hoping he will kill a bird, because he knows there is one there, for he heard it chirp. Such a physician is more mischievous than the disease left to its own natural tendencies.

Please do not understand me as arguing the necessity of an examination in every case of backache or pelvic pain, because I am decidedly of the opinion that not more than half those who complain in this way do really suffer from uterine deviations or displacements. But where, in the estimation of the attending physician, it is necessary, and especially when the patient has been ailing a good while and treatment has proved ineffectual, I say she is a fool who refuses to submit to each and every proper mode of investigation which can lay bare the obscurities and difficulties that embarrass the decision and action of the physician. But he is a greater fool who consents to commence, or continues in care of, a case which he has no reasonable hopes of curing, because he does not understand it. And he cannot understand it, neither can he think he does, because no intelligent, reasoning man will ever consent to rest his judgment upon a ground-work of guesses.

The many painful sensations experienced within the pelvis are but the expressions of some disordered condition. What that condition is, can only, in many instances, be decided by an examination. It may be that the case is one of true prolapsion, and that the pain arises from the pulling or stretching of the nerves caused by the sinking down of the organ ; or, as in many instances, the aches and pains which so resemble those of the true disorder, and which lead astray so many who reason à priori, are purely neuralgic, and entirely unconnected with any deviation or displacement.

The principal reason why so many females fail to receive permanent benefit from their family physician, and finally resort to some empiric, or to some one who makes it his special business to treat this class of diseases, is, that they but half relate their case to him, who above all others should be their confidant in matters of this kind. True they frequently complain, and receive a prescription for their complaints; but it is all done in an incidental sort of way, perhaps while the physician is attending some other member of the family. He prescribes without giving the case much attention, and perchance would never think of it again; but she, in a few days, finding no relief, mentions the fact and receives another prescription, and so on, mentioning her case every time the physician happens to be in the house, but never sending for him for herself, " because she is always complaining." After suffering in this way for months, perhaps years, she finally resolves that, as her physician can do nothing for her, she will consult one who has a reputation for curing complaints like hers. She does so, and the first thing demanded is an unconditional surrender of her fastidious modesty; an examination is at once instituted, which reveals the whole difficulty; a cure is commenced and rapidly completed. Now, if the family physician had been allowed the same privilege and received the same compensation, he would have located the disease and made a cure just as promptly as the stranger did. Some may say, " The physician is as much to blame as the patient; he ought to explain that the case needed investigation before a remedy could be given." I do not think so; the doctor is asked casually if he cannot give something for a weak back; of course he can, and does give something that affords temporary relief. He cannot, neither is it expected that he will, sit down and unravel what, no doubt, he foresees will prove a very knotty case before he prescribes.

I would advise every one who wishes to consult a physician, and especially in regard to diseases of the description we are now considering, and kindred cases, to make a business of it. Go, and in a formal manner relate your case. Do not button-hole a physician in the street or at a neighbor's, but call at his office, or have him call upon you, and whatever method of investigation he may think will best reveal the true nature and indication of the disease, you will do well to submit to in a quiet, sensible, becoming manner. If your physician be a man fit for the high calling of a missionary of health, filled with respect and compassion for the afflicted, you need have no fear of being subjected to any unnecessary or humiliating treatment.

PROLAPSUS UTERI, OR FALLING OF THE WOMB.

DEFINITION. — CAUSES. — For a better understanding of this common disorder, let us for a moment consider the anatomy of the organ. The uterus or womb, in its natural unimpregnated state, in females who have borne children, weighs about one ounce and a half. It is shaped like a pear flattened from before backward, and is from two and a half to three inches long by one and three-quarters wide at the top, and terminates below in the neck or mouth, which is about half an inch across.

It is situated in the pelvic cavity, between the bladder in front and rectum behind, resting upon the upper end of the tube of the vagina, and retained in this position by four ligaments which act as guys. These ligaments are about two and a half inches in length, and their sole duty is to retain the womb upright in the centre of the pelvis.

Prolapsion of the womb, or, as it is commonly called falling of the womb, is in reality a descent or sinking down of the organ. In some instances the displacement is but slight, scarcely noticeable, while in others it is so great as to permit the uterus to protrude through the external parts.

Properly speaking, this is not a disease of the womb, but rather of one of the supporting tissues. It will readily be observed, that although the looseness and extensibility of its connection allow the womb to float about, as it were, in the pelvic cavity, it will be absolutely impossible for it to become displaced, prolapsed, or retroverted while the vagina retains its natural dimensions and the ligaments of which we have spoken are but two and a half inches long. It is quite true, that in many instances of prolapsus uteri, other lesions of the organ, as inflammation, ulceration, etc., are associated with the displacement; but these are not the cause of it. Nor, will their removal replace the womb in its natural position. Women suffering from this complaint can usually state the exact date of its attack, or at least they imagine they can, and will lay it to a fall, to lifting a heavy child, or raising some heavy · weight, or a wrench of the body which caused a sensation of something giving way in the interior of the back. Perhaps it would not be quite true to say that this is all imagination, — that such a thing *never* did happen; still it is very doubtful whether a sudden jerk or strain would cause the womb to prolapse and *stay so.* It requires a long course of preparation so to destroy sufficiently the tone, and relax the walls of the vagina, and other supporting tissues as to allow of a true prolapsion. Menorrhagia, or a debilitating leucorrhœa, if allowed to continue for a considerable time, would almost certainly produce the disorder. Another occasion of it is a too early getting up after delivery or abortion, or a too early resumption of household duties. The stretching, relaxing, and other changes of the supporting parts consequent upon child-bearing, miscarriages, or difficult labors, undoubtedly predispose to all the displacements to which the organ is liable. Another cause, especially in females of a delicate constitution, is the incessant running up and down long flights of stairs, which some are compelled to do. Among the causes we may enumerate tightlacing, the weakening effect of cathartics and other drugs, the injudicious application of the bandage after confinement by thoughtless physicians, or ignorant midwives or nurses, and also large doses of ergot and other allopathic drugs.

SYMPTOMS. — The degree of prolapsion is no guide as to the severity of the case, or to the amount of pain attending it. I have known the slightest possible descent of the womb to produce most intolerable pain through the lower abdomen and back ; while other cases, where the organ has descended so low as to peer out at the genital fissure, produced no pain whatever. The pain in many instances is truly an abdominal neuralgia, excited, quite likely, by the depression.

There is generally more or less bearing down, or dragging sensation in the lower part of the abdomen, drawing from the small of the back and around the loins and hips, pressure low-down toward the lower parts, with a desire to make water, sometimes without ability to do so, "or if it does pass, it is reluctantly, and oftentimes painfully hot, — a sense of faintness, and occasionally a variety of nervous or hysterical feelings and alarms which almost overwhelm the patient. A pressure and feeling about the rectum, resembling a slight tenesmus, sometimes importunately demand the patient's attention, which, if she obeys, almost always end in unavailing efforts. The pain in the back is sometimes extremely distressing, while the patient is on her feet, and gives to her walk the appearance of weakness in her lower extremities. A benumbing sensation shoots down the thighs, especially when the woman first rises upon her feet, or when she changes this position for a horizontal one. In some few instances, the woman is .obliged to throw her body very much in advance, or is obliged to support herself by placing her hands upon her thighs, when she attempts to walk. But all these symptoms subside, almost immediately, if she indulge in a recumbent posture, and this circumstance pretty strongly designates the disease."

Though to seeming certainty, the above symptoms appear to indicate a true case of prolapsion, we are still in doubt as to the true nature and indications of the disease, and must ever remain so, until an examination of the parts involved has been made.

See *"General Remarks"* under head of *"Uterine Displacements."*

TREATMENT. — The majority of females who are troubled with falling of the womb, think it necessary that they should wear some kind of an abdominal supporter. As this is the only opportunity, perhaps, that I shall have, in these pages, to express my opinion in regard to these abominable contrivances, I may as well at once give vent to my ire, and denounce what I consider the most unscientific, anti-surgical, and unphysiological of all the inventions man ever yet devised. The very idea that to diminish the capacity of the abdomen — for this is the direct effect of all utero-abdominal supporters — would, in any way, tend to hold up the womb, is perfectly preposterous.

These belts and pads, however applied, offer no support whatever to the bowels; on the contrary, the squeezing to which they are subjected, forces them and the womb with them, down into the pelvis, where they ought not to be. Do not, I pray you, ever for a moment think of reverting to this expensive and nonsensical gearing, under the impression that it will afford you any relief, as it affords none.

Thank Heaven, that among all the names appended to the unblushing advertisements of utero-abdominal supporters, which are constantly thrust into public notice, you will find not one homœopathist!

Cases of prolapsion are constantly occurring under the observation of physicians, where the patient, suffering intense pain, can be promptly relieved by simply raising the womb up to its place. Now, an instrument that will keep it in place will make that relief permanent. The utero-abdominal supporter will not do it, but a *pessary* will. (A pessary is an instrument made of wood, ivory, silver, caoutchouc, or other material, and is introduced into the vagina affording direct support to the womb.) But it is an unpleasant and disagreeable thing to wear, and the relief so obtained is but palliative. Sometimes, we have to accept it as a necessary evil, and use it, as we would a splint upon a fractured bone, to hold the parts in position while a cure is accomplished.

It is very seldom indeed, however, that the homœopathic physician has to resort to anything but the specific power of his medicines to make prompt and perfect cures.

The remedies are, *Aurum, Belladonna, Calcarea-carb., Nux-vomica, Platina, Sepia.*

Belladonna. — When the following symptoms are present, pressure, as from a load in the lower abdomen, or, as if the contents of the abdomen would fall out; heaviness even in the thighs, crampy pains through the abdomen and pelvis, also extending down to extreme point of spinal column ; great sensibility and irritability; also when accompanied by leucorrhœa and menorrhagia.

Nux-vomica. — For congestion of the womb, with pressure downwards, especially when walking or after walking ; great heat and weight in the womb and vagina ; dragging, aching pain in the back, also from the abdomen down into the thighs. During the menses, abdominal spasms and headache; disposition to miscarriage ; menses too early, and too profuse ; leucorrhœal discharge of fetid, yellow mucus.

Sepia. — Irregular menstruation, too early, too feeble or suppressed; heat in the womb; pains in the back and abdomen, aggravated by walking ; frequent desire to urinate ; contractive, pressive pain in the abdomen, as if everything would be pressed out; colic before the menses ; itching, excoriating leucorrhœa, with a discharge of a yellowish or reddish water, or a mattery, fetid fluid.

The treatment, in most cases, may commence by administering one dose of *Nux-vomica* every four hours, and be continued for one week; during the next week no medicine should be taken, but the week following, the patient may take one dose of *Sepia,* night and morning. If the symptoms indicate *Belladonna,* that remedy should be preferred to *Nux-vomica,* at the beginning, and followed by *Sepia,* or *Calcarea-carb.*

Calcarea-carb. — This is an excellent remedy for persons of a weak or lax muscular system, or of a scrofulous habit, and especially when menstruation is exhaustive, too profuse, and too frequent.

Secale cornutum — is occasionally called for, especially when there is prolonged bearing-down, forcing pains ; profuse menstruation ; depression, lowness of spirits ; deficient contraction after miscarriage.

Other remedies, as *Mercurius, Thuja, Kreosote, Nux-mos, Stannum,* are at times called for.

Too high praise can hardly be awarded to *cold water,* when properly used in this disease. The most convenient and advantageous mode of application is the wet bandage, renewed two or three times a day. Frequent sitting-baths, of short duration, will be of great benefit.

At the commencement of the treatment, it is often necessary for the patient to maintain a recumbent position a large portion of each day, and in all cases it is absolutely necessary to refrain from all active exercise, sweeping, lifting, carrying a child; running up and down stairs is especially objectionable. The patient should strictly avoid taking cathartic or opening medicines.

Diet. — In all diseases which have a tendency to debilitate the patient, a good nourishing diet should always be prescribed. In this disease, a stimulating diet should be avoided. Coffee and tea are strictly prohibited.

In all cases, when possible, consult a homœopathic physician.

LEUCORRHŒA. WHITES.

Definition. — Perhaps there is no term in the whole catalogue of diseases more generally undefined than that of leucorrhœa. It is derived from two Greek words, which, literally translated, signify a white discharge. The general application of the term to all non-sanguineous, vaginal discharges, no matter what their character, and without any definite knowledge of the various diseased conditions which give rise to them, has led to much confusion and loose treatment.

When we come to speak of the causes of this disease, it will be seen how essential it is, in the treatment of the sick, that our conclusions should be based upon correct premises.

In popular phraseology, "Leucorrhœa," "Fluor Albus," "Whites," and "Female Weakness" are synonymous terms, and by them is understood a light, colorless discharge from the female genitals, varying in hue from a whitish or a colorless to a yellowish, light green, or to a slightly red or brownish, varying in consistency from a thin, watery, to a thick, tenacious, ropy substance, and in quantity from a slight increase of the healthy secretion, to several ounces, in the twenty-four hours. — *Copland.*

Leucorrhœa may occur at any period of life, from early infancy to old age, but it is most frequent between the ages of fifteen and forty-five. It seldom continues later than this period, except when the discharge has its origin in some organic disease of the womb. It may occur even before the first menses have made their appearance, especially in scrofulous subjects, and materially interfere with the full and free development of this important function. As a general thing, the leucorrhœal discharge is more copious at the time of the menses than at any other time.

With propriety the disease may be divided into acute and chronic; the acute being nothing more nor less than an attack of inflammation of the mucous membrane, lining the parts, from whatever cause it may arise. A large majority of the mild cases consist simply of a catarrhal inflammation, occasioned by taking cold. The chronic form is but a continuation of the acute; the inflammatory stage of the disease being neglected, or improperly treated, passes into the chronic, with ulceration of the neck of the womb.

CAUSES. — By many, leucorrhœa is looked upon as simply the result of "general debility;" and most patients, suffering from the disease, think, if they could only get something that would strengthen them, they would be cured. Under this delusion many a woman has dallied away valuable time in taking some one of the many strengthening bitters or universal panaceas advertised to rejuvenate waning humanity. This is a mistake: the debility is not the cause of the disease, neither is the discharge; they are simply the outward manifestation, or the result of morbid action going on in some portion of the uterine organism. What that disease is, or has been, which has given rise to these symptoms, can generally be ascertained by the nature of the discharge, and the peculiarities which each particular case presents. For instance, the discharges from the vagina are of three kinds, namely: mucus, purulent or mattery, and watery. And there are morbid conditions capable of producing each of these evacuations. Inflammatory action is not necessary for the secretion of mucus, but it is for the secretion of pus, etc. It is very necessary that we should ascertain definitely the cause of the various forms of the malady, that we may be able to treat it intelligently, and not blindly.

For the most part, leucorrhœa has been looked upon and treated as a vaginal disease. With the very limited knowledge of its

5

pathology which most physicians possess, it is no wonder they have fallen into the habit of treating it upon routine principles, and that the success of that treatment has been anything but flattering. Leucorrhœa, especially the inveterate forms of it, is not a disease of the vagina, but in truth a disease of the womb; and he who attempts to treat it without taking into consideration the condition of this organ, will make a miserable failure, as, indeed, is done every day.

Wherever I have had an opportunity to make a thorough examination of these cases, I have found nine out of every ten to consist of congestion, inflammation, or ulceration of the neck of the womb; and when these were removed, the discharge soon ceased, and "general debility" tarried but a little while.

That leucorrhœa is a hereditary disease can hardly be questioned. Perhaps, however, it would be more proper to say, that many females — those, for instance, possessing a lymphatic, nervous constitution, with soft flesh and pale skin — are *hereditarily predisposed* to uterine affections. With such, a cold, errors in diet, nightly dissipation, tight-lacing, or other indiscretions would result in leucorrhœa, or some kindred disease; while other women might commit the same imprudent acts, but, being differently constituted, would be affected in a totally different manner.

Among the exciting causes of the disease may be enumerated cold, sitting upon very cold seats, upon stones, or upon the ground exposure of the neck and shoulders to cold air, violence, excessive indulgence, tight-lacing, irritation from stimulating injections, inflammation of the rectum, hæmorrhoids, miscarriages, abortions, uterine displacements, ulceration of the womb, tumors of various kinds, purgatives and emmenagogue-medicines, which are intended to promote the menstrual discharge, warm injections, abuse of warm baths, late hours, improper articles of diet, fish, crabs, lobsters, oysters, acid and watery fruits, the excessive use of tea and coffee, depressing passions, chagrin, grief, etc.

Leucorrhœa is quite common in cold, damp climates, and among that class of people who are compelled to live in narrow lanes and alleys, and in basements, where the atmosphere is damp and loaded with noxious matter, exhaled from dirty streets. However, it is by no means confined to this class, and perhaps it is quite as common among the highest strata of society, where over-indulgence in luxurious habits, sleeping upon downy beds, sitting upon soft cushions, living in houses tempered to suit the delicate susceptibilities of these favored ones of earth, have exhausted their vitality and energy, and rendered them capable of raising as luxuriant a crop of uterine diseases as those who live and suffer so far from them at the other extreme of society.

A leucorrhœal discharge is not unfrequently produced in young female children by the presence of pin-worms, which find their way from the rectum to the vagina. In these cases, a removal of the worms is speedily followed by an abatement of the annoyance. It is always well, when little girls are troubled with a discharge or an itching of the parts, to make a close examination of the vagina. I have frequently known children to be kept awake, night after night, from the itching occasioned by two or three little worms just within the lips of the vagina. They can be easily removed with a little piece of cloth, after separating the lips.

SYMPTOMS. — In some rare instances, the vaginal discharge is about the only symptom complained of. As a general thing, however, the disease is attended with quite a long list of aches and pains. In fact, the constitutional symptoms, in many cases, are so plainly marked as to be mistaken for the *cause* of the disease. Not a few will complain of pains here and there, with which they are afflicted; they will refer to the color of their skin, to their bloated faces, to their shortness of breath, to their constipation, to the obtuseness of their moral and intellectual faculties, their general debility, etc., etc., and say not one word about the vaginal discharge.

As before stated, the discharge varies in color, quantity, and consistency. At times it is a white or colorless fluid, scarcely leaving a stain upon the linen, but stiffening it, and coming off, under friction, in the shape of a fine white powder, or of little scales. At other times it is of a yellowish, greenish, or brownish tinge. It may vary in consistency from a thin, watery fluid, to that of milk or cream, or even thicker. In quantity, the discharge varies in different cases, and in the same case under different circumstances. At times, it is a mere exudation. Again, it is so copious that the patient is compelled to protect herself, as during her menstrual flow. Sometimes it is even more severe, and is evacuated in considerable quantities; indeed, there are cases where the leucorrhœal discharge amounts to a perfect flooding. In such cases, it seems as though the secretion accumulated in the cavity of the womb, and, during its accumulation, all the concomitant symptoms, especially the uneasiness and pain in the region of the womb, the aching in the limbs and joints, and the weariness are increased. When the discharge takes place, it affords marked relief.

In mild cases of catarrhal leucorrhœa, the discharge is mild or bland, and does not irritate or excoriate the parts; it therefore causes but little inconvenience, and is often scarcely noticed; but where the discharge depends upon a low grade of inflammation of the neck of the womb, or of the mucous lining of the womb and vagina, or where ulceration or other organic lesions of the uterus or its appendages exist, the discharge becomes acrid, corrosive, fetid, and of a brownish or greenish color, and excoriates the lips and adjoining skin, to the great annoyance of the patient.

If the attacks of leucorrhœa are but slight, and not too frequent, the general health of the female suffers but little. In acute cases, — those arising from colds, — the symptoms are those of catarrhal inflammation; there will be a sense of heat and soreness of the parts, with a feeling of weight or heaviness, or of a bearing-down pain, with languor and a general feeling of weariness. These symptoms are at times accompanied with slight chills, and are followed by pain in the back, quick pulse, thirst, high-colored urine and other febrile symptoms. Soon after these symptoms appear, the discharge manifests itself, and, as it increases, the symptoms abate. If, at this stage of the disease, proper treatment is instituted, there will be very little difficulty in controlling it, while the discharge and other symptoms will gradually subside.

If improperly treated, and especially if astringent injections and cathartic medicines are made use of, the disease will almost certainly pass into the chronic disorder; the languor and debility will increase; the discharge will continue or become more copious; pain and a sense of heaviness in the abdomen will be complained of; digestion will become impaired, nausea, loss of appetite, headache, vertigo, palpitation of the heart, weariness upon the slightest exertion, and a host of dyspeptic symptoms will soon manifest themselves. Ultimately, the disease extends to and into the womb; congestion, inflammation or ulceration takes place; the tissues become relaxed; prolapsus uteri soon follows, with a general increase of constitutional disorder.

After a season the patient grows thin, the pulse quick and small, the tongue coated, and either dry or pasty, often flabby and indented with the teeth; a constant aching is felt in the small of

the back, and especially low-down between the hips; great exhaustion after the slightest exertion, and general debility; eruptions of small, black-headed pimples appear upon the forehead
and face; the face becomes thin, or bloated and pale, the eyes
sunken and surrounded by a dark circle. The intellectual and
moral faculties are always more or less weakened.

Much experience and great care are often necessary to determine the nature of the disease, and its exact location, which give
rise to vaginal discharges. Where the discharge is purulent,
bloody-colored, or fetid, the disease occasioning it is evidently of
a grave nature, and should receive prompt attention from a skilful man.

Simple leucorrhœa is scarcely ever, in itself, a serious disease;
but as there are so many diseases of the uterine organs which
produce a discharge simulating that of leucorrhœa, it is best that
in these cases the patient should be under the care of an intelligent, practical physician. It is a disease, which, at best, is difficult to cure, and the longer it continues, the more obstinate it
becomes; however, a well-directed and persistent course of treatment seldom fails to afford permanent relief.

TREATMENT. — To obtain satisfactory results from remedies in
this disease, it is necessary that the patient, physically and mentally, should be placed in a favorable condition. All the surrounding circumstances which may in any way tend to excite or
aggravate the disease should be promptly removed. Late suppers,
and all dissipations, whether bodily or mentally, must be avoided.
The diet should receive strict attention; the food should be nourishing, as little stimulating as possible, and be taken at regular
intervals. Tea and coffee with all acid and watery fruits must be
abandoned.

Moderate exercise in the open air will be of great assistance in
promoting the cure; care, however, should be exercised to avoid
fatigue. The clothing should be so adjusted as to admit the
freest motion of the body, and about the waist should be especially loose. All exciting and depressing emotions ought, as far
as possible, to be avoided. If the patient resides in a low, marshy,
or damp, unwholesome region, she should, if convenient, be removed to a dry and open country. This, I am aware, in many
cases will be impossible, but whenever it can be done it is strongly
advised.

When these requisitions are complied with, the following remedies will be found efficient in arresting the disorder in by far the great majority of cases : —

Pulsatilla. — When the discharge is thin and acrid, excoriating the parts at times, with swelling of the vulva, or when the discharge is thick, like cream, and attended with crampy or cutting pains in the abdomen. Leucorrhœa, occurring during pregnancy, or during and after menstruation. *Pulsatilla* is especially applicable to females of a mild disposition, with soft, muscular system, light hair, and pale skin, and subject to menstrual derange ments.

Sepia. — Especially indicated in sensitive and delicate females. The discharge is yellowish or greenish, sometimes mixed with matter and blood, and often fetid, more or less acrid, and attended with stitches in the vagina, and with burning pain and soreness of the parts ; leucorrhœal discharge of a yellowish or reddish green water ; itching of the external parts, with redness and soreness ; falling of the womb.

Alumina. — Leucorrhœa after the menses, profuse discharge of transparent mucus during the day, stiffening the linen ; corrosive leucorrhœa, producing heat, soreness, and itching of the vulva ; between and after the menses discharge of bloody water ; profuse leucorrhœa just before and after menstruation.

Calcarea-carb. — Leucorrhœa before the menses, itching, burning leucorrhœa, with milky discharge at intervals, or when urinating. Especially applicable to females of light complexion, with pale skin, lax fibre, and sluggish circulation, and those who are troubled with too frequent and too copious menstruation.

Kreosotum. — Whitish, acrid leucorrhœa, attended with great weakness ; discharge of blood and mucus from the vagina on rising in the morning ; excessive pains in the small of the back ; falling of the womb ; smarting and itching of the external parts.

Nitric Acid. — For fetid, brownish leucorrhœa, mucus and greenish, or flesh-colored leucorrhœa.

Mercurius. — Purulent corrosive leucorrhœa, discharge of flecks of mucus and matter from the vagina, with smarting.

Cocculus. — Watery, bloody leucorrhœal discharge during pregnancy ; scanty menses, with leucorrhœa between the periods ; leucorrhœa instead of the menses.

Conium. — Smarting, burning, acrid, excruciating leucorrhœa, preceded by pinching colic, and lameness in the small of the back, and excessive itching of the external parts.

Sulphur. — In stubborn cases, slimy or yellowish, smarting, excoriating leucorrhœa, preceded by colic.

Silicia. — Leucorrhœa like milk, acrid, excoriating leucorrhœa, with itching of the pudendum.

ADMINISTRATION OF REMEDIES. — Of the selected remedy, give six pills every morning and evening until five doses have been taken. If the case does not improve, give a dose of *Sulphur* and omit four days, when the remedy may be repeated. If this affords no relief, or if the symptoms assume a new form, the remedy may be changed.

Most physicians speak highly of water as a remedial agent in this disease, and experience has taught me that too high encomium can hardly be awarded to it when judiciously employed. Under the head of *causes* of leucorrhœa, it will be remembered, I asserted that the disease was not unfrequently occasioned by the use of water. This is true. Some over-fastidious females are not content with cleaning their external person, but think it necessary to syringe themselves out frequently with tepid or cold water. Vaginal injections during health are not only uncalled for, but are positively injurious. I will explain how they are so. In its natural, healthy condition, the lining membrane of the vagina is kept constantly moistened by a mucous secretion. Now an injection, even of simple water, washes away this secretion, and leaves the surface dry, in a condition easily irritated and prone to disease. Water, however, is not the only injection made use of, and cleanliness not the only pretext for its use. But upon this point we have already spoken in another chapter. The object now is, simply to protest against the use of injections in health.

The use of vaginal injections in leucorrhœa is of great importance. In those cases where the discharge is acrid and excoriates the parts with which it comes in contact, I have found injections of water especially beneficial. The water dilutes the secretion, and thus renders it less irritating; besides, it has a tendency to re-establish a healthful action in the parts. I have known many cases where prompt recovery has speedily followed the use of simple injections of cold water. Cold hip-baths are also beneficial.

I have treated some severe, obstinate cases with injections of a decoction of *Hamamelis Virginica.* It can be obtained from any botanic drug-store, and a tea made with but very little trouble. Before using it, the vagina should be thoroughly washed out with injections of warm water, or warm castile soap-suds. It may be used three or four times a day.

UTERINE TUMORS.

It is unnecessary in a work like this to say anything in regard to the various kinds of tumors, indurations, contractions, and cancers which may affect the womb and vagina. They can be recognized by an experienced physician only.

CHLOROSIS OR GREEN SICKNESS.

DEFINITION. — This disease is almost exclusively peculiar to young ladies. As a general thing, it manifests itself at about the period of puberty, and is characterized by a pale, yellowish-green countenance, deficient warmth, perverted appetite, with occasional nausea or sickness, great physical and mental weakness, impaired digestion, palpitation of the heart, and general derangement of the sexual function.

CAUSES. — It has been asserted by some, and maintained with a good deal of pertinacity, that chlorosis is essentially a disease of the uterine organs, amenorrhœa. I am inclined to think, however, that the uterine disorder is rather a coincident effect of the same diseased condition or state that produces the chlorosis, — perhaps a defective energy of the nervous system, a lack of vital force. However this may be, we are well aware that in chlorotic patients we not only find retained or deranged menstruation, but we also find all the organic functions — those of digestion, assimilation, sanguification, and nutrition — inadequately performed. Therefore, there would be just as much propriety in calling the disease dyspepsia as amenorrhœa.

We find the disease most frequent in young girls about the age of puberty, and especially among those who have led a sedentary life. Those who work in crowded and ill-ventilated manufactories, and who sit constantly at the sewing-machine, or follow any employment which requires a stooping position, and particularly when they have been put to such trades or occupations when very young, before the frame is fully developed.

Those who possess feeble and delicate constitutions, or who reside in damp, unwholesome localities, and those who have insufficient, unwholesome, or innutritious food, as well as those who drink largely of tea, coffee, diluted acids, herb teas, bad wines, and are addicted to too great indulgence in warm bathing, to tight-lacing, excessive sleeping or watching, are *predisposed* to the disease. In fact, anything which tends to debilitate and relax the system *predisposes* to this disease.

The *exciting* causes are, disturbing emotions, troubles of all kinds, and especially unrequited love, or unfortunate and imprudent attachments, homesickness, depression of spirits, long-entertained feelings of anxiety or sadness, etc.

SYMPTOMS. — The groups of symptoms which characterize this disease, do not come upon the patient all at once, but manifest themselves gradually, insidiously, almost insensibly. The patient first complains of general lassitude, and has a great aversion to either mental or physical exertion; she loses her complexion, becomes emaciated, has no appetite, or if any, only for such things as are unwholesome, and has a longing for such substances as chalk, slate-pencils, or charcoal. The tongue is generally coated white, and there is a pasty taste in the mouth, especially on rising in the morning; the breath is offensive. The least bodily exertion causes shortness of breathing, and palpitation of the heart. Sleep is disturbed and unrefreshing. The bowels are constipated; sometimes there is nausea and vomiting; the pulse is small and frequent; there is a desire to be alone. She has frequent fits of weeping, is sad and sighs frequently. The menses are either retarded or scanty, and of a pale color.

If the disease continues, all these symptoms grow worse, and new ones manifest themselves. The countenance becomes more and more pale, and presents, in a marked degree, the characteristic feature of the disease, — the greenish yellow tint. The lips and gums, and whole interior of the mouth, become pale; the eyelids livid and swollen; the white of the eyes looks almost chalky; there is a sad expression about the face; the muscles are flaccid, the extremities cold, and the ankles swollen. The appetite becomes more and more capricious and vitiated; various dyspeptic symptoms become troublesome, — heartburn, sour stomach, pain in the stomach, accompanied with nausea and vomiting, especially in the morning. The patient experiences great difficulty in breathing, and palpitation of the heart upon the least exertion, especially on going up-stairs. The menses, when they have appeared, gradually become scanty, and are attended with an unusual amount of pain; they continue but a short time, are pale and watery, recur at long periods, and finally cease altogether. The abdomen often becomes tense and swollen; so much so, that not unfrequently the patient is accused of being pregnant. The veins of the skin are pale, and never distended as in health; the blood is thin and watery; frequently there is severe pain under the left breast, or pain through the chest, with slight cough and hectic fever; sometimes the cough is attended with expectoration of small clots of blood, — symptoms which have all the appearance of a rapid decline. During the course of the disease, various hysterical complaints manifest themselves.

Chlorosis is not unfrequently mistaken for disease of the heart, or for tubercular consumption.

TREATMENT. — From what has already been said about this disease, you may well infer that it always needs the attention of a skilful physician; for, if neglected or improperly treated, it will *surely* destroy the health and happiness of the patient. We shall here simply draw the attention of the reader to a few prominent remedies appropriate to the disease, and especially those whose early application will tend to arrest its progress.

At the outset, before much else can be done, we must, if possible, ascertain and remove the exciting cause. Perhaps in no other disease does change of occupation, of climate, and of scenery produce such salutary results, as in chlorosis. A complete change should be made in the patient's whole existence. If she leads a sedentary life, change it to an active one; if she is confined to school, and close application to her books, break off her studies; postpone her education; send her to the country; give her plenty of active out-door exercise; institute the frequent use of cold bathing. The mental and moral causes, which perhaps are the most important ones, are the most difficult to remove. New scenery and new friends will go a great way toward erasing old attachments. But for the evils of unrequited love, or for a lacerated heart, the cool advice of the medical man is far exceeded in value by that which gushes forth from the warm heart of a loving mother. "We come to men for philosophy, to women for consolation."

For those who sit from early dawn till late at night in some close, ill-ventilated factory, or those who from choice or necessity work all day long, and especially if their occupation compels a stooping position, an absolute radical change is necessary before medical treatment will be of any avail. A change of employment *must* be made.

From the list of remedies for this disease, we will mention *Pulsatilla, Calcarea-carb., Ferrum, Bryonia, Sepia, Sulphur, Lycopodium,* and *Belladonna.*

Pulsatilla. — Especially for females of a mild, easy disposition, given to sadness and tears. The symptoms indicating this remedy are: sallow complexion, alternating with redness and flushes of heat; frequent palpitation of the heart; great difficulty of breathing, with a sensation of suffocation after the least exertion; weariness, and heaviness of the legs; cold hands and feet; looseness of the bowels; pressure and heaviness in the abdomen; nausea; vomiting; chilliness; swelling of the feet; frequent headache, especially upon one side; buzzing in the ears, with pains which extend to the teeth, and frequently fly from one side of the face to the other; acrid, burning, or thick, painless leucorrhœa, like cream.

Bryonia — May be given in alternation with *Pulsatilla*, when there is frequent congestion of the chest; bleeding from the nose; constipation; coated tongue; flushes of heat; chilliness; cough, with expectoration of clots of dark, coagulated blood, and especially when *Pulsatilla* affords but partial relief.

Ferrum. — Especially when there is great debility; constant desire to sit or lie down; want of appetite; nausea, and hectic cough; dropsical swelling about the eyes; swelling of the extremities; want of vital heat; difficulty of breathing, as if from contraction of the chest; palpitation of the heart; extreme sallowness of the skin; the lips very pale, looking almost bloodless.

Sulphur. — For obstinate cases, and especially, though apparently well indicated, if the above fail to afford relief; also, when the following symptoms occur: pain in the back of the head; throbbing pains in the head; humming in the ears; great depression after talking; difficulty of breathing, with sense of weight in the chest; constant drowsiness in the day-time; pressure in the abdomen; voracious appetite; sour eructations; emaciation; constipation of the bowels, with hard stool; sensitiveness to the open air. Suitable for irritable persons, or those inclined to sadness and crying.

Calcarea-carb. — This remedy is suitable after *Sulphur*, especially when the difficulty of breathing is very great, and there is excessive emaciation; weariness and heaviness of the body; palpitation, nausea, and vomiting; desire for wines, salt things, and dainties.

Lycopodium. — Especially when *Calcarea* affords but partial relief; also, where there is obstinate constipation, extreme languor and cough, with tendency to consumption.

Belladonna. — When there is a pressing or bearing-down pain, as though the internal parts would fall out, with or without leucorrhœa, scanty and painful menses, preceded by colic.

China. — When the disease occurs after a severe fit of sickness, or after hemorrhages.

All patent or forcing medicines, which are recommended to bring on the menses, under the impression that, when they appear, the patient is cured, should be strictly avoided, as numbers of females have ruined their health forever, and shortened their lives considerably, by resorting to such quackery.

ADMINISTRATION OF REMEDIES. — Of the selected remedy, give five or six globules, dry, upon the tongue, once in six hours. As soon as 'improvement sets in, lengthen the interval between the doses to twelve hours ; and, as it continues, to two or three days.

If preferred, you may dissolve twelve globules in six spoonfuls of water, and of the solution take one spoonful at a dose.

DIET AND REGIMEN. — The diet should be perfectly plain, and of the most nutritious kind ; all stimulants and condiments of every description should be strictly avoided. Coffee, green tea, and all spirituous and malt liquors are objectionable.

The directions in regard to mental and physical exercise, together with strict attention to diet, are indispensable to a successful treatment.

CHAPTER III.

PREGNANCY.

GENERAL REMARKS. — The woman whose privilege it is to bear within herself a human being occupies a high position in the scale of humanity; within her has started a ripple, which may continue as a wave upon the great ocean of life, itself producing and reproducing others like unto it, until its ever-widening circles are lost from human gaze in the greater ocean of eternity itself.

It is said, that not a single sound that was ever made has ceased to reverberate, that it continues and continues on, ever ending, ever beginning, always changing, yet immutably the same. It is also said, that every little pebble thrown into the ocean disturbs the whole waters of that mighty deep, that the little circlet of waves thus set in motion is ever widening, ever breaking upon distant shores, ever returning, but never ending. If this be true, that slight impressions, thus made upon material substances, produce such great results, who can estimate the influence of a single individual upon the generation in which he lives, or, in fact, upon all generations which are to follow him.

It is very easy for any of us to look back upon times that are long passed, and pick out prominent characters, the marks of whose influence we can now trace upon individuals, in our own immediate circle of friends. We all, no matter how obscure our positions, in journeying through life, make impressions upon those with whom we come in contact, either for good or evil, which are never effaced; not only that, but these very impressions are transmitted to others, and to others still, and, though originating with us, may affect thousands yet unborn.

Thoughts like these lead one to contemplate the responsibilities of life, and when applied to the mother, existence itself must assume a more sacred import, as she gazes upon her new-born babe, that miniature human being, whose moral and physical form she herself has moulded and stamped with the imprint of her own nature. She, certainly, to a great extent, must be responsible for its future life. Springing from herself, — an almost imperceptible germ, — and growing day by day upon the very life it drew from her, it has attained a body which is a part and parcel of her very self. This close connection could not have, has not existed without their lives being one. The infant's mind, its disposition, its habits, its loves, and its hates, have been formed by the parent in the exact image and likeness of herself, and undoubtedly, were the child kept *exclusively* in the society of its mother, away from the influence of all other persons, it would grow up, at least as far as its general disposition and mental characteristics go, the exact counterpart of her who gave it birth; as it is, even in spite of all surrounding influences, — the society of other children, the precept and example of adults, education at school and in society at large, — we still see prominently exhibited, during the whole period of life, the same habits of thought and action which were peculiar to the parent.

It is true, in the highest possible degree, that the habits of mind, the impulses and emotions of the mother during pregnancy, do have a direct and powerful influence upon her offspring, and to such an extent, too, that it is in the power of the mother to determine, at least in a great measure, what class of passions shall have predominance in the souls of her children.

It would be hardly natural to suppose that a female who, during pregnancy, was habitually irritable, passionate, and sour-tempered, would give birth to an infant, sweet-tempered and amiable; in fact, any of us would about as soon expect to obtain sweet fruit from a sour apple-tree. Some one has said, " Show me a child, and I will tell you the disposition of its parents." Scripture has it, " By their fruits ye shall know them." Any of us meeting a matron of gentle and winning disposition, one in whose mental composition all the faculties are well balanced, and governed by a pure morality, would expect, on being introduced into her family, to find the same commendable qualities exhibited in her children. And why? Because observation has taught us that the general disposition and mental characteristics of the parent descend to the infant.

We naturally suppose, besides, we have ample proof in support of the supposition, that the physical composition of the infant resembles that of the mother, during her pregnancy; we know, that if the parent be suffering from any constitutional disease, it is transmitted to the child. Now is it not just as correct to suppose that habits of mind, the impulses and emotions of the mother, are also transmitted to the infant? Cases are by no means rare where the excessive anxiety or sadness of the mother, during the period of gestation, is shown in the after-life of the child. "By the unalterable decree of the Divinity, impressions indulged by the mother during this period, as they are received by her own highly impressible and delicate organization, are conveyed from each of those organs to the corresponding organs of the child she bears, and she is thus forming for good or for evil, for virtue or for vice, one who is hereafter to be her happiness or her misery, her honor or her reproach." — SMALL.

We can scarcely over-estimate the importance of maintaining a proper state of the mind and feelings during this period; not only intellectually, but morally and socially, the habits and condition of the mind of the mother are important to the future welfare of the child. Violent anger, terror, or jealousy, seldom fails to produce unpleasant effects, and the consequences are sometimes alarming, and even fatal. Abortion is not unfrequently the result of mental excitement.

Most women are very particular to avoid all unsightly objects, for fear of producing some physical deformity in their child. In my estimation, this popular precaution has very little, if anything, in truth to support it. It is against the production of mental, and not physical, deformities in their offspring, that pregnant women should be most on their guard.

Gloomy and harassing thoughts and impressions should be most sedulously guarded against. I am well aware that during this period, owing to the unusual irritability of the system, the difficulty of controlling the feelings becomes greatly enhanced. But then a greater effort should be made to confine the thoughts and attention to the beautiful and true. Every means should be taken to procure a healthy and vigorous tone of mind. Cheerful conversation, pleasant friends, agreeable books, music, household duties, and out-door exercise, should all contribute their share to promote comfort and enjoyment.

There is no period, perhaps, in the whole life of a female when she stands so much in need of sympathy as at this interesting period, and especially the sympathy of him who has sworn to protect and love her, in sickness and in health. She needs the sooth-

ing care of affection, and the strong arm of love, to smooth over
the rough places of life, and to surround her with every comfort;
to encourage her cheerful moods, and banish her gloomy emotions;
in a word to make to her the foreboding path which she has to
travel as smooth and easy as kindness, patience, and affection can.

I do not mean that every nonsensical whim should be gratified, —
that she should be surrounded with all the luxuries that her morbid imagination may suggest; she will have many *longings* for
strange, and sometimes unheard-of things, and will endeavor to
make herself and others believe, what perhaps she has heard from
those who are old enough to know better, that, unless these fancies
are indulged, evil will result to the child. This is sheer nonsense,
and, were a woman's mind engaged as it should be, such things
would never enter her head. To gratify such whims tends only
to increase the trouble. If she finds that, by longing for a thing,
she can have it, she will long for things you little dream of. She
will want strawberries in the winter, and if she cannot have them,
will imagine that her child will be marked with a strawberry upon
its face. The proper treatment, in such a case, is to occupy the
mind with other things, — cheerful employment, entertaining
books, and healthful exercise.

SIGNS OF PREGNANCY.

The diagnosis of early pregnancy is no easy task; it frequently
baffles the most experienced physicians; therefore, great care and
discrimination should be exercised before venturing upon a positive assertion. The general condition of a pregnant woman is
that of a full habit; the pulse is quicker and fuller, and the quantity of circulating fluid is said to be increased. There are various
and well-marked sympathies excited in distant organs. Variations in temper and disposition are of frequent occurrence. The
appetite is often capricious. The skin sometimes becomes sallow
or discolored in patches.

The deviations from the ordinary state of health in some particulars are quite remarkable, and constitute the special signs upon
which our conclusions must be based. Those we shall now notice
in detail.

CESSATION OF MENSTRUATION. — This is one of the earliest and
most unvarying signs of pregnancy. The non-appearance of the
catamenia at the proper time is one of the first circumstances
which leads a female to suspect that she is pregnant; and, if the

second period passes by, and they are still absent, it is deemed conclusive. But, strictly speaking, it is not conclusive, because menstruation may be arrested by various diseases; besides, menstruation may occur for some months after conception, or even monthly during the whole period of gestation. Nevertheless, although exceptions do occur, when menstruation ceases without any perceptible cause, the woman otherwise remaining perfectly well, we take it as pretty good evidence that conception has taken place.

MORNING-SICKNESS. — When combined with other symptoms, this one is of great value, but, when taken alone, there is not much reliance to be placed upon it, because it may be altogether absent, and yet the patient be pregnant; on the other hand, it may be present, as morning-sickness, from various causes, and yet the patient not be pregnant.

This irritability of the stomach, arising from sympathy with the uterus, commences soon after conception, and ceases soon after the third month.

SALIVATION. — Salivation is sometimes present, though not, I take it, as a general thing, as I have never observed it myself. This form of salivation differs from that induced by mercury, in the absence of sponginess and soreness of the gums, and of the peculiar fetor.

ENLARGEMENT OF THE BREASTS. — "About two months after conception, the attention of the female is attracted to the state of the breasts. She feels an uneasy sensation of fulness, with throbbing and tingling pain in their substance, and at the nipples. They increase in size and firmness, and have a peculiar, knotty, glandular feel; the areola — a colored circle round the nipple — darkens, and, after some time, milk is secreted. But it must be recollected that the breasts may enlarge from other causes; this happens with some women at each menstrual period when the catamenia are suspended, or after they cease, and at such times a milky fluid may be secreted." — CHURCHILL.

ENLARGEMENT OF THE ABDOMEN. — The gradual enlargement of the abdomen, taken in connection with the symptoms already enumerated, enables us to estimate with a good deal of certainty the period of pregnancy at the time the examination is made. During the first four months the entire womb is contained within the cavity of the pelvis, and, therefore, cannot be felt through the walls of the abdomen; but, soon after this time, it may be felt,

7

especially in thin females, just above the share-bone; about the fifth month, it reaches midway between the umbilicus and share-bone, and gives a roundness and fulness to the abdomen; about the sixth month, it is as high as the umbilicus, which it protrudes; during the seventh and eighth, it fills the whole abdomen, the intestines having been pushed above and behind it.

Distension of the abdomen, however, sometimes takes place from other causes than pregnancy; therefore, this sign alone is not sufficient to warrant us in pronouncing upon a case.

QUICKENING. — This term is applied to the first movement of the child within the womb, or rather, to the first perception of such movement on the part of the mother. Some women labor under the erroneous belief, that the child is not alive until the fourth month, — the time at about which quickening is first felt. The fact is, however, we have just as much reason to believe that the child is alive at the fourth week as at the fourth month. Indeed, we have no reason to doubt but that the child is actually alive from the *very first moment of conception.* Quickening takes place at about the fourth month after conception, though some women feel it earlier, and others not until the sixth or seventh month. " The sensation is at first like a feeble pulsation, and, though so slight, is often accompanied by sickness of stomach and faintish-ness, or even complete syncope. By degrees it becomes stronger and more frequent, until the movements of the different extremities are distinguishable." — FLETCHER.

PRESERVATION OF HEALTH DURING PREGNANCY.

We have already remarked upon the close connection that exists between the health and happiness of the mother during pregnancy, and the future disposition and well-being of the child. We have shown that the moral and physical condition of the infant *in utero* are intimately connected with the health and regularity of the parent's system; that moral and physical perturbations of any kind not only act injuriously on the female herself, but, through her, upon the infant, and impressions thus made are not transient, but affect the whole future welfare of the child. Therefore, during the interesting period of pregnancy, it becomes the female sedulously to guard against all accidents and circumstances which may have an unfavorable influence upon the delicate organization of the incipient being within her womb. And nothing perhaps can contribute more to the general good health and vigor of both body and mind, and thereby promote the happiness of the individual and the well-being of her offspring, than proper attention to *dress, exercise,* and *diet.* Upon each of these points we shall offer a few remarks.

DRESS. — Lycurgus, the great Spartan lawgiver, once ordained a decree that all pregnant women should wear wide, loose clothing. A similar law also prevailed among the Romans. The dress should be warm, loose, and light, during the whole period of pregnancy; and, at this day, were there such a law, and the proper power to enforce it, you would hear fewer complaints of "bad gettings-up," "fallings," "prolapsuses," "broken breasts," "weaknesses," and many other complaints which do so much to undermine the constitutions of married women. Let out your dresses early; no part of the dress should be tight; even garters should be abandoned; everything should be loose, so as to allow a free circulation of the blood. Tight-lacing is highly injurious: how can it be otherwise? While nature is gradually increasing the capacity of the abdomen to accommodate the steady development of the child, for the woman to compress her chest with stays, or to gird her abdomen about with skirts, it seems to me that any one with common intelligence must know cannot fail to prove injurious both to mother and to child. Special care should be taken that the dress is loose about the breast. This is highly important; for not unfrequently the breasts and nipples are so completely flattened out, by direct pressure, that, after confinement, there is nothing that can properly be called a nipple to be found. More will be said upon this subject, when speaking of the *Preparation of the Breasts.*

EXERCISE. — There is an impression among some people, that a pregnant female should carefully avoid exercise, especially during the early months. This is altogether a mistake. Experience has taught us that it is impossible for a pregnant woman to enjoy good health, unless she daily takes active exercise; passive exercise — such as riding out in a carriage — is not sufficient.

The daily household duties should be continued just the same as usual. Actual fatigue should be induced by continued, moderate action; not, however, to such an extent as to interfere with quiet sleep.

Quick and violent action should be avoided, as lifting of heavy articles, or great exertion in moving them; there should be no exposure to sudden strains, as in lifting or reaching; or to jar or falls, as in jumping, etc.

Those women who are actively employed during the whole period of gestation are seldom annoyed by any of that multitude of bad feelings, which are so apt to attend their state; and this, of itself, is a sufficient reason why a woman should never be idle, — should never have time to picture to herself the reality of any of those mishaps, to which her *particular friends* have informed her that females in her condition are peculiarly liable.

In addition, however, to the usual exercise of accustomed employment, active exercise should be taken in the open air for enjoyment. As the time advances beyond the fifth month, the amount of exercise and work should be gradually diminished.

DIET. — The diet should be well regulated. Whatever tends in any way to disturb the general health should be strictly avoided. Many females, during this period, have a morbid desire for things which, if indulged in, would prove highly injurious; producing dyspeptic and other troublesome symptoms, which, beside being a source of much suffering to the mother, may seriously affect the future health of her offspring. No specific rules can be laid down as to what a pregnant female should eat, because some can digest with ease what would produce in others serious gastric disorders. Each one, by using a little discrimination, can best regulate her own diet. She knows what formerly agreed with her, and will soon find out what agrees with her now. All she needs is the resolution to refrain from partaking of what she has reason to think, or what she has proven to be, injurious to her.

The diet should be simple, purely nutritious, generous, but not excessive; and everything medicinal or stimulating, such as all highly-seasoned food, spirituous, vinous, and fermented liquors, strong teas and coffee, should be for the most part discarded.

It used to be thought necessary, and with some people is still, that the bowels should be kept in a loose state during the whole period of pregnancy, under the absurd notion that it would make labor easier. Just exactly how castor-oil produced this desideratum, I have never seen stated; certainly I cannot think that they should imagine that oil poured into the stomach would lubricate the passage from the womb. However, it matters little: the notion is antediluvian, and long ago exploded: no one thinks of following such advice now, except the ignorant.

DERANGEMENTS DURING PREGNANCY.

Although pregnancy is a perfectly natural and perfectly healthy condition, yet, owing to the increased activity going on in the general system, we not unfrequently meet with disorders which are peculiar to this state. All females are not affected alike; some pass through the whole period of pregnancy without experiencing any trouble whatever, while others are not so fortunate.

We shall briefly treat of some of the deviations from health which are met with during pregnancy.

CONTINUED MENSTRUATION.

This is comparatively of rare occurrence, yet it is occasionally met with, and should receive attention. The following remedies will usually have the desired effect of arresting the discharge : —

Cocculus. — When there is severe spasmodic pain low-down in the abdomen.

Crocus. — When the discharge is dark and copious.

Phosphorus, Platina, and *Sulphur* are also serviceable remedies. For their special indication, and also for other remedies, see " PAINFUL MENSTRUATION."

HEADACHE AND VERTIGO DURING PREGNANCY.

Not unfrequently, pregnant women are very seriously affected with giddiness and pain in the head. With these headaches there is almost always a sense of dulness, and disinclination to active employment; sometimes there are nausea, dimness of sight, sparks before the eyes, palpitations, and nervous tremblings.

These symptoms are generally caused by nervous irritability, and whatever tends to derange the general health of course predisposes the patient to their frequent visitations. To prevent their recurrence, then, it is specially necessary that the patient should follow the strictest hygienic rules in diet and exercise, as well as avoid, to as great an extent as possible, all mental and physical excitement. It is not necessary to take a dose of medicine for every little ache or unpleasant sensation that may occur; the better plan, a great deal, is to abstain a little from the accustomed diet, to take a little more out-door exercise, and avoid excitement of every description.

When, however, this will not avail, and it becomes necessary that medicine of some kind should be taken, you can select from the following remedies : —

HEADACHE. — *Aconitum, Belladonna, Nux-vomica, Platinum, Pulsatilla, Opium.*

VERTIGO. — *Aconitum, Belladonna, Opium, Nux-vomica, Sulphur.*

SPARKS BEFORE THE EYES. — *Aconitum, Belladonna, Pulsatilla, Sulphur.*

SLEEPLESSNESS. — *Coffea, Ignatia, Hyoscyamus, Nux-vomica, Opium.*

SLEEPINESS. — *Opium, Pulsatilla, Nux-vomica, Crocus, Tartar-emetic.*

ADMINISTRATION OF REMEDIES. — After selecting a remedy, dissolve twelve globules, in eight or ten spoonfuls of water, and take of the solution one spoonful every three or four hours.

For congestive headache, or whenever the pain is very severe, a dose may be taken as often as every half-hour, or every hour.

MORNING-SICKNESS.

NAUSEA. — Most pregnant women suffer more or less from nausea and vomiting, especially on rising in the morning, and this is what is termed "morning-sickness." It is not an actual disease, but rather a sympathetic affection, and plainly illustrates the close connection existing, through the nervous system, between the womb and other organs and functions of the body. When morning-sickness occurs at the regular time, and in the usual manner, it is of great value as an evidence of pregnancy; that is, when it is combined with other symptoms which look to the same cause.

This irritability of the stomach may commence immediately after conception, but, as a general thing, it sets in about the fifth or sixth week, and ceases soon after the third month; sometimes, however, it continues, with but slight modification, to the end of pregnancy. The daily attacks, which commonly take place immediately on rising in the morning, last but a short time, — from ten minutes to an hour, — after which the patient completely recovers, and is able to take her breakfast. Occasionally, but not often, they return in the evening.

It is asserted by some physicians, and with a great deal of plausibility, too, that when vomiting is entirely absent, gestation does not proceed with its usual regularity and activity, because irregularities in this particular are most generally followed by deviations of other kinds.

TREATMENT. — Morning-sickness being such a distressing accompaniment of pregnancy, thankful, indeed, might we be, had we the remedies to offer that would afford prompt and permanent relief; but, as long as the *exciting cause* remains, we can at best expect to offer but temporary relief. In prescribing for these cases, it will be found that what will give to one instant relief often entirely fails with another.

Ipecacuanha. — When there is bilious vomiting; nausea and vomiting with uneasiness in the stomach; vomiting of undigested food or of drink; bowels loose or relaxed.

Arsenicum. — Excessive vomiting after eating or drinking, with attacks of fainting, great emaciation.

Nux-vomica. — Nausea and vomiting in the morning; or for nausea which comes on while eating, or immediately after eating; acid and bitter eructations; hiccough; heart-burn; sensation of weight in the pit of the stomach; constipation; irritability.

Pulsatilla. — When *Ipecacuanha* and *Nux-vomica* prove insufficient, and especially if the vomiting comes on in the evening, or at night; depraved appetite; longing for acids, beer, wine, etc.; diarrhœa, alternating with constipation.

Sepia. — Especially for vomiting of milky mucus.

When other remedies fail, recourse may be had to *Natrum-muriaticum, Phosphorus, Petroleum,* or *Sulphur.* Also *Kreosote,* in some cases.

ADMINISTRATION OF REMEDIES. — Dissolve of the selected remedy twelve globules in eight spoonfuls of water, and take, for severe cases, one spoonful every three hours. For ordinary cases, however, it will be as well to take six pills, dry upon the tongue, night and morning.

CONSTIPATION DURING PREGNANCY.

Pregnancy is frequently accompanied by a sluggish condition of the bowels, and especially in persons of a naturally costive habit. Constipation long continued gives rise to a train of symptoms, which sometimes prove excessively annoying; the appetite fails; digestion becomes difficult; the sleep disturbed; and the patient generally nervous and irritable.

The only treatment usually necessary to be instituted is a change in diet; a little more of vegetables or fruit; a little more exercise in the open air; a good drink of fresh cold water on rising in the morning, and perhaps two or three times during the day. The avoidance of all indigestible food, coffee, and other stimulating liquids, is often sufficient to remove the difficulty.

When, however, other assistance is required, you can resort to *Nux-vomica*, and take one dose night and morning, for three or four days. *Nux-vomica* is especially called for when there is dull headache; heat in the abdomen; frequent but ineffectual desire for stool. This remedy may also be given in alternation with *Opium*, especially where *Nux* alone fails to have the desired effect. Should both these remedies fail, and where constipation has continued for a long time, *Lycopodium* or *Sulphur* may be taken. *Bryonia*, *Alumina*, and *Sepia* are often of service.

If constipation has lasted three or four days, and there be frequent urging to stool, but inability to expel the fæces, on account of their hardness and size, recourse must be had to the injection of tepid water.

Under no circumstances should cathartics be given, even those of the mildest description. See " CONSTIPATION."

DIARRHŒA DURING PREGNANCY.

Diarrhœa in a pregnant female is an untoward symptom, and should be changed as soon as possible; otherwise, the health of the patient may suffer severely.

Sometimes there is a simple looseness of the bowels, which can scarcely be called a diarrhœa, where the movements are more loose and frequent than is natural, but not otherwise much altered in their appearance. The appetite remains good, and there is no general disturbance of the system. Such cases can generally be controlled by a proper regulation of the diet, by avoiding such articles as have a tendency to produce relaxation.

The form of diarrhœa, however, which is foreboding of evil, and which prompt measures should be taken to remove, is where the stools are liquid, dark-colored, and very offensive; the tongue is coated; there is little or no appetite; bad breath, and disagreeable taste in the mouth. In cases like this, the diet must be restricted to very small quantities of the very mildest kind of food. See " DIARRHŒA."

The remedies called for, are, *Lycopodium*, *Chamomilla*, *Sulphur*, *Pulsatilla*, *Dulcamara*, *Antimonium-crudum*, *Rheum*, and *Calcarea*.

Chamomilla. — Should there be violent colic, with yellow, greenish stools, or resembling stirred eggs.

Pulsatilla. — When the stools are watery or greenish, preceded by colic, with slimy, bitter taste in the mouth.

Dulcamara. — When the diarrhœa results from a cold, or getting wet.

For further particular indications, and for administration of remedies, see " DIARRHŒA."

PRURITIS. ITCHING DURING PREGNANCY.

Itching of the private parts. Perhaps there is no disorder to which a pregnant female is liable that causes her so much annoyance and trouble as this. It takes away all rest and sleep, and thus sometimes produces the most extreme debility. Some of the cases are, to an unimaginable degree, distressing; the itching is all the time not only constant and severe, but also attended with frequent exacerbations, during which it is so intolerable that the sufferer cannot resist the impulse to scratch, even to the extent of seriously wounding the surface.*

There is some little doubt as to what is the cause of this disease; in most cases, however, it arises from a vitiated condition of the mucous secretion of the parts. In others, it depends upon an aphthous eruption, or an eruption resembling the thrush of infants, when it is accompanied by a burning heat, with dryness, redness, and perhaps some swelling. This affection is not confined to the pregnant state alone, but may occur at any time.

The chief remedies for this disease are, Conium, Kreosote, Bryonia, Arsenicum, Rhus, Pulsatilla, Silica, Sulphur, Lycopodium, and Graphitis.

After selecting a remedy from the above list, dissolve six pills in six spoonfuls of water, and take one spoonful of the solution, every four hours; or the remedies may be taken dry upon the tongue, six globules at a dose. Give each remedy used a fair trial of three or four days before selecting another. Use the remedies in their order as given.

A very efficacious and simple external application is made by dissolving one ounce of borax — biborate of soda — in a pint of rose-water, or, if rose-water is not as handy, soft rain-water. The parts affected may be washed several times a day with this solution ; where the disease extends too far within the body to be reached otherwise, a female syringe may be used for applying the wash.

This wash, in connection with the internal use of one of the above-named remedies, will very speedily cure the disease in almost every case.

The utmost cleanliness should be observed, especially in those cases where the irritation is produced by a constant oozing of a thin, watery secretion. This secretion, when allowed to accumulate, produces an itching which is absolutely intolerable.

DYSPEPSIA. HEART BURN, DURING PREGNANCY.

Pregnant females often experience a great deal of discomfort from heartburn and acid stomach. This is often very distressing, and frequently commences immediately after impregnation. *Nux-vomica* and *Pulsatilla* are the principal remedies to regulate these derangements. Sometimes a slice of lemon, sugared and kept in the mouth, is salutary. Also frequent but small quantities of lemonade give relief. One drop of sulphuric acid in a tumbler of water is highly recommended. Pleasant sour or sub-acid apples and other fruits are often very effectual in this complaint.

Physicians now-a-days strongly object against using magnesia, limewater, charcoal, chalk, prepared oyster-shells, and the like, as they often produce other diseases, besides at best but absorbing the acid present in the stomach, and exerting no power in preventing its new formation. See " DYSPEPSIA."

HYSTERIC FITS—FAINTING—DURING PREGNANCY.

Females of a nervous, hysterical, or delicate constitution are frequently, especially during the early months of pregnancy, attacked with fainting, hysteric spells. They are occasioned by want of sleep, excessive fatigue, tight-lacing, warm-rooms, or a disordered digestion. The attack is usually preceded by a constriction about the throat, by sobbing, or repeated attempts at swallowing. Then the patient rolls about from side to side, or sometimes lies perfectly still and motionless for some little time; then the sobbing becomes violent, or the patient bursts out into tears, and the paroxysm then terminates.

Generally the attack passes over in a short time, without any bad consequences. These attacks do not, as far as I have observed, interfere with the progress of gestation; cases of premature labor, however, are recorded as having taken place during these paroxysms.

The speediest means of reviving a patient from fainting is to dash cold water upon the face, admit plenty of fresh air into the room, or hold some volatile alkali to the nostrils. When the paroxysm is over, a single dose of *Chamomilla* or *Coffea* — six globules — may be given, and the patient permitted to go to sleep. On waking, she will find herself quite restored.

Where these derangements of the nervous system arise from a disordered digestion, of course our attention must be turned to the state of the stomach, and *Nux-vomica* or *Pulsatilla* will be found our best correctives. *Nux-vomica* is best suited to choleric, petulant, peevish females; while the amiable ones, with mild, easy dispositions, will be soonest reached with *Pulsatilla*.

Chamomilla. — When the disorder arises from a fit of anger.

Belladonna and Aconitum. — When there is much congestion to the head.

ADMINISTRATION OF REMEDIES. — The remedies may be either given dry or in solution. When given dry, put six globules upon the tongue, and let them dissolve. When given in solution, dissolve six pills in six spoonfuls of water, and give one spoonful at a dose. The doses may be repeated every two, three, or four hours, according to the severity of the case. See "HYSTERICS."

PALPITATION OF THE HEART DURING PREGNANCY.

Palpitation of the heart not unfrequently causes pregnant women a great deal of annoyance, and sometimes serious alarm. Should this affection occur, for the first time, during pregnancy, there is no fear of its being connected with organic disease of the heart, and it should therefore cause no uneasiness. It is, however, very distressing, and, on that account, requires medical treatment.

TREATMENT. — The principal remedies are *Coffea, Ignatia, Chamomilla, Nux-vomica, Pulsatilla, Belladonna, Nux-moschata*.

When caused by anger, *Chamomilla*; by fear, *Veratrum*; by joy, *Coffea*; by sudden fright, *Opium*. For nervous persons, *Ignatia, Coffea, Chamomilla*. For plethoric persons, *Aconitum* and *Belladonna*.

ADMINISTRATION OF REMEDIES. — Dissolve of the selected remedy twelve globules in twelve spoonfuls of water, and take one

spoonful of the solution every hour, or oftener if the severity of
the case demands it.

TOOTHACHE DURING PREGNANCY.

This is one of the most common, as well as one of the most
troublesome, disorders a pregnant woman has to contend with.
From other aches and pains she seeks relief, but from this there is
no escape ; she suffers on patiently, to the end, under the impression, which is so prevalent, that the extracting of teeth is liable
to induce a miscarriage, or mark the child. Of course, this impression has not originated with herself; it is the common belief
of all women, and was furnished to them by physicians. It is
astonishing, yet, nevertheless, true, that many eminent medical
men still cherish this opinion. I am well aware that there are
plenty of cases upon record where abortion has followed the removal of a tooth ; but where is the proof that the simple shock
of extraction was the sole cause of the mishap ? There is just as
much proof that the nervous irritation occasioned by decayed
teeth, is the cause of abortions, as there is, that it is all the result
of the shock of extraction ; in fact, there is a great deal more.
I unhesitatingly recommend the removal of decayed teeth the
moment they induce pain and suffering which a few doses of some
simple remedy will not allay. It is seldom, however, that a
homœopathic physician meets with a case that cannot readily be
controlled by medicine. I have yet to see the first case where
abortion has been the result of such practice. On the other hand,
cases are not wanting, in every physician's practice, where abortion is the result of nervous irritation, and this irritation, the
result of decayed or ulcerated teeth.

It has always been my custom, when pregnant women have applied to me for the cure of toothache, to send them at once to a
competent dentist and let him examine their teeth. If they are
not too far decayed, I recommend filling, but if they are past
saving, then the sooner they are removed, the better. I have no
more hesitation about their being extracted during pregnancy,
than at any other time.

Sometimes the pain, though caused by, is not situated in, the
decayed tooth. This is a sympathetic pain ; great care should
therefore be taken that the wrong tooth is not removed. The
dentist, by gently rapping, with a small metallic instrument, upon

each suspected tooth, will be able to detect the real offender, — the real *cause* of the pain, — as this tooth will be found more sensitive than the others, and, when extracted, perhaps ulcerated at the root.

Not unfrequently women, whose teeth are quite sound, suffer severely from flying pains through the face, and all the teeth, during pregnancy. These pains are of a neuralgic character, and can be easily removed by taking the proper remedies. Sometimes the severest form of toothache is instantly relieved by one of the following remedies: *Sepia, Chamomilla, Pulsatilla, Coffea, Belladonna, Nux-vomica, Staphysagria, Mercury, Sulphur.*

When the pain is erratic, flying from one tooth to another, *Pulsatilla.*

For toothache, in carious teeth, *Antimonium-c., Chamomilla, Nux-vomica, Staphysagria, Mercurius, Sulphur.*

For violent pain, which comes on in paroxysms, *Coffea, Chamomilla, Belladonna.*

For nervous toothache, *Ignatia, Coffea, Chamomilla, Hyoscyamus, Belladonna, Sepia.*

ADMINISTRATION OF REMEDIES. — After selecting a remedy, give, according to the severity of the pain, at intervals of from one to six hours, a dose of six globules, until better. See " TOOTHACHE."

NEURALGIA DURING PREGNANCY.

When we take into consideration the increased irritability of the nervous system, caused by the new action which is going on within the womb during pregnancy, we need not be at all surprised to find neuralgic disorders, throughout the whole system, during this period. And we do find them, in the abdomen, under the short ribs, near the hips; in the region of the bladder; in the muscles of the limbs, back, and head, — all originating from one and the same cause, excessive nervous sensibility.

One of the following remedies will generally afford prompt relief: *Belladonna, Aconitum, Coffea, Chamomilla, Bryonia.*

Give, according to the severity of the pain, at intervals of from one to four hours, a dose of six globules. See " NEURALGIA."

PAINS IN THE BACK AND SIDE DURING PREGNANCY.

Not unfrequently, women suffer considerably from pain in the small and lower part of the back during pregnancy. Occasionally there is a deep-seated pain or aching in the right side just under the ribs. The patient also feels a sensation of heat in the affected part. Benefit will be derived from the following remedies:

For pain in the back, *Bryonia, Rhus, Belladonna, Pulsatilla, Nux-vomica.*

For pain in the side, *Aconitum, Chamomilla, Mercurius, Pulsatilla, Sulphur.* If the pain is attended with much heat, *Mercurius* and *Aconitum.*

When the pains are of a dull, heavy character, *Chamomilla.*

ADMINISTRATION OF REMEDIES. — The same as for Neuralgia.

CRAMPS DURING PREGNANCY.

Cramps in the calves of the legs, hips, feet, back, or abdomen, often torment pregnant women day and night, and very much fatigue them, by depriving them of rest and sleep.

The best remedies for cramps in the limbs are, *Veratrum, Colocynthus, Chamomilla, Nux-vomica,* or *Sulphur.*

For cramps in the abdomen, *Nux-vomica, Belladonna, Colocynth,* or *Pulsatilla.*

For cramps in the back, *Ignatia,* or *Rhus.*

ADMINISTRATION OF REMEDIES. — The same as for Neuralgia.

VARICOSE VEINS, OR SWELLING OF THE VEINS, DURING PREGNANCY.

Some women suffer a great deal during pregnancy from distention of the superficial veins of the lower extremities. This distention is caused by the pressure of the enlarged uterus upon the veins within the abdomen and pelvis, thus preventing a free return of the blood upward.

Usually the swelling commences at the ankles, and gradually extends upwards toward the thigh. Not unfrequently the swelling is confined to the leg below the knee, the veins of the calf of the leg alone presenting any unnatural appearance. Both limbs may be involved, or the disease may be confined to one.

When the disease first commences, the veins beneath the surface assume a reddish hue; but, as the distention increases and the vessels become knotted and swollen, they change to a dark blue or leaden color. The swelling decreases, when the patient is lying down, or when the limb is kept in an elevated position, as, lying upon a chair when the patient is sitting; but when the patient is compelled to be constantly or a large part of the time upon her feet, or when the limb is allowed to hang down, the distention is very much increased, and the disease aggravated.

This condition of the veins, at first, is not painful, and becomes so only from actual distension of the vessels. Sometimes the swelling is so great, that the veins actually burst, and large quantities of blood are discharged, either externally or effused beneath the skin.

After delivery, the pressure of the pregnant uterus on the large veins of the abdomen and pelvis being removed, the swelling disappears, and the veins resume their natural size.

TREATMENT. — At the commencement of the difficulty, and in cases where the swelling is not extensive, nor the pain very severe, frequent bathing with cold water, or diluted alcohol, will afford relief. But when the veins are large and painful, or when they are knotted, the leg requires the careful application of a bandage, and rest in the recumbent posture. Few women are able, without some instruction, to apply a bandage as it should be; therefore, in all severe cases, it is advisable to consult a medical man, as serious consequences often follow neglect or wrong treatment.

Persons who are constantly on their feet, should constantly wear the bandage, or a laced stocking. The stocking or bandage should be applied in the morning on rising, at which time there is the least swelling, beginning at the toes, and progressing upwards, with a moderate and equal pressure. At night, on retiring, the bandage should be removed, and the whole limb freely bathed, and rubbed *upward*, with cold water, or water and alcohol, or a weak solution of *Tinct. Arnica.*

Varicose veins, though occurring more frequently during pregnancy, are not by any means confined to this state; they may take place at any time in the female, and are not unfrequently met with in the male sex. When occurring under any other circumstances than those which we have just been considering, they are indicative of constitutional debility.

The remedies which are most to be depended on in this affection, are, *Arnica, Hamamelis-virg., Pulsatilla, Nux-vomica, Arsenicum, Lachesis, Lycopodium, Carbo-veg.* .

Nux-vomica. — When the disease is attended with hæmorrhoids, constipation, frequent bearing-down pains, enlargement of the abdomen, and irritable temper.

Pulsatilla. — This is the principal remedy for varices, especially when there is much swelling of the veins, and of the whole limb, with severe pain, and more or less inflammation, or when they are of a bluish or livid color, which is imparted to the whole limb. Should *Pulsatilla* give some relief, while the swelling and discoloration continue the same, *Lachesis* may be substituted. In some cases, especially where the occupation of the patient compels her to be constantly upon her feet, *Arnica*, given in alternation with *Pulsatilla*, proves very efficacious.

Arsenicum.—When the swelling is of a livid color, and attended with a good deal of burning pain, when the burning continues after the administration of *Arsenicum*, give *Carbo-veg.*

Lycopodium.—For inveterate cases, after the failure of other remedies.

ADMINISTRATION OF REMEDIES.—Of any of the indicated remedies, dissolve twelve globules in twelve teaspoonfuls of water; give a spoonful every four hours for two days; if no relief is obtained, select another remedy.

HÆMORRHOIDS OR PILES, DURING PREGNANCY.

Pregnant women are very often subject to piles. Many have supposed this disorder to originate from obstructed circulation; but the fact seems to be that the most frequent cause which operates in its production, is habitual constipation of the bowels. If this is avoided, by the means already pointed out in the article on constipation, much suffering and inconvenience will be avoided; but, if it is permitted to exist, and temporary relief only sought by an occasional cathartic, the disease will continue throughout the whole period of pregnancy, and perhaps throughout the remainder of the patient's life. The inexperienced can scarcely imagine the amount of suffering some females undergo from piles; and the suffering is constant day in and day out. Various external applications have been devised for their removal; even the knife has been resorted to. Against all these we would warn you, as they are not only exceedingly painful, but, during pregnancy especially, highly dangerous.

It is very important that a pregnant woman, and especially if it be her first pregnancy, should pay strict attention to the state of her bowels, not allowing either constipation or diarrhœa to become seated, as *early* attention to either of these derangements will cause their prompt removal.

TREATMENT.—The appropriate remedies for this disease are,

Pulsatilla, Nux-vomica, Ignatia, Opium, Sulphur. Also, *Arseni-cum, Carbo-veg., Belladonna, Hepar-sulphur, Natrum-muriat., Ha-mamelis-virg.*

Pulsatilla. — When blood and mucus are discharged with the fæces, with painful pressure on the tumors, pains in the back, pale countenance. Where this remedy proves insufficient, follow it with Sulphur.

Nux-vomica and *Sulphur* are the principal remedies against hæmorrhoids; *Nux-vomica*, especially, when there is a burning, pricking pain in the tumors; also, when there is a discharge of light blood after each evacuation of the bowels, and a constant disposition to evacuate. This remedy may be given at night, and *Sulphur* in the morning. *Sulphur* is well adapted for all forms of piles, and like *Nux,* is especially called for when there is that constant, ineffectual inclination to stool. It is also serviceable when there is considerable protrusion of the tumors, so much so that it is difficult to replace them. Also, when there are violent, shooting pains in the back.

When these two remedies, after two or three days' trial, fail to afford relief, recourse should be had to *Ignatia,* especially if the pains, like violent stitches, shoot upward, or where, after the evac-uation, there is painful contraction and soreness, or the rectum protrudes at each evacuation.

ADMINISTRATION OF REMEDIES. — Take ten globules, dry, upon the tongue, night and morning; or, in severe cases, the remedy may be repeated every hour until relief is afforded.

In addition to the internal administration of remedies, much benefit may be obtained from a proper use of cold water. When the piles do not bleed, cold applications, either as sitz-baths, com-presses, or injections, are of great benefit. As evil results some-times follow the sudden suppression of the discharge, it is not advisable to use cold water where there is much if any bleeding. When, however, the bleeding is profuse, to such an extent as to cause alarm, cold applications are the best styptic. Warm water or steam is preferable when the tumors do not bleed, or when, from any cause, the bleeding has ceased and there is considerable pain.

When, after each evacuation, the bowels, or a small tumor, pro-trude, causing great pain, relief may be obtained by gently press-ing them up again with the ball of the finger. Injections of cold water, when judiciously administered, are of the greatest value; but more harm than good is so often done from the carelessness

9

of introducing the injecting tube, that I seldom recommend
them.

DIET.—As the use of condiments and stimulants of every description tends to produce gastric and intestinal derangements, it
is advisable that, in this disease, they be dispensed with, and the
patient confine herself strictly to the homœopathic rules of diet.
Meat diet should be avoided as much as possible; some physicians even recommend their patients, suffering from this complaint, to eat nothing, for a few days, except bread and water.

JAUNDICE. ICTERUS, DURING PREGNANCY.

In another place we have spoken at large upon this disease.
We shall here merely mention it, because, occasionally, toward
the end of pregnancy, it sometimes occurs, caused partially, perhaps, by mechanical pressure of the distended uterus upon the
gall-duct, and partially by the sympathetic action going on in the
liver, in common with the other digestive organs.

The symptoms of this disorder are, constipation, with whitish,
almost colorless stools, urine of an orange color, and dry skin,
with slight remittent or intermittent fever.

The remedies which may be taken for its removal are, Mercurius, Hepar-sulphur, China, Lachesis, Sulphur, Nux-vomica,
Chamomilla.

To commence, take *Mercurius*, six globules every three hours
for three days; follow this with *Hepar-Sulphur* or *Lachesis*, two
doses of six globules, daily, night and morning.

When the disorder arises from a fit of pain, take *Chamomilla*
and *Nux-vomica* in alternation about four times a day; dose, the
same as *Mercurius*.

INCONTINENCE OF URINE DURING PREGNANCY.

This disorder consists in a partial or total inability to retain the
urine. There is a frequent desire to urinate, which is sometimes
so violent, that a few drops of urine will escape before the patient
can reach the vessel. This is one of the most annoying complaints
that ever befalls a pregnant woman, not only on account of the
urgent necessity that she should be at all times where she can relieve herself the moment the inclination overtakes her, but in
spite of all her care and watchfulness, there will be a constant
dribbling of urine, which irritates the parts with which it comes

in contact, producing a degree of inconvenience and actual suffering, which the inexperienced can scarcely imagine.

The principal remedies for this difficulty are, *Pulsatilla*, *Sepia*, *Belladonna*, *Causticum*, *Hyoscyamus*.

ADMINISTRATION. — Of the chosen remedy, take six pills, dry, upon the tongue, once in three or four hours. Commence with *Pulsatilla;* should this prove insufficient, next try *Sepia*.

DYSURY AND STRANGURY DURING PREGNANCY.

DEFINITION. — These two affections are equally frequent among pregnant females. Dysury means, simply, difficulty in passing the urine. When there are frequent, painful urgings to discharge urine, and it passes off only by drops, or in very small quantities, the disease is called strangury.

The causes which give rise to these diseases are the pressure of the extended uterus upon the bladder and urethra; the irritation of the mucous membrane excited by this pressure; spasms at the neck of the bladder; excesses in eating or drinking; exposure to cold, etc.

Sometimes the pressure of the enlarged uterus upon the urethra, or passage from the bladder, is so great, as to cut off completely the discharge of urine, when the patient soon begins to suffer from distension of the bladder.

This state of things should not be allowed to continue for any length of time, or serious consequences may be the result. If, by placing herself in different positions, especially by reclining, in order to relieve the neck of the bladder from the mechanical pressure of the womb, the patient cannot relieve herself, a physician should be sent for, as it may be necessary to draw off the accumulated fluid with a catheter. A *catheter* is a small, silver tube about the size of a goose-quill, and its introduction causes no pain whatever.

TREATMENT. — The best remedy for this trouble is *Pulsatilla;* one dose of six globules may be taken every two hours. If *Pulsatilla* is not sufficient, *Nux-vomica*, will be of service. *Aconitum*, *Belladonna*, *Cocculus*, *Cantharides*, *Phosphoric acid*, and *Sulphur* are also valuable remedies, any of which may be taken the same as *Pulsatilla*.

When this disease arises from cold, or from spasmodic contraction at the neck of the bladder, in addition to the above treat-

ment, great relief will sometimes be afforded by drinking freely
of linseed tea, watermelon-seed tea, pumpkin-seed tea, gum ara-
bic water, barley-water, or something else of the same nature.
This dilutes the urine, and thus renders it less irritating to the
bladder and urethra.

FLOODING DURING PREGNACNY.

DEFINITION. — During the last month of pregnancy, and at the
commencement of labor, patients are sometimes subject to severe
attacks of flooding.   This arises from the detachment of a portion
of the after-birth from the surface of the womb, produced, perhaps, by a violent shock, such as a fall, or a blow; or by great
exertion ; violent straining at stool; lifting heavy weights; a
hearty fit of laughing.

SYMPTOMS. — The exciting cause may be immediately followed
by the discharge, or the patient may simply complain of local or
general uneasiness, dull pain and aching in the back, weight in
the abdomen, and faintness.   At length, with or without pain, the
discharge commences, varying in amount from a few ounces to a
quantity sufficient to endanger the patient's safety.

TREATMENT. — In the first place, the patient should be placed in
bed, on a hard mattress, and very lightly covered with bed-clothes ;
the temperature of the room should be reduced very low, and
nothing but cold drinks allowed ; her mind should be kept free
from care, and the greatest quietness preserved in the whole
house.

Internally, administer the *Tincture* of *Cinnamon*; put two or
three drops into half a tumbler of water, stir it up well, and
give a teaspoonful every half hour, or oftener, according to the
urgency of the case, until you can procure the services of a homœopathic physician.   If the Tincture of Cinnamon is not at
hand, a small piece of common cinnamon may be chewed.

I shall not offer any extended remarks upon this disorder, because I do not deem it safe for such a case to be left in the hands
of a layman.   A physician should be immediately sent for, as the
case may prove serious, although at the commencement no alarming symptoms are present.

The following are the principal remedies which we make use of
for flowing : *Arnica, Ipecacuanha, Belladonna, China, Bryonia,
Chamomilla, Platinum, Hyoscyamus, Ferrum.*

MISCARRIAGE, OR ABORTION.

DEFINITION. — By miscarriage, or abortion, is understood the expulsion of the fœtus from the womb before the sixth month; subsequent to this period, it is called premature labor.

Although the expulsive action of the uterus may be exerted at any period of gestation, it is much more common for it to occur at or before the third month, owing, no doubt, to the frail connection existing between the embryo and the womb at this early period. It is also more liable to occur at the beginning of each. month, corresponding to a menstrual period.

A woman who has suffered one abortion, is more liable to a similar mishap, in subsequent pregnancies, than she who has not suffered in this way. There are many women who cannot carry their gestation out beyond a certain time; they always miscarry; why it is so, perhaps it would be difficult, satisfactorily, to explain. This accident is always an untoward event, although it cannot be considered dangerous, unless it be accompanied by great flowing; and even then it is rarely fatal. In the first two months the embryo, which is yet very small, is often expelled without pain, or any considerable flowing; miscarriages at a later period, however, are much more serious.

CAUSES. — Could you but know the extreme tenuity of the membranes and vessels which connect the early embryo with the womb, you would rather be amazed at its power to live on to its term, than surprised at its occasional death and expulsion. The union between the embryo and the womb is *very* slight; it may be overcome by a sudden jar; the after-birth may become slightly detached, and a few drops of blood become insinuated between it and the womb, thus little by little separating the connection sufficiently to destroy the life of the new being.

Among the more common causes of abortion are enumerated blows; falls; violent concussions; excessive or sudden exertions; straining; lifting heavy weights; in ascending stairs; running; dancing; riding over rough roads in a carriage; riding on horseback; severe coughing; mental emotions of anger, joy, sorrow; good or bad news suddenly told, which may excite contractions of the uterus, and expel the fœtus from the womb. Lastly, the action of drugs, which some women are constantly taking for some imaginary trouble; emetics, cathartics, herb-teas, patent medicine, etc.; the application of blisters, mustard plasters, etc.

Although we frequently see miscarriages arising from some

trivial, almost insignificant, cause, such as laughing, coughing, or sneezing, and we do not so much wonder at it, when we come to consider the extreme delicacy of the existing link between the mother and child ; yet it does excite our amazement to see with what tenacity the embryo is sometimes retained by persons of delicate constitution, and under very trying circumstances. Thus a woman, far gone in consumption, conceives, completes the term of gestation, and is delivered of an apparently healthy child. And Dr. Mauriceau mentions a case of a woman who fell from a window in the third story of a house, in the seventh month of pregnancy, and broke one of the bones of her arm, dislocated her wrist, and bruised herself very much, yet she fulfilled the period of pregnancy, and was delivered of a living child. Dr. Davis also relates a case of a lady who was thrown from her horse, when three or four months pregnant, and much bruised, yet without interruption to the gestation. — *Churchill.*

SYMPTOMS. — In enumerating the causes of abortion, we should have also mentioned, as a prominent one, the death of the fœtus. Abundant observation has proven that the child is liable to disease as well before as after its birth. Numerous cases of measles, small-pox, intermittent fever, pleurisy, and a host of other diseases, are on record as having existed in newborn children.

When the embryo dies, it ceases to possess any vital property, becomes the same as a foreign body, and, by its irritation, a new action is awakened in the womb; its muscular fibres begin to contract, and its whole contents are soon expelled. The signs of the death of the fœtus are, the cessations of its movements ; the flaccidity of the abdomen ; a sensation of dead weight and cold-ness in the abdomen ; the shrinking of the before well-developed breasts ; the general health becomes bad ; fœtid breath, and re-peated chills. It is a difficult matter, however, even for the most experienced, to decide positively whether the fœtus is dead or not.

The symptoms of abortion vary with each given case ; most patients, however, at first, complain of a sense of uneasiness and weariness, with aching or pain in the back ; this is soon followed by regular labor-pains, the pain sometimes being even as great as those of labor at full term. Occasionally, especially in those females who have repeatedly miscarried, the fœtus slips out of the womb readily, producing hardly any pain, with little or no hemorrhage,. and occasioning the patient but slight inconvenience, while in other cases very alarming hemorrhage may precede or accompany the accident, and reduce the patient to the lowest possible ebb.

The progress of different cases is so dissimilar, that, without making an article too lengthy for such a work as this, it would be impossible to make mention of one half the variety of phases that these cases sometimes present. Whenever possible, a miscarriage should be placed under the care of a competent physician.

TREATMENT. — The first thing to be attended to, on taking charge of a case of abortion, is to see that the patient is placed in a proper position.

In those cases where, from experience, the woman knows that a miscarriage will not to her prove a very serious affair, it will be only necessary for her to maintain a recumbent position for a few days. It will not be necessary even to go to bed; she can lie upon a sofa through the day, and take the remedies below recommended. But, as, in the vast majority of cases, an abortion is far more serious than a natural labor, it is important that each case be treated with great care and consideration, that all complications may be warded off, and the patient secured a good getting-up, free from the provoking sequels which are so apt to follow these mishaps.

When a female is threatened with a miscarriage, she should immediately assume the recumbent posture; she had better at once go to bed; the bed-clothes should be light, the room cool; all causes of excitement should be removed, and every means be employed to insure perfect tranquillity of mind. By this means, with the remedies recommended further on, it may be possible to arrest the flow, even after the patient has lost many ounces of blood, and instead of her miscarrying, pregnancy go on as regularly as though no accident had happened.

Should the hemorrhage be severe, producing great paleness and weakness, the pillows should all be taken from under the patient's head, allowing her to lie upon an absolute horizontal plane. Indeed, we not unfrequently elevate the feet of the bedstead four or five inches, by placing some blocks under the lower bed-posts. Cold should be applied to the vulva by means of a cloth dipped in cold water, and suddenly applied.

After the fœtus has been expelled, the hemorrhage generally stops, and no further treatment is necessary but the one which we have recommended in ordinary confinement. If anything, the patient should keep her bed longer than the usual time, in order to give the uterus a chance to recover from the shock, and regain its natural strength. By this means we shall avoid that large class of disorders dependent upon weakness, such as " fallings," "leucorrhœa," etc. The diet, for some days, should be bland and unstimulating. See " FLOODING."

The following remedies will be found of service for this disorder : —

Arnica. — If the attack has been brought on by a fall, blow, violent concussion, overlifting, misstep, or walking, or great physical exertion of any kind.

Cinnamon. — Should *Arnica* fail, have recourse to this valuable remedy next.

Secale. — After the miscarriage has occurred, and particularly in females who have miscarried more than once, or in those who have a weak and debilitated constitution, where there is a discharge of dark, liquid blood, and the pains are but slight.

Chamomilla. — When there is excessive restlessness, severe pain in the back and loins ; also periodical pains, resembling those of labor, and each pain is followed by a discharge of dark-colored blood.

Cinchona. — For weak and exhausted persons ; also when there is spasmodic pain in the uterus, or a bearing-down sensation, with a considerable discharge of blood at intervals. This is a most valuable remedy, in restoring the exhausted energies of the patient, after the hemorrhage has ceased.

Hyoscyamus. — For miscarriages, attended with spasms or convulsions of the whole body.

Ipecacuanha. — In alternation with *Secale*, if, with flooding, there is nausea, and cramps ; profuse and continuous discharge of bright red blood ; disposition to faint, whenever the head is raised ; chills and heat ; cutting pain in the umbilical region. Should *Ipecac* fail, recourse may be had to *Platinum.*

Platinum. — When there is a discharge of dark, thick, or clotted blood, attended with pressing or bearing-down pains.

Belladonna. — This is a valuable remedy at the commencement, perhaps more so than any other that we possess, especially when there are great pains in the loins and the entire abdomen ; severe bearing-down, as though the intestines would be pressed out ; pain in the back, as though it were dislocated or broken ; profuse discharge of blood.

Crocus. — This remedy is especially indicated when there is a discharge of dark, clotted blood, with a sensation of fluttering or moving about in the umbilical region; an increased discharge of blood on the slightest movement. If other remedies fail, this sometimes will help.

Nux-vomica and Bryonia — May be given alone or in alternation, for the following symptoms: Severe, burning or wrenching pain in the loins; painful pressure downward, with a mucous discharge. Also, in cases attended with obstinate constipation.

In cases where there is a disposition to miscarriage, or if the patient have previously miscarried, as she approaches again the same period, she should lie on the sofa or bed the greater part of the day, taking an occasional dose of *Sabina*, until the period has passed.

As preventives of this disorder, the principal remedies, are, *Sabina, Secale, Lycopodium, Calcaria, Sepia.*

ADMINISTRATION OF REMEDIES. — Dissolve of the selected remedy twelve globules in twelve spoonfuls of water, give one spoonful every fifteen or twenty minutes in severe cases; in milder ones, every two hours, and in the early stages when no danger is apprehended, every six or twelve hours. If relief is not obtained in five or six hours, another remedy should be selected.

PREMATURE BIRTH.

A birth, occurring after the sixth and during the eighth month, is called a premature birth. It should be treated just the same as a labor at full term. See "LABOR."

PREPARATION OF THE BREASTS.

Young mothers not unfrequently find great difficulty in nursing their children, owing to some organic defect, or incapacity of the nipple, — the result of carelessness in early life. Not a few instances have come under my observation, of young mothers, whose nipples were almost entirely obliterated, from compression during their girlhood. Perhaps this compression has not been relaxed during married life; and, when, after confinement, the nipple is to be put to its natural use, there is nothing deserving the name of nipple to be found. This direct pressure from tight dresses, stays, etc., affects not the nipple alone, but the secreting structure of the breast itself is permanently injured, and the important function of lactation never attains that state of perfection which it otherwise would. The suffering resulting from this state of things, to both mother and child, is by no means trifling.

10

The breasts of a pregnant woman should be specially guarded, as pressure, however slight, when constant, will have an injurious effect; they should be effectually protected from any influence which will have the least tendency to prevent their free development. In most cases, if the nipples are effectually protected from compression by the clothing, they will stand out prominently from the breasts; sometimes, however, in spite of all the care and attention which you can give them, they will not project sufficiently from the breasts, so as to be grasped easily by the mouth of the child. In such cases, it becomes necessary to draw them out occasionally. This may be done with a common breast-pump or pipe. The suction at first should be moderate, not sufficient to give pain; after the pump is removed, a ring of beeswax, or the common glass nipple-shield should be applied and constantly worn. In moderate cases the glass nipple-shield will be sufficient, without the pump. This process of drawing the nipple may be commenced any time before confinement, — if the case be a bad one, the earlier the better, — and continued up to the time of delivery, once, twice, or more times daily, if necessary.

If this operation excites, as it sometimes will, abdominal pains, it should be immediately suspended, and not again resorted to; because, if persisted in, it may produce abortion.

Dr. Tracy, in the " Mother and her Offspring," — a work from which we have before quoted, — suggests the following method for keeping the nipples permanently prominent after they have once been drawn out: —

"It consists in the winding of a bit of woollen thread or yarn two or three times around the base of the nipple, previously drawn out sufficiently, and tying it moderately tight, but not so tight as to interfere with the free circulation of the blood. By this means, a young mother, a patient of mine, was enabled to keep her nipples sufficiently prominent. I can see no objection to its being put in operation at any time desirable before confinement, and should strongly recommend its trial in cases where the nipple, after having been drawn, if left to itself, soon sinks back again, and becomes imbedded within the breast."

Retraction, or a want of sufficient development, is not the only difficulty to which the nipple is subject. The most common affections, perhaps, to which these parts are liable, are excoriations, cracks, inflammation, scaly eruptions, and small abscesses. Inflammation, excoriations, etc., arise from the extreme sensitiveness of the skin, occasioned by the nipple being kept folded down upon the breast by the clothing; in this way, the skin around the base of the nipple, being folded upon itself, becomes very delicate and thin, unfitted for the purpose to which it was designed. The result is, as we would naturally expect it would be, as soon as the child begins to nurse, the skin becomes irritated and inflamed; cracks, fissures, or abscesses form, and the mother is subjected to untold misery, every time the child is put to the breast.

Now, the main object to be attained in preparing the breasts, during early pregnancy, for their future important function, is to thicken and toughen the skin upon and at the base of the nipple. We shall give some of the most important means, which have been recommended at different times, to accomplish this desirable end. For several weeks previous to delivery, the entire breast and chest should be bathed in cold water daily, and afterwards well dried and rubbed with coarse towels. Some recommend bathing the nipple and breast with brandy, twice a day, for several weeks anterior to confinement; others prefer a decoction of green tea, a decoction of oak bark, or of pomegranate. For my own part, I should recommend, in preference, the cold water treatment, or simply rubbing the parts upon all sides, and in every direction, with the palm of the dry hand. This rubbing should be commenced soon after the commencement of pregnancy, and repeated two or three times a day, till the time of confinement.

Should there be tenderness, soreness, or slight excoriation, the parts may be bathed in a weak solution of Arnica. See " SORE NIPPLES."

FALSE PAINS.

DEFINITION — CAUSES. — Previous to delivery — sometimes but a few days — sometimes a week or even longer, — women are not unfrequently very much annoyed with what is termed *spurious*, or *false pains*. These pains sometimes so closely resemble true labor-pains, that it is exceedingly difficult to discriminate the one from the other. From this close resemblance arise what are termed " false alarms," the patient and nurse, anticipating that delivery is about to come on, make all the needful preparations, perhaps send and arouse the doctor from his warm couch, to walk a mile or two in the middle of the night, perhaps through a heavy shower or a drizzling rain, to be met at the door with, " La! doctor, the pains have all blown over. It was nothing but a ' false alarm.' "

Now, in view of all this, it becomes quite essential that both patient and nurse should fully understand the difference between true and false pains. False pains chiefly differ from labor-pains, " in the irregularity of their occurrence ; in being unconnected with uterine contraction, and chiefly confined to the abdomen, with sensibility to touch and movement, and in not increasing in intensity as they return." — LAURIE. True labor-pains commence low down, and are first felt in the *back*, extending gradually to the front, recurring with regularity, and increasing in intensity with each return.

Spurious pains arise from various causes, such as over-fatigue, indigestion, cold, mental emotions, constipation, errors in diet, and are, no doubt, occasionally excited by the motions of the child.

TREATMENT. — As these spurious pains, when they come on early in pregnancy, are liable to bring on premature labor, or; when at the full term of gestation, occasion great distress and loss of rest, it is always desirable to relieve them as speedily as possible. This may generally be done by one of the following remedies : *Bryonia, Nux-vomica, Pulsatilla, Dulcamara, Aconite.*

Bryonia deserves a preference when the following symptoms are present : The pains in the abdomen and about the loins, resembling a dragging weight.

Pulsatilla. — Similar pains in the abdomen and loins, with a feeling of stiffness or lameness ; also when the pains arise from indigestion, eating rich or fat food.

Nux-vomica. — When there is a pain as if from a bruise, in the region of the bladder ; constant but inefficient urging to stool, and when the exciting cause appears to be constipation. The pains occur chiefly at night.

Dulcamara. — When the pains are sharp and violent in their character, or when they arise from taking cold, — the effect of a chill. The pains are mostly confined to the small of the back, resembling labor-pains.

Aconitum. — Especially in women of a plethoric constitution, with a full, bounding pulse ; head hot ; skin dry. *Belladonna* may be given in alternation with *Aconite*, especially when the head is hot, and the feet are cold.

ADMINISTRATION OF REMEDIES. — Dissolve, of the selected remedy, twelve globules, in twelve teaspoonfuls of water ; give one spoonful of the solution, in severe cases, every half-hour ; in others, every three or four hours, until relief is afforded, or another remedy selected.

CHAPTER IV.

PARTURITION OR CONFINEMENT.

GENERAL REMARKS. — It is not my desire, nor my intention, to make practical accoucheurs of all the women who may read this work. No; whatever woman's rights may be, this, certainly, is not her mission. Her sympathizing nature is too much alive to all the humane charities, and prompt gushings of a tender heart, to stand by unmoved, and, with a calm and calculating eye, gaze upon the writhing agonies of a difficult or complicated case of labor, and there, amid the importunities of anxious friends and the shrieks and groans of her suffering patient, calmly to weigh the relative value of all the therapeutic agents that may be brought to bear upon the case. No; the heart-promptings of her allied nature are better adapted to soothe, sympathize with, and encourage her suffering sister than, with unflinching resolve and steady hand, to perform the agonizing though necessary operation which alone, perhaps, can terminate the patient's sufferings, and save the life of both herself and child.

It is with no selfish hand, that I thus strike from the list of woman's rights the duty of an accoucheur. Heaven forbid that I should infringe upon her province, or turn a single honest penny from her needy purse. And, were it not for a duty which every physician owes the community at large, I should have passed the subject by unnoticed.

Do not infer from what has been said, that I would wish to withhold all information in regard to the necessary attendance upon a parturient woman. By no means. On the contrary, I consider it quite essential that every female should be more or less conversant with the necessary duties and cares of a lying-in chamber, for the purpose, not of making them poor physicians, but of fitting them to be competent, efficient, and trustworthy nurses, so qualified, that, in cases of emergency, they may render intelligent assistance.

It not unfrequently happens, especially in quick cases, that your medical attendant cannot be had just at the moment when you wish him; how important, then, it is, to know yourself, and feel assured that those around you are alike informed, what is necessary to be done for the safety of yourself and child!

Though, as I have before observed, labor is a perfectly natural process, and the majority of cases would terminate favorably, with none present but an ordinary nurse, yet events *may* occur, which would call for prompt interference, and such interference as none but a well-educated and qualified medical man could afford.

The passage of large ships over the bar, and into the harbor of New York is a perfectly natural process; but who would think of trusting a merchantman, heavily laden with a precious cargo, or a packet-ship freighted with human souls, to be brought into port by any but a well-informed and skilful pilot. Although he may stand at the wheel and have apparently little or nothing to do, yet it is he that knows the channel, — where the water is deepest; he has in his mind's eye every sand-bar and rock; he knows what currents are capable of deviating the ship from its proper course; in one word, he is aware where every danger lies, and by guarding against them in season, can avoid them, with so little effort on his part, that you would almost think he was not aware of their existence. Now, the duties of a pilot and those of a physician are, in many respects, very similar. The well-informed accoucheur, appreciating the responsibilities of his position, is ever on the alert that his precious charge is not exposed to any of the accidents that threaten the new life in its passage from the womb to the fond, expectant mother's arms. The many little points at which he renders you valuable assistance may seem trivial, or even of no account; yet he knows best, and is well acquainted with all the liabilities to which you are exposed, and by his timely though slight interference, *you* escape evils, without so much as being aware of their existence. And, in those cases where danger is obvious to all, he may sit by your bedside, *apparently* unoccupied and unconcerned; but you can rest assured that he is most deeply interested, and the consciousness of the abilities he commands will sustain his self-possession, and enable him to render the proper attentions at the proper time. He well knows that nothing is gained by pointing out to others, or to you, the dangers that lie on this side, or the other; if he can avoid them quietly, as they appear, without causing alarm to the patient or her friends, so much the more does he show his abilities and fitness for his calling.

Labor being a perfectly natural process, and its termination almost uniformly a happy one, it is somewhat strange, that so much fear and foreboding should possess the minds of all parturient women. It is not so much the actual pain and suffering, which they must necessarily pass through, as it is the anticipation of an unfavorable termination. I doubt not, that if you could insure a woman that her labor would terminate favorably, and without being followed by any unfavorable consequences, the anticipation of a few hours' suffering would give her but little uneasiness. Unfortunately, physicians cannot give absolute insurance of safety, although we can positively assure them that there is no more cause for apprehension of evil in regard to their coming confinement, than there is in regard to an excursion by boat or carriage. In either case, an unhappy termination may take place, but the probabilities are, that, with proper care, the result will be happy.

If you have lived prudently and maintained proper habits during pregnancy, and have proper attendants during confinement, you need apprehend no danger, but may confidently expect your sickness to terminate without any untoward circumstance either to yourself or child.

LABOR. CHILDBIRTH.

Natural labor generally takes place at the end of the ninth month after conception, or two hundred and eighty days from the commencement of pregnancy.

It is not absolutely certain, however, that confinement will take place at this precise period, for it not unfrequently happens that pregnancy is protracted to the two hundred and ninetieth day or even later. Neither is a premature confinement an uncommon thing, especially in first cases. Still, in the great majority of instances, you can confidently expect your sickness at the expiration of the two hundred and eighty days.

The commencement of actual labor is usually preceded by some of the following premonitory symptoms: — agitation, nervous trembling, lowness of spirits; irritability of the bladder, with frequent inclination to urinate; nausea and vomiting, flying pains through the abdomen, followed by an increased mucous discharge or flow, sometimes streaked with blood.

Your preparation for confinement, if not already completed, should now be attended to without delay. Your medical attendant should be notified as soon as possible, so as to give him an opportunity to arrange his other engagements and be in readiness. Your female friend and nurse should be sent for. Do not alarm the whole neighborhood and call in all the old women of your acquaintance; for their presence is not only useless, but often highly injurious. Nothing annoys a physician more, when entering a lying-in chamber, than to find several women standing around whose services will not be needed.

The physician, nurse, and one female attendant are all that are necessary or desirable.

Your room should be put in perfect order. The clothing for yourself and infant should be in readiness, arranged in the order in which the articles will be wanted, and placed in some convenient position, where, when wanted, they can be had without trouble.

You should also have convenient a pair of sharp scissors, and a couple of short pieces of strong, round cotton cord.

The occurrence of true labor-pains may soon be looked for, after the premonitory symptoms above described. The pains usually commence in the back; sometimes they are first felt at the lower and front part of the abdomen, and extend to the loins and lower part of the back. They are not constant, but periodical or intermittent, coming on at regular intervals of longer or shorter duration.

At the commencement, they are not actual pains, but rather a feeling of uneasiness. When active pains first begin, they are slight and of short duration, lasting but a few moments and with intervals of rest lasting from half an hour to an hour or more. By degrees they become more and more frequent, gradually increasing in intensity, until labor is completed, which usually takes from four to six hours.

The duration of labor, however, is variable; some labors are short, and others long, from reasons entirely beyond our knowledge. Anxiety on account of the length of labor should never be indulged. If the position of the child is right, protracted labors are no more dangerous than short ones. First labors are generally longer than subsequent ones.

As soon as true labor-pains manifest themselves, your medical attendant should be called in, and you should now submit yourself entirely to his directions, confidently expecting that he will render you all possible aid at the proper time. "His attention will probably be first turned to inquiries whether your general system is in a healthy condition. He may be able to satisfy himself upon this point without doing or saying anything that shall indicate that his thoughts are turned in that direction; but he may have occasion to feel your pulse, to look at your tongue, and examine the state of your skin. He may also have occasion to ask a variety of questions relative to yourself.

" It is desirable that he should know as much about the state of your general system as it is possible for him to know, in order that he may adopt such measures as are best calculated to preserve you from a protracted and severely painful labor, and to render your confinement as speedy and easy as possible.

" It will probably not be long after his arrival before it will be advisable for him to make an *examination*, in order that he may know positively the state of the womb, and the actual position of the child. It is important that this should be attended to at an early period, otherwise he may be the means of much evil by remaining an idle spectator while he ought to be acting. When he deems it the proper time, he will inform you himself, or will do so through your female attendant.

" In order for this examination, you will lie upon your *left side*, with your back very near that side of the bed which will be at your left hand, as you stand at the foot, facing the head-board.

" Your limbs should be flexed forward upon your body, with the knees bent about as they naturally are when you lie upon your side, and a small pillow should be rolled up and put between your knees. While occupying this position, with the bed-clothing spread over you, the examination may be conveniently made without any exposure of your person.

" Now, although he may learn by this means that you will have no further occasion for his presence for some hours, perhaps, and, therefore, that he may with safety be absent upon other duties that may be pressing upon him, and although he may thus relieve you also from the tedium of waiting hour after hour, in constant and anxious expectation of the completion of your labor, and from no trifling amount of solicitude on account of the delay, yet, if the parts were never found deformed nor dry, painful and unyielding; if the top of the child's head was always the part which presented itself first at the mouth of the womb; if the head were never too large to pass between the bones of the pelvis; if neither an ear, nor the face, nor the chin, nor the back of the neck, nor a shoulder, nor an elbow, nor a hand, nor the back, nor a side, nor the belly, nor a hip, nor a knee, or other part except the top of the head, were ever the first to present itself there; if the umbilical cord never came down first; if the placenta were never attached over the mouth of the womb, or very near it, so as to produce dangerous flooding in the early stages of labor, besides various other *ifs*, — this examination might not be considered as absolutely necessary.

" But, inasmuch as each and all of the things above-named, as well as many others, are to be looked out for, and guarded against or removed, or their evil consequences mitigated, each at the proper moment, and in the proper way, I think you cannot fail to agree with me, when I say, that no false modesty should be allowed to throw the least obstacle in the way of the due performance of the one, or of your securing to yourself the services of the other." *

After this examination is completed, if everything is found in favorable condition, your medical attendant will have little to do except to watch closely the progress of the case.

To do this, it may be necessary for him to repeat the examination occasionally till delivery is accomplished. During these examinations, he will frequently be able to render you great assistance, relieving your suffering and accelerating your delivery.

He may not deem it necessary to sit constantly by your side; he may even retire to another room, giving you permission to walk about, or sit up in an easy-chair. It is not necessary for you to lie in bed, nor for him to remain constantly with you, until near the close of labor.

It is unnecessary for me to say more upon this subject here, as I have anticipated you are under the care of a well-educated and qualified medical man. Nothing further that I can here say, will be of any use to you, but that you submit yourself entirely to his direction, confidently expecting that he will do everything for you to facilitate your labor, and render you as comfortable as possible.

LABOR, IN THE ABSENCE OF A PHYSICIAN.

Now, as I have already said, you may be disappointed in securing the services of your medical attendant; you may have

---

* " The Mother and her Offspring," by Dr. Tracy, — an invaluable little book, and one which should be in the hands of every mother.

neglected to notify him in time; or he may be away from home, or engaged with another case, or, from some other cause, he may necessarily be absent.

I will endeavor to explain, in as brief a manner as possible, what is to be done in such an emergency.

I have already pointed out to you the manner in which the pains will, in all probability, commence. At first, though slight, they may interfere with your moving around, or even sitting up. It is best, however, that you should keep about as long as possible, or until you feel a disposition to bear down with every pain, or until the pains become so severe that you feel as though you must lie down.

At about this stage of the proceedings it is customary to " make the bed," which is done by placing a square of oiled silk over the mattress to protect it at that part of the bed which will be occupied by your hips; over this are placed the under-blanket and sheet, and upon them two or three sheets, folded square, on which your hips are to be placed.

These folded sheets will absorb most of the discharges, and can afterwards be removed, without any trouble, leaving the dry bed-linen beneath. The oiled silk is allowed to remain for some time longer.

When you go to bed, your night-dress should be drawn up, underneath you, beyond the hips, to escape soiling. You can assume any position that best suits your convenience or inclination. As soon as there is an involuntary disposition to bear down, you can encourage it by active efforts of your own.

It is useless for you to press or bear down, unless there is present an involuntary disposition to do so, for such efforts will greatly fatigue you, and not expedite the labor at all. Wait until you are *obliged* to make such efforts, and then they will be useful.

Between the pains, gain all the rest you can ; if possible, take a short nap; but when they commence, assist them with a right good-will.

When the pains become severe, you will feel a disposition to hold your breath, and to press with your feet against the bedstead, or any other solid substance. This you may do, and at the same time, you can take hold of the hands of your attendants and pull. Or, what perhaps is better, you can pull upon a sheet, one corner of which has been fastened to the opposite corner of your bedstead.

During these efforts, the pain in your back, which at times

becomes very severe, may be relieved by one of your attendants pressing hard with the palm of her hand against the lower part of your back. At about this period, the sack of waters will become ruptured, after which labor will progress with more rapidity.

Should you become uncomfortably warm, the bedclothes must be lightened, and your room *all* the time should be pleasantly cool and fresh.

Food cannot be taken at an advanced period of the labor, but warm drinks, such as whey, gruel, or tea, you can have as often as you desire.

It is not unfrequently the case, that, as labor draws near to a close, the patient becomes despondent, and feels as though she would never live to see the end of it. Do not indulge any such feelings, if you can possibly help it. Show your friends that your courage is not to be daunted; for, rest assured, that, whether you encourage it or not, a buoyant hope will soon take possession of your flagging spirits ; and, as the pains increase and come in quick succession, are more severe and of longer duration, you will have a renewed feeling of strength and of ability to help yourself, by holding your breath, bearing down, pushing with your feet, and pulling with your hands.

These last throes of nature, as she ushers a new being upon earth, you may thank your stars, are of short duration. During the last few pains, your best position will be upon your left side, and near the edge of the bed, with your knees drawn up and a pillow between them. This will be more convenient for your attendant, who, as the last pain, which will be a long and a hard one, expels the head of the child, will receive it upon her hand, and support it, so that its weight will not be sustained by the neck.

As soon as the head is born, there will be a short interval of rest before the recurrence of another pain. Your attendant should now ascertain whether the umbilical cord is wound around the child's neck; if it is, she should endeavor to loosen it, and slip it down over the shoulders, or up over the head. If this cannot be done easily, she should desist from further attempts, resting satisfied with having unwound or loosened it sufficiently to prevent compression of its vessels.

The interval of rest coming between the pain which expels the head and that which expels the rest of the child, varies in different subjects ; in some cases it is scarcely perceptible, while in others it is so long, that your attendant may have fears for the safety of your child, especially when she observes that it does not breathe. The child is exposed to little danger, however, while in this position; it never yet has breathed, and there is no immediate necessity for its doing so now, as the circulation is still carried on through the umbilical cord and the placenta.

As soon as the pain which is to expel the body of the child commences, your attendant will raise the head upon her open hand and convey it downward from your person just fast enough to make room for the advancing body. She need make no drawing upon it whatever, but simply receive it and carry it away from the discharges, and place it in such a position that it will rest easy and have no obstruction in the way of its breathing. If the child is healthy, and has not suffered long from pressure, it will cry as soon as born, and that first cry will be sweet music to your now happy ear.

As soon as respiration is fully established, a ligature should be placed around the umbilical cord, at about two inches from the navel, and a second one a few inches further on, and the cord divided between the two with a pair of scissors. The ligatures with which the cord is tied should be of strong, round cotton cord; they should be drawn tightly and tied in a hard knot.

Some excessively nice people think all this should be done beneath the bedclothes; but, for my part, I prefer to have the child in such a position that I can see what I am going to do. There is no occasion for the least exposure of your person, provided you lie in the position which I have above recommended.

As soon as the cord is tied and cut, the child should be wrapped up closely in a soft flannel blanket, which has previously been well warmed, and then be removed.

Immediately after the birth, the binder should be applied. This may consist of a folded towel, or other broad bandage, placed around the whole abdomen and extending down over the hips. It should be pinned firmly, but not too tight. Be careful to have it smooth, so as to give an even support to the whole surface of the abdomen.

Some physicians and many nurses discard the use of this bandage altogether, or use it, if at all, only for the first few days. I believe it is deserving of rather more attention than is usually paid to it, especially in feeble women, and, also, when the patient suffers faintness immediately after delivery. If properly applied at first, it is very useful in maintaining a certain degree of contraction of the uterus, and giving support to the abdominal walls. It also assists in promoting a return to the natural condition of the abdomen, preventing that loose, flabby state of the abdominal walls which so frequently follows confinement. I recommend that it should be worn several weeks after getting up. I think it has a happy effect in preserving the natural form and dimensions, especially of women who have several children in a few years.

This band may be worn either over or beneath the ordinary under-garment, as most agreeable. Care should be taken that it is kept well over the hips, for if allowed to work up, and get like a string around the waist, it will do more harm than good.

When the bandage is applied, you can rest awhile before any attempt is made to extract the after-birth. In the course of fifteen or twenty minutes, your pain will commence for the expulsion of the placenta, or after-birth. These pains will not be severe. When they come on, you can bear down the same as during labor; usually the second or third pain, sometimes even the first, brings it away. If it does not come away spontaneously, let your assistant make slight traction upon the cord; great care and gentleness must be used, and if it does not come away readily, let it remain. It will be better to let it alone where it is than to run any risk from its forcible removal, by any but the most experienced accoucheur.

After the placenta has been expelled, or withdrawn, or if neither is accomplished, in the course of half an hour a warm napkin should be applied to the external parts, and the binder tightened, if necessary. The soiled sheet beneath you may now be removed, and your night-dress drawn down, but no further change should be made for at least three or four hours, as it is most important that you should avoid all exertion and excitement at this time. Have the room darkened, and, if possible, take a good nap; at least avoid talking and all other excitement.

As soon as you are rested and feel able to be dressed, which will not be under three or four hours, the napkin should be removed, and such parts of your person as require it may be washed with soft, warm water, to which a few drops of the Tincture of Arnica has been added, and another napkin applied. This operation should be performed every few hours for the first few days, and as often afterwards as may seem necessary for your comfort. The Arnica need not be added to the water except during the first few days, or only as long as any soreness remains.

Such of your clothing as has become soiled should now be exchanged for other garments which have been well aired and warmed. If you are still to occupy the same bed upon which you have been confined, the clothing may all be changed upon one side while you are upon the other; then allow your attendants to move you to that side. Do not attempt to help yourself even if you feel perfectly able to do so; above all, do not attempt to rise up; but retain your recumbent position for at least five or six days.

TREATMENT AFTER DELIVERY.

LIGHT, TEMPERATURE, AND VENTILATION OF THE ROOM.

LIGHT. — Many persons still labor under the antiquated idea, that it is necessary to keep a lying-chamber constantly darkened, and, in support of the notion, they advance many reasons; the principal one of which is, that the infant's eyes are delicate, too sensitive to bear the ordinary light of day. I often wonder they do not attempt to dilute the atmosphere we breathe, under the pretence that the air supplied by nature is too strong for the infantile lungs. What a wonderful improvement some people would make upon all created things, if they only had the power! For the first two or three days it is well enough that the light should be somewhat modified, but the idea of keeping the room constantly darkened for weeks is simply ridiculous; nay worse, it is injurious to the health of both mother and child. You will not find a *tidy, modern* nurse committing this error. She knows that light is just as necessary for growth and health as air. Besides, her room is always neat and clean; she is not afraid of the light's exposing any of her negligence, and, therefore, has no occasion to darken the room upon the absurd pretence of protecting the child's eyes.

VENTILATION is another important point which is too frequently neglected. Why, not many years ago, it used to be the custom to exclude, as far as possible, every breath of fresh air from the sick-room, to put sand-bags under the chinks of the doors, to nail the window round with lasting, and, by every other possible precaution to oblige the inmates to breathe over and over again the vitiated atmosphere!

Is it any wonder that women in those days suffered more from puerperal diseases than they do now? This pernicious habit, I am happy to say, is fast falling into disuse, and as people become more and more enlightened, new facts take the place of old fallacies, so that, at this time, almost every one recognizes and acknowledges the important truth, that nothing is so essential to good health as plenty of pure, fresh air. Nevertheless, physicians yet have frequent cause for complaint upon this point. Not a few will insist upon keeping windows down and doors closed, for fear of colds and fevers. A surer method of exciting fever could scarcely be adopted. Confine a patient in a vitiated atmosphere, exclude the light, feed her upon water-gruel and slops, and then why wonder that she has fever, or that she does not have a good getting up ?

Do not fall into either of the errors, but be careful that your room is *constantly well lighted and well ventilated.*

The TEMPERATURE should be such as will be most agreeable to your feelings, and this you will find to range somewhere between 67° and 73°. It is not only desirable that your room should be of a proper temperature, well lighted, and well ventilated, but it should also, for the first week at least, be kept entirely free from every kind of excitement; visitors and children should be excluded. In fact, no persons should be admitted, except those whose duty it is to care for you.

If your labor has been at all difficult, and is followed by general soreness, it will be advisable for you to take an occasional dose of Arnica, say six pills once in three hours. In case there is much or any local pain or soreness, you will obtain relief from the external application of a lotion prepared by mixing about thirty drops of the Tincture of Arnica in half a pint of water.

Should you be troubled with any nervous excitement, which sometimes follows delivery, you can take an occasional dose of Coffea; in case this fails to afford relief, you can have recourse to Aconitum, especially should there be any febrile symptom present.

AFTER-PAINS.

After the expulsion of the child and after-birth, the uterine contractions still continue with more or less force until the uterus has resumed the condition which it had before conception. The effect of these contractions or *after-pains,* as they are called, is to expel the clots of blood and shreds of membranes which may have remained within the cavity of the womb.

As a general thing, they commence within half an hour after delivery, and ordinarily cease within thirty or forty hours, though they may continue longer. They vary a good deal in their frequency, their severity, and their duration; usually they are not accompanied with bearing-down efforts.

Their operation is, within certain limits, undoubtedly salutary. They prevent flooding, diminish the size of the uterus, and expel its contents. Nevertheless, when they occur in an aggravated form, and are unduly protracted, which frequently happens in females of excitable, nervous sensibility, they should be subdued as speedily as possible.

In many instances, the employment of *Arnica*, after delivery, is sufficient to prevent the excessive development of these pains. But, should the pain become severe, the Arnica having proved insufficient, and you should feel nervous and excitable, with great restlessness and tossing about, you should take a dose of *Chamomilla*, — six pills, — and follow it in about an hour with *Nux-vomica*.

Pulsatilla is indicated in persons of a mild and gentle disposition, when the pains do not return very frequently, but are protracted and continue for several days.

Coffea. — When the pains are intense, or when they are followed by coldness and rigidity of the body.

Secale. — For pains of a violent description, occurring in females who have borne many children.

ADMINISTRATION OF REMEDIES. — With regard to the dose, you may dissolve a few globules in a wineglass full of water, and take a table-spoonful of the solution from one, two, or three hours apart, according to the severity of the pains; in many cases a single dose will be sufficient. In all cases discontinue the remedies as soon as relief is obtained.

As a general rule, females do not suffer from after-pains subsequent to the first confinement. Exceptions do, however, occasionally occur.

FLOODING AFTER DELIVERY.

A certain amount of blood is always lost after delivery; nor is this injurious; and it is only when it is so great as to produce an impression upon the constitution and the pulse, that it is to be considered as "flooding." Of course, in all cases it escapes from the mouths of the vessels which have failed to contract after the separation of the after-birth.

I presume one of the most frequent causes of hemorrhage after delivery is mental excitement.

The congratulations of friends over her safe delivery will often excite emotions in the fatigued patient, which may prove highly injurious. Worriment from hearing other children in the house crying; depression of spirits, on finding out that her own child is of a different sex from what she anticipated and wanted; in fact, excitement of any kind, be it either of a sad or joyous nature, is almost always certain to produce an injurious result.

It is, therefore, necessary that all excitement should be religiously avoided, and sleep — that great restorer of health and strength — should be courted.

TREATMENT. — The first object is to obtain a firm and steady contraction of the uterus. This can be effected by one of the following remedies:

Belladonna. — When there is pressure, as if everything would fall out from the private parts.

Chamomilla. — When the hemorrhage is accompanied with pains similar to those of labor.

China. — In severe cases, attended with giddiness, loss of consciousness; sudden weariness; faintness; coldness of the extremities; paleness of the face; or if accompanied by colic. It may be given in alternation with *Ipecac.*

Ipecacuanha. — When flooding is very copious and long continued, with cutting pains through the abdomen; chills, with flushes of heat rising into the head; great weakness.

Platina. — Where the discharge is thick and dark, and the flooding has been produced by violent mental emotions.

Pulsatilla. — When the discharge is clotted and appears at intervals, ceases and reappears, *Pulsatilla* may be followed by *Crocus* or *Sabina.*

A drop of the tincture of cinnamon in a tumbler half full of water, giving a teaspoonful every few minutes, will often prove serviceable when other remedies fail.

Cold water is a valuable auxiliary, and in all severe cases should be freely used. Cloths dipped in the coldest water should be applied to the abdomen and genitals, and renewed every few minutes, or pounded ice, if necessary, may be put in bags, and applied in the same manner. Cold drinks are also of service.

ADMINISTRATION OF REMEDIES. — Of the selected remedy, put one drop of the dilution, or twelve globules, in twelve teaspoonfuls of water, and take one spoonful every half hour; in extreme cases every fifteen minutes, until relief is afforded. Lengthen the intervals as improvement becomes manifest, and if no improvement take place in from three to six hours, change the remedy.

DURATION OF CONFINEMENT.

It will be advisable for you to lie quietly in bed for six or eight days after delivery. The length of time, however, will, in a great measure, depend upon circumstances; many women are better able to stand upon their feet within six days than others are within three weeks. Should your general health be poor, your strength gone, or should the flow or discharge called "*Lochia*" be profuse, as it sometimes is, amounting to a hemorrhage, and producing great debility, you will be compelled to remain in a horizontal position for a greater length of time.

And although you may feel strong and perfectly able to get up and help yourself, you had better spend the most of your time, for at least nine or ten days, in bed, or in a recumbent position. This is the safest way; besides, it is always better to err upon the safe side, and remain in bed a little longer than is absolutely necessary, than to get up a little too soon, and, by the indiscretion, bring on local displacements, or other serious diseases, from which it may take you years to recover.

After the first nine or ten days, you may get up, and be seated in an easy-chair for a short time every day; do not, however, attempt to stand upon your feet, or walk about, under twelve or fifteen days at least. After this period, if you feel pretty strong, you can walk about your room; but do not leave it before the expiration of the second week, and do not by any means attempt to go up or down stairs until the end of the third week after delivery.

No woman should consider herself entirely recovered from her confinement until after the expiration of the sixth week.

DIET AND REGIMEN DURING CONFINEMENT.

By a strict and well-regulated regimen during confinement, you will be able to ward off a great many accidents. As I have, in a previous article said, the first thing to be attended to after confinement, or rather, after you have had a short season of rest, is a proper regulation of your clothing; everything about yourself or bed, that has been soiled, or is at all damp, should be removed, and replaced by other articles which are dry. Care should be taken to dry and warm thoroughly every article, before using it.

Great care should be taken, that the utmost cleanliness is preserved; such parts of your person as require it, should be washed at first with soft, warm water; afterwards the temperature of the water may gradually be reduced; never, however, using it entirely cold.

This operation, for the first few days, should be repeated every few hours, and, afterwards, as often as seems necessary for your comfort. The linen should be changed at least every twenty-four hours.

Your food should be of easy digestion, moderate in quantity, and not stimulating. It is not necessary, nor advisable that you should starve yourself; only be careful that your diet is light, and easy of digestion. It may consist of gruel, panada, farina, toast, bread, black tea, broths, and other articles of a similar kind. After the third day, or when the milk-fever has passed, you can have a little soup two or three times a day, and gradually the quantity and kind of nourishment may be increased, until the usual mode of life can be resumed again without danger.

Ales, wines, coffee, and stimulating drinks of every description, which are used to promote the secretion of milk, should be avoided as injurious. Most of these preparations predispose to fevers, and not unfrequently give rise to night-sweats. Coffee especially deranges the nervous system of both mother and child, and produces numerous diseases of the digestive organs.

As a drink, you can use black tea, cold water, either pure or with a little strawberry or raspberry syrup. A few drops of claret, added to cold water, make a good beverage for a woman in confinement. Some women are very fond of alkathrepta, or prepared cocoa; you can procure it at any homœopathic pharmacy.

As regards the restoration of strength after confinement, I do not know as any particular treatment is required; if no serious derangement has taken place, it will soon return, by means of a good diet, and rest. Sleep is an excellent restorer of strength in cases of confinement; should nervous excitement prevent you from getting as much sleep as you need, take an occasional dose of *Coffea*. Should this fail, and there be any febrile symptoms present, *Aconite* will usually suffice. *Chamomilla* is another excellent remedy, especially when there is great restlessness and tossing about.

If you should have become exhausted and debilitated by night-sweats or flooding, you may dissolve six or eight globules of *China* in twelve teaspoonfuls of water, and take a spoonful of the solution once in three hours.

DISEASES FOLLOWING PARTURITION.—THE LOCHIA.

The discharge of blood which accompanies delivery, continues for several days afterwards, doubtless from the mouths of the vessels exposed by the separation of the after-birth. After three or four days the character of this discharge changes, and, instead of continuing a mere escape of blood, it takes on the character of a secretion. This discharge is called the "lochia." For the first three or four days, it continues of a 'red color, but much thinner and more watery than blood ; it then sometimes becomes thick and yellow, but more frequently it maintains its watery consistence, and changes its color successively to greenish, yellowish, and lastly to that of soiled water.

The duration of the lochial discharge varies considerably in different females ; in some it is thin and scanty, and ceases in a few days, while in others it continues for several weeks, and sometimes so profuse as to almost amount to a hemorrhage.

As this secretion is necessary to health, its *sudden* suppression is generally attended with evil results. Frequent washings, with soft, warm water, should be practised as long as it continues.

SUPPRESSION OF THE LOCHIA.

The causes of lochial suppression may be either exposure to cold, errors in diet, or sudden mental emotions.

The symptoms accompanying a suppression are generally chilliness, fever, thirst, headache ; sometimes delirium, pain in the back, limbs, etc.

TREATMENT. — *Bryonia* will be of service when there is accompanying the suppression, fulness and heaviness of the head, with pressure in the temples, throbbing headache, pain and aching in the small of the back, and scanty discharge of urine.

Pulsatilla, — however, is the principal remedy for sudden suppression, either from mental emotions, exposure to dampness, or any accidental cause, particularly if it is followed by fever and headache, coldness of the feet, and frequent desire to pass water.

Dulcamara and *Puls.* — may be taken in alternation, when suppression arises from exposure to cold. If suppression is followed

by diarrhœa and colic, take *Chamomilla.* For suppression consequent upon some mental emotion, take *Platinum.* Warm com-
presses around the abdomen, warm hip and foot baths are also of
service.

ADMINISTRATION OF REMEDIES.—*Pulsatilla* and *Dulcamara* may
be taken in alternation every two hours. Of the selected remedy,
dissolve twelve globules in twelve spoonfuls of water, and take one
dessert spoonful of the solution from one to four hours apart, according to the necessity of the case.

EXCESSIVE OR PROTRACTED LOCHIA.

When the lochial discharge is too profuse, or continues too long,
one of the following remedies should be taken.

Aconitum. — Is indicated in profuse lochial discharges of a deep
red color. Should *Aconitum* prove insufficient, take *Calcaria-
carb ;* but, as a general thing, *Aconitum* will be found sufficient
to check it in two or three days, without the assistance of other
remedies.

China and *Ipecac.* — In alternation, if the discharge takes place
in paroxysms, with nausea, vertigo, fainting, coldness of the extremities, paleness of the face, and debility.

Crocus.—Is indicated, when the discharge is dark-colored, black,
and of a viscid or sticky consistancy, with a feeling in the abdomen as of something alive.

Rhus. — In cases where the lochia return after they once had
ceased.

ADMINISTRATION OF REMEDIES. — Of the selected remedy, take
six globules once in four hours until better.

Tepid hip-baths are valuable assistants, and in all severe or
obstinate cases should be freely made use of.

Complete rest and good nourishment are indispensable to correct this disorder.

MILK FEVER.

About the third or fourth day after confinement, you may expect your breasts to become distended with milk, and at the same
time you may experience something of a chill, followed by more
or less fever and headache. This is called the " Milk Fever."

The appearance of this secretion, though ushered in with some
fever, is seldom attended with sufficient disturbance to call for

medicinal interference, especially when you nurse your own infant, and the milk can be drawn off as soon it commences to flow.

If, however, you do not nurse your infant, this fever may become complicated with other ailments, which it is necessary to prevent. For this purpose, you will find one of the following remedies sufficient: *Aconite, Arnica, Bryonia, Belladonna,* and *Pulsatilla.*

Aconitum. — If there is much fever, with hot skin; violent thirst; breasts hard and knotted; restlessness; anxiety, and discouragement.

Arnica. — Given internally, and applied externally to the breast as a lotion, will be found useful, when there is much distension, hardness, and soreness.

Bryonia. — When the symptoms have been partially removed by Aconitum, or if the breasts become much distended by the milk, with painful oppression of the chest.

Belladonna. — After, or in alternation with *Bryonia,* when the latter has not been sufficient to remove the symptoms entirely; or if head-symptoms set in, with stupefying headache, glistening eyes, delirium, etc.

Pulsatilla. — Will be found particularly useful in severe cases, especially when caused by taking cold, and when there is great distension of the breasts, with soreness and rheumatic pains, extending to the muscles of the chest, abdomen, and arms. " A timely administration of this remedy will, in many instances, prevent a threatened attack of childbed-fever." — HERING.

External applications are of little use during a milk-fever, except perhaps the Arnica Lotion, which I have already mentioned. Some physicians recommend bathing the breasts with hot lard, to which has been added a little *diluted* Arnica Tincture, and then covering them with raw cotton. The milk should be drawn out as soon as possible, either by the child or by the nurse.

ADMINISTRATION OF REMEDIES. — Dissolve twelve globules of the selected remedy in twelve teaspoonfuls of water, and of the solution take one spoonful every two hours in severe cases; in other cases, every four hours, until relief is obtained.

DIET. — During the continuance of this fever, none but the lightest articles of diet should be partaken of; such, for instance, as gruel, boiled rice, toast, toasted crackers, and black tea, homœopathic cocoa, or other equally light articles.

SUPPRESSED SECRETION OF MILK.

The secretion of milk being a natural function, its sudden suppression not unfrequently produces sudden disorders, such as internal or local congestion and inflammation, determination of blood to the head, chest, or abdomen, and the usual train of symptoms which constitute childbed-fever. The circumstances which cause a suppression of the milk are numerous and variable. The most prominent, however, are, exposure to cold or dampness; errors in diet; sudden mental emotions; nervous excitement, and diseases in other parts of the system.

The evil effects of a suppression of this secretion are frequently of so serious a nature, that the slightest diminution in the quantity of milk should excite your apprehension, and place you upon your guard; for in the great majority of cases, when this difficulty first manifests itself, the administration of *Pulsatilla* will be found sufficient to check the disorder at the outset, and restore the flow of milk. If any unpleasant symptoms still remain, after the use of *Pulsatilla*, they will, in most cases, yield to the administration of *Calcaria-carb*.

Should active febrile symptoms set in, such as hot, dry skin, thirst, etc., take *Aconite* at short intervals until a favorable impression is made. Where there is excessive restlessness, in addition to the fever, you may alternate *Aconite* and *Coffea*, especially if mental emotion has caused the trouble, and there is great nervous excitement.

Belladonna and *Bryonia* will be found serviceable when there is congestion of the head or lungs, with fever, pain and aching in the limbs, and especially if these symptoms have been preceded by a chill.

ADMINISTRATION OF REMEDIES.—Dissolve of the selected remedy twelve globules in twelve teaspoonfuls of water, and of the solution take one dessert-spoonful every two or four hours, according to the necessity of the case, for twelve hours. If the symptoms become favorable, diminish the frequency of the doses, but continue the remedy for twenty-four or forty-eight hours longer. Should any unpleasant symptoms remain, take *Zincum-metal.*, four doses. at intervals of twelve hours between the doses.

EXCESSIVE SECRETION OF MILK.

It sometimes happens that the secretion of milk is too abundant causing painful distension of the breasts, and involuntary emis sions of milk, which is not unfrequently followed by emaciation and debility, or nervous and inflammatory disorders. In cases of this description, *Calcaria-carb.* or *Phosphorus* will generally afford relief.

At first, commence by dissolving six globules of *Calcaria* in twelve spoonfuls of water, and take one spoonful every six hours. If this does not afford relief, prepare and take *Phosphorus* in the same manner.

When febrile and congestive symptoms arise from over-distention of the breasts, *Aconitum* and *Belladonna*, will be found useful. They may be taken singly, or in alternation from three to six hours apart, according to the urgency of the case.

For involuntary emissions of milk, where the parts are kept constantly wet, and thus rendered more liable than usual to cold, upon slight exposure, either one of the foregoing remedies may be taken, as above, a dose every twelve hours; or, for debilitated persons, from loss of fluids or other causes, *China*; and, for females of mild, easy dispositions, *Pulsatilla* may be administered in the same manner.

In all cases, it is advisable to make an external application of cotton-batting. This tends to reduce the swelling and mitigates the pain.

CONSTIPATION AFTER CONFINEMENT.

It appears to be natural for the bowels to remain inactive for several days after delivery, the secretion from the intestinal tube being wholly or partially suspended; and this is not to be wondered at, when we take into account the great changes going on at this time within the female organism, whereby a great quantity of liquids is discharged from the womb and breasts. This, together with the vicarious action of the skin, demonstrating itself by the increased perspiration, amply compensates for the temporary inactivity of the alimentary canal, and, by this provision of nature, the balance of the system is kept up.

We cannot, therefore, too strongly condemn the use of aperient medicines in these cases: they only tend to promote irritation, which is indeed but the stepping-stone to inflammation. And, besides, the relaxation thus produced always interferes with the proper secretion of the milk.

It used to be the custom, and in fact is yet, to a great extent, among the Allopathic fraternity, to give a mild cathartic on the

13

second or third day after delivery. This was considered essential, and by many physicians recommended, even though the bowels had previously been freely moved. Of what use it was, I am certain I do not know.

I once inquired of an elderly physician why he always gave Castor-oil the second day after delivery. He hesitated a moment, and then said, "It is to carry off the impurities." A very definite answer ; but, nevertheless, I ventured to suggest that perhaps it was a provision of nature, that the bowels should remain in a torpid state for the first five or six days, so as not to disturb the patient, but to let her remain quiet until the uterine organs had time to regain their natural form and position.

He did not coincide with my view, but thought constipation a state which, if not speedily removed, would be productive of many evils.

I have frequently known cathartics, when administered under these circumstances, to produce most serious results ; but I have yet to see the first case where any trouble has arisen from this temporary inactivity of the bowels. In fact, I am of the decided opinion, that this state of torpidity is a wise provision of nature for woman's comfort, safety, and convenience. I would, therefore, advise you to eschew Castor-oil, and all other cathartics, no matter how innocent they may appear, or how strongly they may come recommended.

As a general thing, upon the fifth or sixth day the bowels will move spontaneously ; however, should they fail to do so, and you complain of uneasiness or pain in the bowels, with fulness of the head, a dose or two of *Bryonia* — six pills — will usually afford relief. Should this prove insufficient, take one dose of *Nux-vom.* in the afternoon, and a dose of Sulphur the following morning.

In obstinate cases, which, by the by, are seldom met with, where it appears necessary to afford mechanical assistance, you can make use of an injection of lukewarm water, to which you have added a little Linseed-oil.

DIARRHŒA AFTER CONFINEMENT.

Diarrhœa during confinement is to be looked upon as a highly dangerous condition, and prompt means should be at once taken for its speedy removal.

As a general thing, the occasion of this disorder is cold, errors in diet, or the abuse of aperient medicines.

When arising from checked perspiration, produced by chills, from exposure to cold or dampness, *Dulcamara* will be the appropriate remedy.

Rheum or *Antimonium-c.* — For watery and exceeding offensive evacuations, *Rheum*, especially if the stools smell sour and fetid, with much pain and straining after each evacuation. *Antimonium*, when the tongue is coated white, and there are frequent, bitter eructations, diarrhœa worse during the night and early in the morning.

Hyoscyamus. — For painful and almost involuntary evacuations.

Pulsatilla. — When the diarrhœa occurs mostly at night, and is accompanied by inefficient straining; the evacuations are small, sometimes only a little mucus passing, accompanied with severe pain in the anus.

For obstinate and protracted cases, when the evacuations are watery, painful, and almost involuntary, take *Phosphorus.* Should this fail to afford relief, try *Phos-ac.*

For diarrhœa with clay-whitish, curdled, or sour-smelling, musty evacuations, accompanied with nursing sore-mouth, take *Nux-v.* and *Hepar-s.* in alternation every three hours.

ADMINISTRATION OF REMEDIES. — Of the selected remedy, dissolve twelve globules in twelve spoonfuls of water, and take one dessert spoonful from two to four hours apart. Or, if you prefer it, you can take the medicines dry, upon the tongue; five or six pills for a dose.

RETENTION OF URINE, OR PAINFUL URINATION, DURING CONFINEMENT.

It not unfrequently happens, especially after severe labor, that the neck of the bladder and the whole tract of the urethra become extremely sensitive, causing painful emissions, and sometimes even entire retention of urine.

This sensitiveness arises from the great amount and long-continued pressure to which the parts have been subjected.

Retention of the urine, when it lasts for any considerable length of time, is to be held as a dangerous affection, because, if relief be not obtained, if the pressure on the inner surface of the bladder be not relieved by the removal of the accumulated water, inflammation must necessarily follow. Fortunately, complete retention is seldom met with, and the painful and difficult emissions of

urine which are frequent, as a general thing, yield readily to the
following treatment : —

Arnica should be the first remedy taken, as it is especially
indicated in cases like the present, where the difficulty arises
from mechanical injuries. Should *Arnica* fail, and there be considerable fever, with burning heat in the region of the bladder,
take Aconitum.

Belladonna. — When there are darting and pricking pains, extending from the lower part of the back to the bladder; also
when there is great agitation and colicky pain.

Camphor is indicated when the retention arises from spasmodic
contraction of the neck of the bladder.

Nux-vomica and *Pulsatilla* are also valuable remedies. *Nux-
vomica*, especially, if there is also constipation.

The application of warm fomentations to the parts will sometimes prove of valuable assistance, or sitting over a pan which
contains warm water will often have the desired effect.

ADMINISTRATION OF REMEDIES. — Of the selected remedy, dissolve six pills in twelve teaspoonfuls of water, and take one teaspoonful of the solution every two hours until relief is obtained,
and if no relief is afforded in the course of eight or ten hours,
proceed to select from the other remedies.

SORE NIPPLES.

This frequent and exceedingly annoying complaint, may, in the
large majority of cases, be prevented, if proper care of the breasts
is taken previous to confinement. Of this we have spoken at
large, under the head of "PREPARATION OF THE BREASTS," which
see.

There appears to be a constitutional tenderness of the skin in
some females, which predisposes it, upon the slightest occasion, to
the development of cracks and sores of a most distressing nature,
and which at times prove most obstinate to heal. Wherever a
tendency of this kind exists, the utmost care should be taken to
avoid the least irritation or abrasion of the skin, either by your
clothing, by the shield, if you use one, or by the breast-pump.
When a shield is made use of, it should with care be frequently
removed, and the parts bathed with a weak lotion of Arnica Tinct.
or brandy and cold water. This will obviate the otherwise certain result of tenderness and consequent excoriation.

Do not spare any pains or labor to do all in your power to prevent a siege of sore nipples; for I can assure you, if women are to be believed, the pain is of a most intolerable kind; and, indeed, we cannot doubt it, if one may judge from the appearance of the mother, down whose cheeks tears are seen to flow as she submits to the torture of nursing her infant.

There is no doubt that many cases of broken breasts owe their origin to the reluctance of the mother to encounter the pangs of suckling her infant while these cracks and fissures remain uncured.

The most frequent form of sore nipples consists of a long, narrow ulcer, about as wide as a horse-hair, and varying in length from the sixteenth of an inch to the whole circumference of the nipple.

The chief difficulty in healing sores of this nature, you will readily observe, arises from their being constantly torn open afresh by the efforts of the child in nursing. It is, therefore, very important, especially where the fissures are deep and gape open, that some means should be devised to keep the edges pressed together. This can be accomplished with a narrow bit of adhesive plaster, or you can spread some adhesive salve upon a piece of narrow ribbon; the latter, on account of its pliability, I have found to answer the purpose better than the common adhesive plaster. I have also used arnicated collodion, in the same manner, with great success. This, as well as the other application, will admit of the child's nursing without tearing the fissures open afresh.

In all cases, as soon as the child has left the breast, the nipple should be washed with *cold* water, to which a few drops of Arnica Tinct. have been added, and should then be thoroughly dried. Then, taking the nipple between the thumb and first two fingers, gently compress it. This is done for the purpose of disgorging the small vessels that have become distended by the suction of the child. As soon as you have rendered the nipple soft and flexible, cover it over thickly with powdered wheaten starch or Gum-Arabic. " Pulverized white sugar," says Dr. Hering, " makes an excellent application."

Should the above precautionary treatment, which we have advised, prove inefficient, and the nipple crack become sore and refuse to heal in spite of all your care and attention, you will then have to resort to the administration of internal remedies, to counteract or remove the constitutional taint to which the disease generally owes its origin.

Sulphur seems particularly indicated for most cases of this affection, especially when the nipples are sore and chapped, with deep fissures around the base, which bleed and burn like fire. When these fissures are large, bleed easily, and prove obstinate to heal, you will generally find them to contain little granulations of proud flesh. To all such cases I apply burnt Alum, or pulverized Tobacco-ashes, and then join together the edges as before directed. In cases where *Sulphur* fails to afford relief, *Calcarea-carb.* will generally prove beneficial.

All cases of sore nipples, however, do not present themselves in the form above-described; sometimes the nipple becomes abraded or excoriated, and even suppuration occasionally takes place. I saw a case, not long ago, where from abrasion of the end of the nipple inflammation was excited, which extended down the milk-tubes into the substance of the gland; suppuration followed, and apparently the whole interior of the nipple sloughed out, leaving a large cavity. This case was readily cured by the administration of *Mercurius* and *Silicea.* The only external application made use of was pulverized white sugar, with which the cavity was occasionally sprinkled.

In all cases of soreness or excoriation of the surface, the parts should be frequently laved with Arnica Lotion, and *Arnica* at the same time should be taken internally.

Chamomilla will be found of service when the nipples are very much swollen and inflamed.

Nux-vomica. — For soreness of the nipples with excoriation of the adjacent parts.

For obstinate and severe cases of every description, where the above remedies fail to answer the purpose, recourse may be had to one of the following medicines: — *Mercurius, Silicea, Lycopodium, Graphitis, Sepia.*

A very important point in the successful treatment of these cases is, to keep the parts perfectly dry. This can best be accomplished, as I have already said, by wrapping the nipple in pulverized Starch or Gum-Arabic.

Numerous domestic remedies, in the form of powders, salves, and lotions, have been used with various results. Borax, dissolved in mucilage of Slippery Elm, makes a pleasant wash, which often proves of service. Powdered Potter's Clay, sprinkled upon the parts, often acts like a charm. Reliance, however, cannot be placed upon any form of treatment, especially in severe cases, except the internal administration of appropriate remedies. In all cases where external applications of any description have been made use of, the nipple should be carefully cleansed with a little warm milk and water before presenting it to the child.

ADMINISTRATION OF REMEDIES. — After selecting a remedy, dissolve twelve globules in twelve teaspoonfuls of water, and take one spoonful of the solution every six hours. If you prefer the dry pills to the infusion, you can take six pills dry, upon the tongue, three times a day, morning, noon, and night.

GATHERED BREASTS, OR BROKEN BREASTS.

To make perfectly plain and intelligible the nature and importance of this disorder, I shall first give a brief anatomical description of the female breast.

Beneath the skin on the front of the chest, there lies — one on each side — a large secretory organ, called the mammary gland. It is composed of milk-tubes, nerves, arteries, veins, and lymphatics, the whole being inclosed by a fibrous investment, which also sends out prolongations through the glands, dividing it into numerous lobes. Between these frequent membranous divisions, especially near the skin, exist numerous small cells in which fat is deposited, giving to the surface its beautiful, soft, smooth, hemispherical form.

The nipple is but a bundle of milk-tubes, nerves, and blood-vessels, gathered together, and covered with a thin derm or skin.

The milk-ducts or tubes, resembling little canals, vary from ten to fifteen in number. When distended, they are about the size of a small goose-quill. Starting from the extremity of the nipple, they enter the breast, soon become divided and subdivided, becoming finer and finer as they go inward, until each minute tube terminates in a small hollow globule or granule, about the size of a mustard-seed, from the inner surface of which the milk is secreted. The number of these little granules it would be impossible to count.

If you should take a small syringe and inject each of these ten or fifteen distinct milk-tubes from the nipple with different colored substances, thus, filling one canal with yellow, another with green, a third with violet, and so on, until the whole breast was completely distended, you would see no amalgamation of colors, no uniting or coalescing of tubes, but each injection would follow its own canal, through all its divisions and subdivisions, to its granular termination.

Thus, you observe, we can trace the course of each milk-tube from its exit at the nipple, through all its divisions and divergences, to the actual minute milk-producing granule, just as we can trace a river upon the map, from the broad Atlantic where it empties, to the very springlets among the distant mountains where it has its origin.

The quantity of milk that a given gland will produce at one time does not so much depend upon the size of the organ as upon its secretory power. With a breast-pump some women can draw out a half pint from one breast at one sitting; not that it was actually all present in the breast when she began, but was secreted, as it were, upon demand, — the flow of milk only ceasing when the secretory power of the gland becomes exhausted, and then a period of rest is demanded. To carry on this process of milk-secretion, it is necessary that the organ should be supplied with a large amount of blood and nerve-power. Accordingly, we find numerous branches from large arteries distributed throughout the breast, while, by a great number of nerve-fibres, it is intimately connected with the two great nervous systems.

During lactation, the breasts are in a high state of activity, which, together with their intimate connection with the rest of the system, renders them exceedingly liable to partake of any disorder, either physical or mental, which happens to affect a woman while nursing. Thus we shall find ague in the breast, as it is called, arising from cold, from a chill, from fright, anger, fear, grief, etc.

Gathered breasts not unfrequently arise from a too tardy application of the child to the breasts, or from sudden cessation of suckling, occasioned either by the death of the child, or an unwillingness on the part of the mother to encounter the pangs of nursing the infant, consequent upon sore nipples.

When the breasts become distended with milk, and all of their little milk-tubes are filled and crowded against one another, you will often find it incompressible, and its sensibility so greatly increased, that the least handling produces great pain. Now, unless this tension is speedily reduced, as a natural consequence, inflammation must necessarily follow, or fever soon arises, ushered in by rigors, or severe chills.

The treatment is, of course, to take out the milk: as soon as this is done, the breasts become cool and flaccid, and the freest handling produces no pain.

Do not let the breasts become distended; apply the child often, — as often as the breasts become filled, if it is every five minutes in the day.

The principal remedy for this affection, especially at the commencement, when the breasts become distended, hard, and feel heavy, with shooting pains, dry skin, thirst, and other febrile symptoms, is

Bryonia, either alone, or in alternation with *Belladonna*.

Belladonna is especially indicated, where, in addition to the above symptoms, there is redness of the skin, resembling erysipelas, with shooting, tearing pain through the breasts, and headache. These two remedies will generally, if the breasts are kept well drawn, be sufficient to effect a cure. When, however, some degree of hardness still remains, *Mercurius* should be taken, and repeated once in six hours.

Where hard lumps or cakes are felt deep down in the breast, you must, by some means or other, soften them, and extract the milk.

These lumps, or cakes, as they are commonly called, are caused by the milk-tubes becoming clogged up, or rather, they become distended, and crowd against each other, until they are so compressed, that the flow of milk is obstructed, and thus one division of the gland becomes caked, while the rest remain open.

Nurses make use of all sorts of embrocations and hot applications to scatter the cakes, which simply means, to soften and relax these particular tubes, so that the milk can flow. And this *must* be done, or inflammation, followed by suppuration, will be the result.

A chill acts in the same manner, or at least is productive of the same results; the breast increases in size from congestion of its blood-vessels, and consequent obstruction of the milk-tubes, and the result, if not prevented by prompt interference, as before, will be inflammation and suppuration.

When the breasts become swollen and very tender, *Belladonna* or *Bryonia* should be taken, at least once in two hours; and, at the same time, you will do well to apply externally flannel cloths wrung out from hot brandy.

Should the swelling and tenderness subside, but still there remain lumps or cakes in the breast, you will find relief from applying a plaster made of beeswax and sweet-oil.

The great art in preventing gathered breasts is to keep the

14

breasts well drawn ; if the child is unable to do it, then you must resort to nipple glasses, the breast-pump, or, what is better than either, the lips of the nurse, or some other adult person.

You will seldom find a nurse who will acknowledge that ever such a thing as a broken breast did occur to a patient of whom she had the entire charge ; but all such assertions it is as well to take with a few grains of allowance, for in spite of all precautions the breast will sometimes gather and break.

When suppuration is about to take place, you will know it by the throbbing pain in the breast, accompained by chills ; then take *Hepar-Sulphur* until it breaks.

In the early stages of this disorder, it is best to abstain from applying warm poultices, as it has a tendency to involve a still larger part of the breast within the suppurative sphere. But as soon as the gathering points, or when it becomes evident that it must soon break, it should be hurried along as fast as possible ; and if you employ a physician, he will at this period undoubtedly lance it.   Ground flax-seed makes the best poultice ; it should be applied warm, and changed about once in three hours.

When the abscess has opened, and the matter has been discharged, the breast should be compressed, either by strips of adhesive plaster, or a bandage.   This, you will find, will facilitate the process of healing.

Where there is a profuse discharge of matter, you should take *Phosphorus* alone, or in alternation with *Hepar-sulphur.*

Silicea. — In cases where the discharge becomes watery, fetid, and when it proceeds from several openings, which do not seem disposed to heal.

Sulphur.—In obstinate cases, and where there is a profuse discharge of matter.

Arnica. — In all cases where the disease arises from external injuries.

Should the above remedies fail to produce a cure, you can have recourse to *Graphitis* or *Calcarea-carb.*

During all the time that the breasts have been gathering, and still after the abscess has broken, the infant should be permitted to nurse ; for you must recollect, that milk is secreted by that portion of the gland which is not involved in the abscess, and it must be withdrawn.   If the infant cannot, or refuse to do it, you must resort to artificial means.

ADMINISTRATION OF REMEDIES. — When, at the commencement

of an attack, you wish to take *Belladonna* or *Bryonia*, you should dissolve twelve globules in twelve teaspoonfuls of water, and take of the solution one spoonful every hour; if taken together, alternate them every hour. But the other remedies, after preparing them in the same manner, need not be taken oftener than once in three hours,—from that to six hours; and along toward the close of the case, when you are taking *Sulpur* or *Calcarea*, one dose, night and morning, will be sufficient.

DIET. — The diet should be plain and nourishing, but not stimulating.

CHILD-BED FEVER, OR PUERPERAL PERITONITIS.

I shall not enter into any detailed description of this disease, because I do not deem it safe for any but an experienced physician to attempt its treatment. I shall, therefore, but briefly give its nature and characteristic symptoms, together with such remedial means as will be adapted to the premonitory symptoms and first stages of an attack.

DEFINITION. — Child-bed fever, or Puerperal Peritonitis, as it is technically called by physicians, is an inflammation of the peritoneum, or serous membrane lining the abdomen and covering the bowels. It is not unfrequently complicated with inflammation of the womb and its appendages.

CAUSES. — Among the exciting causes of this disease, may be enumerated, violence during delivery, taking cold, diarrhœa, irritation of the bowels, induced by cathartic medicines, severe mental emotions, suppressed secretion of milk, and so on.

SYMPTOMS. — Child-bed fever is generally "preceded or attended by shivering, and sickness or vomiting, and is marked by pain in the belly, which is sometimes very extended, though in other cases it is at first confined to one spot. The abdomen very soon becomes swelled and tense, and the tension rapidly increases. The pulse is frequent, small, and sharp; the skin hot; the tongue either clean, or white and dry; the patient thirsty; she vomits frequently, and the milk and lochia usually are obstructed. These symptoms often come on very acutely, but they may also approach insidiously. But whether the early symptoms come on rapidly or slowly, they soon increase; the belly becomes as large as before delivery, and is often so tender that the weight of the bedclothes can scarcely be endured; the patient also feels much pain when she turns; the respiration becomes difficult, and sometimes a cough comes on which aggravates the distress; or it appears from the first to be attended with pain in the side, as a prominent symptom. Sometimes the patient has a great inclination to belch, which always gives pain. The bowels are either costive, or the patient purges bilious or dark-colored fæces. These symptoms are more or less acute, according to the extent to which the peritoneum is affected. They are, at first, milder and more protracted in those cases where the inflammation begins in the uterus, and in such the pain is not very great nor very extensive for some time. In fatal cases, the swelling and tension of the belly increase; the vomiting continues; the pulse becomes very frequent and irregular; . . . the extremities become cold, and the pain ceases rather suddenly. The patient has unrefreshing slumber, and sometimes delirium, but she may remain sensible to the last." — *Gardiner's Medical Dictionary*.

TREATMENT. — *Aconitum* in the majority of cases is the first remedy called for, especially if the disease commences with a chill, and is succeeded by a dry, hot skin, thirst, clean tongue, accelerated pulse, and attended with anxiety, forebodings of evil, etc.

Belladonna. — Especially should there be deep-seated pains in the abdomen, with dragging downwards; throbbing pains in the head; face at times flushed and full; glassy appearance of the eyes; delirium; spasmodic eructations, mostly bitter; retention of urine; distention or excessive tenderness of the abdomen, sometimes with shooting and digging pains; painful pressure on the genital organs.

Bryonia. — Sensitiveness of the abdomen; constipation, with shooting pain in the abdomen; high fever, with great thirst. This remedy may be given in alternation with *Aconitum*.

Pulsatilla. — In patients of a mild and gentle disposition, where the attack is mild in the beginning; great pressure downwards, with frequent inclination to pass water; suppression of the lochia; tendency to diarrhœa.

Other remedies applicable to this disease are, — *Apis-mel*, *Arnica*, *Arsenicum*, *Chamomilla*, *Hyoscyamus*, *Nux-vomica*, *Rhus*, and *Sulphur*.

ADMINISTRATION AND DOSE. — In the commencement of an attack of child-bed fever, your safest treatment will be to give Aconitum and Belladonna in alternation, every one, two, three, or four hours, according to the urgency of the symptoms. Give ten or twelve globules at a dose.

I have given you the above outline of symptoms and brief form of treatment, for the purpose of enabling you to prescribe for a patient, in cases of emergency, or until the services of a physician can be obtained.

MILK-LEG, OR CRURAL PHLEBITIS.

DEFINITION. — Milk-leg is the common name given to a peculiar form of disease which sometimes affects women during confinement. As the name implies, it was once supposed that the milk had fallen into the woman's leg. I cannot say that physicians ever took this view of the disorder, but certainly the people did, and it is no uncommon occurrence to meet with persons who still insist that the milk has gone into the leg, because the limb is swollen and looks white; and, besides, the milk has in part or all disappeared from the breast. All the reasoning in the world will not make them believe otherwise. But it is the sheerest nonsense to say the milk has fallen into the woman's leg; for such a thing is impossible.

Physicians, now, who know anything about the disease, call it Crural Phlebitis, which name signifies an inflammation of the veins of the leg; and this is the true seat and nature of the disease. The swelling of the limb is due to the effusion of lymph and serum from the blood into the cellular tissue.

CAUSES. — The exciting cause is generally the impression of cold.

SYMPTOMS. — The ordinary premonitory symptoms of an attack of this disease often resemble and are not unfrequently mistaken for after-pains. There is uneasiness or pain in the lower part of the abdomen, extending along the brim of the pelvis through the hips. The patient is irritable, depressed, and complains of great weakness.

Often, however, there will be no precursory symptoms, — the patient being suddenly seized with pain in the groin or calf of the leg, and not unfrequently the patient will complain of pain in the hip-joint, calling it neuralgia, or rheumatism.

As soon as the inflammation is fairly set in, the region about the groin becomes tumefied; and in a short time — twenty-four or forty-eight hours — the thigh becomes swollen, tense, white, and shiny. The swelling, which sometimes increases the limb to the size of a man's body, or an elephant's leg, may be confined to the thigh, or it may extend down to the foot.

When the pain commences in the calf of the leg, the swelling is first observed there, and gradually extends itself up the leg and thigh. The temperature of the limb is generally increased, although, in some cases, it falls below the natural standard.

Along the course of the inflamed vein, although there is great tenderness, there is neither redness nor other discoloration. In most cases, the vein may be traced from the groin down the thigh, feeling hard, and rolling under the finger like a cord.

" Either leg may be affected, though the left appears to be more frequently attacked, and it not unfrequently happens that the sound leg participates in the disease before the disease is perfectly removed, and then the disease runs a similar course a second time."

TREATMENT. — The treatment of this disease should be undertaken only by an experienced physician. I shall simply enumerate a few remedies, which may be employed at the commencement of an attack.

Aconitum. — If the disease has an acute character, with high fever, heat all over, and violent pains.

Arnica. — If phlebitis sets in after tedious labor, or after an injury.

Belladonna. — This seems to be the better remedy in the commencement of most cases, especially when there are sharp, stitching pains, as with knives ; heaviness in the thighs and lower part of the abdomen ; creeping in the limbs ; violent fever, with burning thirst ; great sensitiveness to touch or motion.

Bryonia. — When there are drawing or lancinating pains from the hip to the foot, with copious sweat, and excessive tenderness to touch or motion.

Pulsatilla. — If *Belladonna* or *Bryonia* effected no improvement. Other remedies recommended for this disease, are *Rhus, Sulphur, Nux-vomica, Arsenicum.*

ADMINISTRATION OF REMEDIES. — Of the selected remedy, take six globules, dry, upon the tongue, once in two hours.

NURSING SORE MOUTH.

In this disease, the soft part and sometimes the whole interior of the mouth becomes very red, and so sensitive and tender, as to render it almost impossible for the patient to partake of any solid food whatever. This is quite a different disease from what is generally called canker sore mouth. In some females, it appears to be constitutional.

As I have before remarked, the breasts are intimately connected with the whole nervous system ; you will not be surprised, therefore, to learn, that this form of sore mouth arises from the peculiar irritation which the act of nursing produces upon the digestive organs.

If not properly treated, it sometimes becomes so severe, and is attended with so much suffering and debility, that the weaning of the child becomes absolutely necessary.

The weaning of the infant has a magical effect upon this disease, —the whole of it vanishing as soon as nursing is discontinued.

In the majority of cases, this disease can readily be controlled by some one of the following remedies : —

Mercurius. — This is a prominent remedy, and may be given in alternation with *Nux-vomica* or *China.* With *China,* especially, when there is great debility and exhaustion.

Should this fail, *Borax* may be used ; and in severe and obstinate cases, recourse must be had to *Nitric-acid* or *Sulphur.*

Sometimes an exhausting diarrhœa accompanies Nursing Sore Mouth. When such is the case, and evacuations are sour, curdled, or musty, *Nux-vomica* and *Hepar-sulphur* may be taken in alternation.

ADMINISTRATION OF REMEDIES. — Of the remedy chosen, dissolve twelve globules in twelve teaspoonfuls of water, and take one spoonful of the solution at a dose, from four to six hours apart. Or, if you prefer, you can take a dose of six globules dry, upon the tongue, once in from six to eight hours. The repetition of the dose should be governed by the severity of the case.

When *Nux-vomica* and *China* are used in alternation, the dose may be repeated as often as once in four hours. When taking *Sulphur,* a dose night and morning will be sufficient.

DIET AND REGIMEN. — The diet of a woman suffering from Nursing Sore Mouth should be generous and nourishing, but not flatulent. Whatever articles of food are found to disagree should be strictly avoided.

Exercise in the open air will be found beneficial.

PERSPIRATION AFTER DELIVERY.

The increased perspiration, which takes place immediately after delivery, and continues for several days, acts, as I have before remarked, as a substitute for the suspended mucous secretion, and consequent inactivity, of the alimentary canal.

Therefore its sudden suppression from exposure to cold, or a sudden chill, is unavoidably followed by some injurious result, not unfrequently gathered breasts, diarrhœa, or child-bed fever.

When sudden exposure to cold, especially dampness, has caused the suppressed action of the skin, *Dulcamara* will be found the most efficient remedy to bring about a renewed action. A dose of six pills may be taken every four hours, until four doses have been taken, when the interval between the doses may be lengthened. Should this remedy fail, and there be great excitability and restlessness, with colic and relaxation of the bowels, take *Chamomilla* and *Mercurius* in alternation.

Belladonna. — Should lateral headache occur, with pain in the back of the neck.

Bryonia. — Will be found serviceable when the suppression is followed by chills, or severe pain in the head and limbs. Should there be much fever, *Aconitum* may be given in alternation with *Bryonia.* In some cases *Sulphur* or *Nux-vomica* may be called for.

ADMINISTRATION OF REMEDIES. — The remedy can be taken dry, or in solution. When taken dry, place six globules upon the tongue, and let them dissolve. When in solution, dissolve twelve globules in twelve spoonfuls of water, and take one spoonful every three or four hours, according to the urgency of the case.

EXCESSIVE PERSPIRATION AFTER DELIVERY.

Excessive perspiration, besides causing great debility, predisposes to other disorders, by the high susceptibility of taking cold which it occasions. As a general thing, a few doses of *China* will be all that is necessary for its removal, unless it be occasioned by the too high temperature at which your room is kept, in which case, the remedy is obvious. When it still remains, after the proper regulation of the temperature of your room, and the removal of all superfluous clothing, you can take an occasional dose of *Sambucus*.

Sulph.-acid. — Especially, when the perspiration is profuse while lying still, but diminished by moving about.

ADMINISTRATION. — Of *China* take six globules, dry, upon the tongue, once in every three hours. The other remedies the same.

CHAPTER V.

THE INFANT.

LET us now return to the infant, which, you will remember, we left wrapped in a warm flannel blanket and laid one side while the bandage was being applied and the mother otherwise cared for or attended to. If the infant appears feeble, and its respiration not well established, the skin having a leaden hue instead of the healthy pink or rose color, it should be permitted to remain undisturbed, for some little time, until it is better able to undergo the fatigue of being washed and dressed. But if it appears strong and cries lustily, it may be washed and dressed as soon as convenient. Some people use cold water to wash the child with, even for the first ablution, under the absurd impression that this early introduction to the vicissitudes of temperature will invigorate and harden the child, and thus make it less liable to the injurious effects of sudden atmospheric changes. I hope Providence has endowed you with more sense than to imagine that any such happy results follow this barbarous practice.

For the whole period of its uterine existence, the infant has experienced a uniform temperature of 98°; now to wash it with, or to put it into, a basin of cold water, must give it a shock which cannot fail to prove highly injurious. I would about as soon think of putting the child into a kettle of boiling hot water. In my estimation, the temperature of the water in which the child is first washed should be as high as 90° at least; and this, you will observe, is still eight degrees below the temperature to which, till within a short time, it has been accustomed. It is not necessary that you should stand with a thermometer in one hand and a kettle of hot water in the other, and thus temper your bath to the fraction of a degree. All that is necessary is, to be certain that the water is *warm* and *soft*, instead of cold and hard.

The white, caseous substance, which, to a greater or less extent, covers the body of every new-born infant, and which sometimes adheres with great tenacity, can best be removed by rubbing those parts to which it adheres, freely with hog's lard, or sweet oil, until the two substances become thoroughly mixed, and then wash with soap and water.

Owing to the extreme sensibility of the infant skin, you should use none but the finest quality of white soap, and a soft flannel wash-cloth. This is important, for a slight abrasion of the cuticle, or even the least irritation, may cause troublesome sores. After the child has been well washed, it should be wiped *perfectly* dry with a soft, fine napkin.

It is the custom, as soon as the child is washed and dried, to dust it over with some kind of powder, especially about the neck, armpits, and joints, or wherever the skin is folded upon itself. I would advise you to get along without this if you possibly can, because the powders that are sold for this purpose are most of them highly injurious, and if your child is properly washed and dried, you will have but little call for them. If, however, you think you must use something of the kind, pulverized starch is the best.

Both the washing and dressing of infants should be done as expeditiously as possible, and with the greatest care, so as neither to hurt nor fatigue them.

The author of "Letters to a Mother on the watchful Care of her Infant," in speaking of the daily washing of children, makes the following remarks, which are no less philosophical than practically true : —

"During this daily process of washing, which should not be done languidly, but briskly and expeditiously, the mind of the little infant should be amused and excited. In this manner the time of dressing, instead of being dreaded as a period of daily suffering, instead of being painful, and one continued fit of crying, will become a recreation and an amusement.

"In this, treat your infant, even your little infant, as a sensitive and intelligent creature. Let everything which *must* be done be made a source, not of pain, but of pleasure, and it will then become a source of health, and that both of body and of mind, — a source of exercise to the one, and of early discipline to the other. Even at this tender age, the little creature may be taught to be patient, and even gay, under suffering. Let it be remembered, that every act of the nurse toward the little infant is productive of good or evil upon its character as well as health. Even the acts of washing and clothing may be made to discipline and improve the temper, or to try and impair it, and may, therefore, be very influential on its happiness in future life. For thus it may be taught to endure affliction with patience, and even cheerfulness, instead of fretfulness and repining at every infliction upon the body and health of the little child.

" The parent and the nurse should, therefore, endeavor each to throw her own mind into her duties toward the tender offspring. And in her intention of controlling her infant's temper, let her not forget that the first step is to control her own. How often have I observed that an unhappy mother is the parent of unhappy children."

DRESSING THE NAVEL.

Most nurses and many physicians have fanciful notions in regard to dressing the navel. Some think nothing will do but a piece of scorched linen ; others want a flannel, either scorched or well besmeared with grease. I am acquainted with one old nurse, who always keeps a box of powdered cobweb, a little of which she sprinkles over the navel before doing it up with a piece of scorched linen. Now this is all useless ; the simplest way is the best, and that is, to take a folded piece of soft, plain cotton or linen cloth, about six inches long and three wide ; cut a hole in the centre, and pass the cord through. The cord should then be laid up toward the child's breast, and the lower end of the linen or muslin folded up over it. Over this place a compress, made of several thicknesses of soft muslin, about the size of a silver dollar, or perhaps a little larger. The whole is to be kept in place by the belly-band, which should always be made of a strip of fine flannel of four or six inches width. This band should be applied smoothly, so as to give even support to the whole abdomen ; pin it just tight enough to keep it in place. For the first few days, the condition of the navel-cord should be carefully examined, to see that the child's movements have not disturbed it nor caused it to bleed. In the course of six or seven days it will become separated from the child, when you can remove it. The parts are now to be carefully washed and the compress re-applied. If the parts around the navel are not properly washed and dried, and perhaps dusted with a little starch-powder once or twice a day, they are apt to become red and sore. In case of soreness, or inflammation of the umbilicus or navel, after the falling off of the ligature, or even before, you had better give an occasional dose of *Sulphur*. Should this fail to accomplish a cure, or produce no amelioration, you should then exhibit *Silicea*; a dose — two globules — night and morning.

In case there is an evident tendency to rupture of the navel, after the ligature has dropped off, great care should be taken to apply a proper bandage, and this bandage should be worn some time after the cure, as a precautionary measure against its return. See "UMBILICAL HERNIA."

CLOTHING OF INFANTS.

I presume it will be entirely useless for me to say one word in regard to the infant's dress. Fashion dictates here, as well as almost everywhere else, frequently to the detriment of the child, and always to the great inconvenience of the mother. But this has ever been the case, and I presume always will be. However, I would have you remember, that the power of generating heat at this early period is very feeble indeed, and the child up to this time has been confined in a temperature of 98°, and at the same time most perfectly protected from the possibility of atmospheric changes.

You will, therefore, see the necessity of clothing the infant warmly. Flannel should always be worn next to the skin for various reasons: first, it is warmer, — being a bad conductor of heat, — and, what is very important, it is much lighter than cotton goods; besides, it is a bad conductor of electricity. The flannel should of course be light, soft, and of the finest texture.

In my opinion, if your child's clothing were all made of this material, it would be far preferable to any other; you would then have a *warm, light* dress; whereas, should you use cotton, it will require a much greater weight than of flannel to obtain the same amount of warmth. Besides, cotton or linen goods do not produce upon the skin that healthy degree of friction which flannel does.

No doubt you will object to flannel frocks, and say they do not look as pretty as nice tucked and ruffled muslin ones do. Well, I will not say they do; but then I think health and comfort should be consulted in preference to appearances.

Another important item in an infant's dress is looseness; the clothes should be so adjusted as to admit of the freest motion of the chest and limbs. The imperfectly developed organization of the child, you will bear in mind, is liable to compressions and distortions from the most trivial causes; many of the bones are as yet but mere ligaments, and as easily bent as the twig of a tree ; the ribs, from the slightest pressure, may become crowded from their natural position, making the child pigeon-breasted, or deformed in other ways.

In a few preliminary remarks to the chapter upon " Diseases of the Respiratory Organs," I have spoken more at large upon the subject of dress. In order that you may be fully assured, however, that the views above expressed are those entertained by the best medical authorities, permit me here to present you with a few extracts from the writings of celebrated physicians : —

" The essentials of the clothing of children," remarks Willis, a writer of the last century, " are *lightness, simplicity*, and *looseness*. By its being as *light*, as is consistent with due warmth, it will neither encumber the child, nor cause any waste of its powers ; in consequence of its *simplicity*, it will be readily and easily put on, so as to prevent many cries and tears, while, by its *looseness*, it will leave full room for the growth and due and regular expansion of the entire form,—a matter of *infinite* importance for the securing of health and comfort in after life."

" With regard to clothing " says Dr. Tracy, in his " Mother and her Offspring," " I do not wish to dictate to your taste, further than is necessary to secure to your child that which shall be *warm, light*, and *loose*." . . . . " It must at once be evident to you that *short sleeves and low-necked dresses* are *never* to be named as suitable for children."

" To leave," says Dr. Condie, " the neck, shoulders, and arms of a child nearly or quite bare, however warmly the rest of the body may be clad, is a *sure* means of endangering its comfort and health ; violent attacks of croup, bronchitis, or even inflammation of the lungs, are often induced by this irrational custom, and it is not improbable that the foundation of pulmonary consumption is often thus laid during childhood. It is an important precaution, therefore, to have the dress worn by children so constructed as to protect the neck, breast, and shoulders, and with sleeves long enough to reach the wrist."—*Diseases of Children.*

" It is certainly a most inconsistent practice, to expose the breast and arms during the weak and tender age of infancy and

childhood, and yet to deem it necessary to keep these parts carefully covered *after* the system has acquired firmness and its full
power of vital resistance, by a more mature age.

"Croup, inflammation of the lungs, catarrh, and general fevers
in cold seasons of the year, and bowel-complaints in the summer,
are often the consequence of this irrational custom, and the foundation of pulmonary consumption is often thus laid during the
first years of life. It ought, therefore, to be *immediately* abandoned as one of decidedly injurious tendency." — EBERLE.

"The young of our species, like those of all animals, require
the aid of external warmth to keep up the requisite amount of
animal heat.

"This important principle in physiology and its hygienic deduc-
tions, are not appreciated, certainly not enforced, by physicians as
they ought to be." — BELL.

"I believe, myself, from what I have seen in this city (Philadelphia) during the last 'eleven years, that the most frightful
causes of bronchitis, and also of pneumonia, inflammation of
the lungs, croup and angina in early life, is the style of dress
almost universally used for young children." — MEIGS, "*Diseases
of Children.*"

But "Fashion," as Dr. Dewees, in his work on "Diseases of
Children," truly says, "has exerted a baneful influence over the
best feelings of the mother, for she has become willing to sacrifice
the health and well-being of her offspring to its shrine. The
preposterous and unsightly exposure of the arms and limbs of
children cannot be too loudly reprehended, since it has neither
convenience nor beauty to recommend it, yet it is attended by the
most serious and manifest injury to the child."

Dr. Meigs, in his work on "Diseases of Children," remarks:
"How constantly do we see the strong and fully developed man
comfortably enveloped in a warm, long-sleeved, flannel shirt,
woollen or cotton drawers, and cloth pantaloons, vest and coat, in
the same room and in the same temperature with the little, puny,
pale, and half-naked child. But it is almost impossible to make
people understand that children need as much clothing as themselves. They always insist upon it that as the child passes the
greater part of the day in the house, it cannot require as much
clothing as the adult who is obliged to go out and face the weather ;
forgetting or refusing to see, that the former wears less than one-
half, or probably not more than a fourth as much covering as the

latter, and that the adult, when in the house and in the same rooms as the child, finds his one-half, or three-fourths warmer clothing not at all superabundant or oppressive."

Such is the testimony of medical writers as to the danger to children, from wearing short-sleeved and low-necked dresses; and every work upon infantile diseases, that I have looked over, bears testimony to the truth of the above quotations.

In our first debut upon the stage of life, fashion assumes supreme command, and her mandate regulates our every article of dress, in every act throughout the drama; our costume is changed to suit her imperious will, and, finally, when we make our exit, she dictates the cut and color of our burial-dress.

It almost seems that vanity, as Dr. Dewees intimates, is a stronger passion, at this day, in our country, than parental love, at least with a majority of our mothers.

The above remarks in regard to dress are particularly applicable to the first stages of childhood, but they apply with gradually decreasing force to the growth of the child up to adult age, " *because the power of generating animal heat is lowest at the time of birth, and gradually increases with the advancing age of the individual till past the period of childhood.*"

See GENERAL REMARKS on Diseases of the air passages and lungs.

APPARENT DEATH, OR ASPHYXIA.

It sometimes happens, after severe or protracted labor, that the new-born infant presents all the appearance of being dead; it does not breathe; the blood does not seem to circulate, and there is no apparent motion. This may be termed the first danger to which the infant is subject on its entrance into this world of trouble and tribulation.

Cases of this kind demand the immediate and energetic attention of the physician and nurse; for, if means be not speedily taken to revive it, the child may never recover from this suspension of vitality.

The first thing to be done is to place the child in such a position that there will be no impediment to the circulation through the cord, then wrap the body and limbs in warm flannel cloths, and rub the hands and feet with soft, warm flannels, or with — what perhaps is better — the warm, naked hand. Ordinarily, this will be sufficient to reëstablish the circulation; the pulsation in

the cord will soon manifest itself, the action of the heart will become apparent, breathing will soon follow, and nothing more will
be required. When the infant has fully recovered, the cord may
be tied and divided.

Now and then cases do occur, however, which do not yield so
readily, but we must not be easily discouraged in our efforts, for
infants have been restored after laboring with them for two or
three hours; we should, therefore, persevere, as our efforts may
ultimately prove successful.

If, after rubbing the infant with warm flannels, the naked hand,
or with some stimulant, for five or ten minutes, still no pulsation can be felt in the cord, the cord should be tied and cut, and
the infant be immersed in a warm bath. While in the bath, continue the friction of the skin; rub and press the chest; also dip your
hand in *cold* water or spirits, and rub the breast; or, as recommended
by some physicians, let a stream of cold water from the spout of
a teapot fall upon the chest from a height of two or three feet.

If, in the course of ten or fifteen minutes, there is no sign of
returning animation, or when there is but feeble pulsation of the
cord, the limbs relaxed, or if the face is purple and swollen, place
one or two globules of *Tartar Emetic* upon the tongue, or dissolve
six globules in six spoonfuls of water, and moisten the tongue
with a few drops of the solution. If this produces no change in
ten or fifteen minutes, prepare and give *Opium* in the same manner.

When everything else fails, artificial inflation of the lungs
should be tried. This may be done, by placing the mouth over
the child's mouth and blowing gently, so as to inflate the lungs,
at the same time closing the child's nostrils between the finger and
thumb, so as to prevent the air from passing out through the nose.
After the lungs are filled, the chest should be compressed gently
with the hands. Care must be taken not to force too much air into
the child's lungs, lest you injure them.

SWELLING AND ELONGATION OF THE HEAD.

It is quite common for the head of the infant to be swollen and
elongated immediately after birth, and especially when the labor
has been difficult or protracted; sometimes the head is so drawn
out or swollen as to be shockingly deformed; and to the unini-
tiated, its appearance not unfrequently causes great alarm. In
most cases this is but a trifling affection, and generally disappears

of its own accord. In case the swelling be extensive, or does not disappear in a day or two, repeated washings with cold water or a weak solution of Tincture of Arnica, — three or four drops to a teacup of water, — will hasten its removal, and at the same time perhaps it may be advisable to give a dose or two of *Arnica* internally. Should this fail to remove it in the course of a day or two, you can give *Rhus*.

Should the fontanel be long in closing, — that is, the uniting of the bones on top of the head, — you had better give *Calcarea-carb.*, two globules once in six days.

SWELLING OF THE INFANT'S BREASTS.

Sometimes at birth, or immediately after, the breasts of infants are found inflamed and swollen. I do not know what causes it; but certainly it is only a simple inflammation of the gland, and as such it should be treated. Our first endeavor should be to reduce the swelling; and to accomplish this, we generally cover the breast with a piece of lint, or soft linen, dipped in sweet oil. This is all the application that I have ever found it necessary to make. Sometimes, when there has been considerable inflammation I have deemed it advisable to give a little *Belladonna*, or or *Chamomilla*, or both, — a dose of two pellets, once in three hours.

Authors speak of a propensity on the part of the nurses to squeeze the breasts, under the absurd impression that there is milk, or some matter, in them, which should be pressed out. I never have had the misfortune to meet with such ignoramuses; but, nevertheless, I can easily conceive how they might do a great amount of injury by exciting an inflammation which would end in the suppuration and disorganization of the whole breast, and thereby in females destroy its usefulness forever.

THE MECONIUM, OR FIRST DISCHARGE FROM THE BOWELS.

The first evacuation from the infant's bowels consists of a dark, bottle-green substance called the meconium. Nurses are never contented until the infant has had a free evacuation of the bowels; and, to make sure of an early movement, they upon its first arrival give the little stranger a good dose of some laxative trash. I have often wondered, if an infant had the use of its reasoning faculties, what would be its first impression of the inhabitants of

16

this world, where the ladies in attendance, without even saying "By your leave, sir," just open its mouth, and force down a téaspoonful of molasses, or perhaps the same quantity of some nauseous compound. It must think it had come into a strange land.

Now, this does seem to me to be the most absurd thing in all the world. Suppose the large intestines are full of meconium; have not they been in the same condition for a long time? What is the great haste to get rid of it? Will it kill the child if it remains there a few hours longer? Nature, who is wise in all her dealings, will take just as good care of the bowels as of the brain or lungs. In fact, she has already made provision for the expulsion of this bugbear, in the kind and quality of the milk secreted in the mother's breast. But, it is a fact, that some people, in their self-conceit, imagine themselves wiser than their Creator, and, at the very threshold of life, commence marring the truly beautiful frame of God's image.

Although it may seem perfectly rational that the early contents of the bowels, called the meconium, should be purged off, you should never forget that nature has made wise provisions for this very want.

As soon as the mother feels herself sufficiently recovered to permit it, the infant should be placed to the breast, where it will obtain just the quantity and quality of *medicine* necessary for its welfare.

The generally received opinion, I am aware, is, that at this early period there is no milk secreted; and this is true; but every physician knows, and it is high time that mothers and nurses were aware of the fact, also, that there is secreted within the mother's breast, long before the birth of the infant, a fluid, technically called *colostrum*, exactly fitted for, and containing the properties to produce, just the necessary amount of mechanical action in the alimentary canal, to assist in the expulsion of the meconium. All artificial assistance is, therefore, entirely superfluous.

If the mother is able to nurse her child, absolutely nothing should be allowed to enter its mouth, for the first few days, at least, but what it gets from her, except perhaps a little cool water, which all children should have.

The colostrum furnished by the breast does not act like physic, producing a succession of stools, but more slowly, so that it may take two or three days for all the meconium to pass away; but when the work thus is once done, it is well done.

Mothers need be under no apprehension, should a temporary delay occur in the passing of the meconium; far greater evil results from the violent method taken for its expulsion, than could possibly occur from its continuance in the alimentary canal, for a longer period than natural.

Should, however, an unusually long period elapse, and the child appear costive, uneasy, and restless, a few teaspoonfuls of warm sugar and water may be given to it, which will generally have the desired effect. Sometimes it may be necessary to give a dose or two of *Nux-vomica* or *Bryonia*.

NURSING.

" That every healthy and well-organized woman should support her child from the natural secretion of her own bosom is the dictate both of nature and of reason."

In support of the above assertion we present the following quotations from several authors, with whose opinions, in this respect, we perfectly coincide : —

" In all cases where the mother can nurse her child with safety, she should certainly do so, as the mortality among infants thus nourished is far less than among those who are brought up by hand." — GUERNSEY.

" It is difficult to estimate the evil which may result from depriving the infant of this its natural nourishment, as no artificial food, however carefully prepared, can fully supply its place." — SMALL.

" Reasons of the most urgent nature only should prevent a mother from suckling her infant." — PULTE.

" No mother should deprive her infant of the nourishment which nature seems to have destined for it, except in case of absolute necessity. No animal refuses to nurse its young ; it is only among the human species that we find mothers cruel enough to deprive a new-born infant of its natural food. If this is done from wilful neglect or indifference, mothers often pay dearly for such violations of nature's laws." — JAHR.

" The first nourishment which the child should receive, when there is no *insurmountable* obstacle to it, should be drawn from the mother's breast ; for that which is therein contained is prepared to answer to the demands of the infant's digestive organs, and no nourishment supplied by art can answer equally well.

"If the baby is allowed to nurse as soon as it seems hungry, and the mother has obtained rest, there will be no need of giving any other laxative or cathartic, such as molasses, castor-oil, etc. ; for nature has made all the provision in this direction which is necessary. For the last twelve years I have not given, in a single instance, any form of laxative medicine to new-born infants, aside from that nourishment provided in the mother's breast; and I am satisfied that children do much better without than with such articles as are frequently given to them to move the bowels. The nearer we follow nature, the better. If the infant is fed a few times before nursing, it often loses the faculty of nursing, and it is in such cases, exceedingly difficult to induce it to nurse." — DR. JOHN ELLIS.

Nor does the child alone suffer from its not being allowed to nurse. "As a further inducement," says Conquest, in his " Outlines of Midwifery," "it should be remembered, that medical men concur in their opinion, that very rarely does a constitution suffer from secreting milk, whilst the health of many women is most materially improved by the performance of the duties of a nurse.

"Unless very peculiar, urgent reasons prohibit, a mother should support her infant upon the milk she herself secretes. It is the dictate of nature, of common sense, and of reason. Were it otherwise, it is not probable that so abundant a supply of suitable food would be provided to meet the wants of an infant when it enters upon a new course of existence.

"But few mothers, comparatively, are to be found, who, if willing, would not be able to support their infants, at least for a few months. Parental affection and occasional self-denial would be abundantly recompensed by blooming and vigorous children.

"By this commendable practice, nursing, the patient is generally preserved from fever, from inflamed and broken breasts, and from the distressing and alarming consequences resulting from these complaints." — *Conquest.*

REGIMEN DURING NURSING.

It is of the utmost importance that nothing should occur to the nursing mother, that may interfere with or arrest the secretion of milk, or alter and diminish its nutritive qualities. The importance of this will be realized, when we remember how far short we fall of supplying anything to take its place ; of all the aliments that human ingenuity has ever yet concocted, none has been found which can fully supply to the tender infant the want of this natural secretion.

Nature always provides for her new-born, and the fountain of life which she has opened within the mother's bosom would ever give forth a bounteous supply of pure and healthy nourishment, were it not for our follies, sins, and fashionable dissipations.

Mental and moral emotions, improper diet and irregular habits have a decided and deleterious effect upon both the quantity and quality of the milk. This is a point which it seems almost super-fluous to discuss, but, nevertheless, in the face of all the proofs which can be brought in support of this fact, there are still in existence persons who wholly ignore the idea that mental emotions affect, in any way whatever, the lacteal secretion, and very much doubt that errors in diet ever produce any very marked changes in the quality of the milk.

Now, it is a well-attested fact, substantiated by incontrovertible evidence, — a fact, too, admitting of the easiest demonstration, — that errors in diet and irregular habits may and do change the milk of the mother, from a source of nourishment, into a most injurious substance to the infant. Who has not seen children suffer from indigestion, attended with vomiting, colic, and diarrhœa, in consequence of the mother's having indulged in a very rich diet ? Some nursing mothers cannot partake in the least of fruit or vegetables, without the nurslings suffering, to a greater or less extent, in consequence. I would not be understood to assert that all nurses should abstain from fruits and vegetables, or even live on a very simple diet, because we not unfrequently meet with women who live upon the richest kind of diet, and eat abundantly of all kinds of fruits and vegetables, without the infant's suffering in the least ; but these are exceptions to the general rule.

We all know that butter, made from the milk of a cow which has fed upon garlic, will contain more or less the flavor of this plant. Dr. Draper, in his work on " Physiology," says, " There are many facts, which show that the identical fat, occurring in the food, is actually delivered by the mammary gland with many of its qualities *unchanged*. Thus, if by chance cows should eat the tender shoots of pine trees, or wild onions, or other strong-smelling herbs, the milk is at once contaminated with the special flavor of their oils. The same, too, takes place where turnips are introduced into their diet. If half the allowance of hay for a cow is replaced by an equivalent quantity of linseed-cake, rich in oil, the cow maintains herself in good condition, but the milk produces a butter more than usually soft, and tainted with a peculiar flavor derived from the linseed oil."

The worst case of colic, I think, that I ever saw in an infant, was produced by the nurse eating unripe fruit.   I am acquainted with a lady, who cannot eat the least thing that is at all sour, or acid, but that her nursing infant is sure to have an attack of colic.

Almost every one can call to mind similar instances. It is, therefore, unnecessary for me to bring forward further evidence in support of the assertion that errors in diet do materially affect the quality of the milk.

It therefore follows from what has been said, that it behoves a nursing mother to be especially careful in the choice of her nourishment in order to impart to the milk such properties only as will make it a wholesome and nutritive agent.   Plain, wholesome food, as a general thing, will produce wholesome milk, while a diet of highly seasoned and fancifully cooked dishes, served perhaps at irregular hours, and accompanied with tea or coffee, is almost certain to impart something to the milk which will prove injurious to the child.

If, after a proper regulation of the diet, the milk still proves unwholesome, you may rest assured that there is some constitutional difficulty resting with the mother, which will have to be removed by internal medication.

The diet should be simple and nourishing, not too rich nor too stimulating; bread, fruit, and vegetables may be freely used, while meats should be partaken of in moderation.

The mother's own wishes will generally point out what kind of food is most wholesome for herself and child.   A little experience will soon teach her what does, and what does not, agree with her infant, and if she be a true mother, she will be willing to sacrifice some of her choice dishes, her coffee and tea and any other little luxuries, which she finds to disagree with her child. Regularity in eating is of the utmost importance.

As I have already observed that a stimulating diet is, under no circumstances, advisable, it may be well here to make a few remarks upon the popular beverages, such as ale, porter, and the like, so extensively made use of for the purpose of increasing the flow of milk.

It has been asserted, " that no idea can be more erroneous than that women, during the nursing period, stand in need of stimulants to support their strength and increase the flow of milk." When you come to look into the subject a little, you will find that this is true.

A great ado was made not long ago by the citizens of New York, because the dairymen from the country and suburbs of the city insisted upon supplying them with swill-milk, or milk secreted by a cow constantly fed upon swill. Now, if people are so opposed to using swill-milk themselves, why will they insist upon manufacturing it for their children? I take it, that no one doubts but what swill-milk is unwholesome. In the first place, as it has already been asserted, the milk contains more or less of the properties of the substance from which it is manufactured. Now, if you manufacture milk by passing swill through a cow, — the udder acting simply as a filter, — you, of course, get more or less of the properties of the swill, whatever they may be. In the second place, a cow fed upon swill soon becomes diseased, and of course gives diseased milk. You will now readily observe, that the milk which you get, in addition to containing more or less of its original properties, as affected by swill, is still further contaminated by being drawn from a sick cow.

Now, it is just the same with a nursing woman fed upon ale and porter; not to so great an extent, it is true, because her diet is not exclusively confined to one unwholesome article, but the milk which she produces is unhealthy, and therefore not a proper nourishment for the infant. Drugs enter largely into the composition of all malt liquors, wines, and brandies, and to a far greater extent, too, than is generally supposed. Milk, impregnated with either of these drugged articles, can scarcely fail to engender obstinate and formidable chronic diseases both to mother and child.

Dr. William B. Carpenter, the most celebrated English physiologist, in his prize essay on the " Use and Abuse of Alcoholic Liquors," says : — " The regular administration of alcohol, with the professed object of supporting the system under the demand occasioned by the flow of milk, is ' a mockery, a delusion, and a snare.' For alcohol affords no single element of the secretion, and is much more likely to impair than to improve the quality of the milk."

In regard to the use of fermented liquors, after detailing a case in which the use of a single glass of wine, or a tumbler of porter, per day, was followed by a speedy and great improvement in the condition both of mother and child, he says: "But it may be questioned whether the practice *is in the end* desirable, or whether it is not, like the same practice under other circumstances already adverted to, really detrimental, by causing lactation to be persevered in, without apparent injury at the time, by females whose bodily vigor is not adequate to sustain it.

"Such certainly appeared to be the case in the instance just referred to; for the system remained in a very depressed state for some time after the conclusion of the first lactation, and on subsequent occasions it hás been found absolutely necessary to discontinue nursing at a very early period of the infant's life, owing to the inadequacy of the milk for its nutrition, and the obvious inability of the mother to bear the drain. Hence it may be affirmed with tolerable certainty that the first lactation, although not prolonged beyond the usual period, and apparently well sustained by the mother, was really injurious to her, and the inability to furnish what was required without the stimulus of alcoholic liquor was nature's warning, which ought not to have been disregarded."

Considering, then, that lactation may be put an end to at any period, should it prove injurious to the mother, the writer is disposed to give his full assent to the dictum of Dr. Macnish, "that if a woman cannot afford the necessary supply without these indulgences, she should give over the infant to some one who can, and drop nursing altogether. The only cases," continues Dr. M., "in which a moderate portion of malt liquor is justifiable, are when the milk is deficient, and the nurse is averse or unable to put another in her place. Here, of two evils we choose the least, and rather give the infant milk of an inferior quality than endanger its health by weaning it prematurely, or stinting it of its accustomed nourishment." Now, upon this the writer would remark, "that a judicious system of feeding gradually introduced from a very early period in the life of a child, will generally be preferable to an imperfect supply of poor milk from the mother, and that, if the mother be so foolish as to persevere in nursing her infant when nature has warned her of her incapacity of doing so, it is the duty of the medical man to set before her, as strongly as possible, the risk — the almost absolute certainty — of future prejudice to herself. The evils which proceed from lactation, protracted beyond the ability of the system to sustain it, may be to a certain degree kept in check by the use of alcoholic stimulants; but the writer is convinced from observation of the above and similar cases, that its manifestation is only postponed. Under no circumstances, therefore, can he consider that the habitual or even occasional use of alcoholic liquors, during lactation, is necessary or beneficial."

Dr. Condie, in his work on "Diseases of Children," says: "The only drink of a nurse should be water, simply water. All fermented and distilled liquors, as well as strong tea and coffee, she should strictly abstain from. Never was there a more absurd or pernicious notion than that wine, ale, or porter, is necessary to a female while giving suck, in order to keep up her strength, or to increase the quantity and improve the nutritious properties of her milk; so far from producing these effects, such drinks, when taken in any quantity, invariably disturb more or less the health of the stomach, and tend to impair the quality, and diminish the quantity of the nourishment, furnished by her to the infant."

Another medical writer, speaking in regard to the use of such beverages, says: "The constitution of each is stimulated by them, beyond what nature ever intended it should be. The laws which govern the animal economy are positively infringed, and it is impossible that mother or infant should escape the penalty of that infringement. Both will suffer to a certainty, in some shape or other, if not immediately, at some future period. . . . . . Thousands of infants are annually cut off by convulsions, etc., from the effect of these beverages acting upon them through the mother."

Professor Small, of Philadelphia, while remarking upon this subject, says: "The relief afforded by such stimulants, if indeed it can be called relief, is of very short duration; it is *invariably* followed by a greater degree of weakness and depression, demanding a repetition of the same, or of more powerful stimulants, which destroy the tone of the stomach, deteriorate the quality of the milk, rendering it altogether unsuited to the delicate organism of the tender infant."

From the preceding argumentative facts, we therefore conclude, that, when the mother does not furnish a sufficient supply of milk for the wants of her child, instead of resorting to ale or porter, to alcoholic or fermented drinks, of any description, it is better that a wet nurse be obtained, or the child be immediately weaned.

17

MENTAL EMOTIONS, AFFECTING THE MILK.

It is just as important that a nursing mother should pay strict attention to the state of her mind, as it is that she should pay strict attention to her diet and general health. No other secretion so evidently exhibits the influence of the depressing emotions as that of the breast.

The infant's stomach is a very delicate apparatus for testing the quality of the milk, far exceeding anything which the chemist can devise. How a mental emotion can affect the quality of the milk, perhaps it would be difficult to demonstrate, and what that change in the character of the milk consists in, no examination of its physical properties by the chemists can detect; but, nevertheless, we are well aware that after severe fits of anger, some change takes place in the milk, which alters it from a healthy nutritive agent to an irritating substance, which produces griping in the infant, and a diarrhœa of green stools.

Inasmuch, therefore, as the quality of the milk is very liable to be injuriously affected by any sudden or unpleasant excitement of the feelings, or other causes producing a constant and continued state of unhappiness, it is desirable that the most assiduous care should be taken to keep the mind in as quiet and happy a state as possible. It may not be possible for nursing mothers to avoid all occasions of getting angry or sad, but it certainly is possible to avoid all violent and artificial excitement. All serious business, exciting amusement, novel-reading, theatre-going can and ought to be strictly avoided.

Grief, of course, is an emotion which we cannot entirely control, and it is not an uncommon occurrence for the loss of a relative or friend to have such a depressing effect upon a nursing mother as to cause an almost total suppression of milk.

It is not unfrequent, either, for a child to suffer from griping pains, and green, frothy stools, while sick with some other disease, and yet there be no connection between the two complaints. We, as physicians, can readily understand it, but the mother little apprehends that it is all owing to her own anxiety.

Terror, which is sudden, and great fear instantly stop the secretion of milk.

Sir Astley Cooper, in his work upon the breast, says: "The secretion of milk proceeds best in a tranquil state of mind, and with a cheerful temper; then the milk is regularly abundant, and agrees well with the child. On the contrary, a fretful temper lessens the quantity of milk, makes it thin and serous, and causes it to disturb the child's bowels, producing intestinal fever and much griping."

The necessity, therefore, is plain, that if you would have healthy, quiet, and good-natured children, you should always yourself be calm, cheerful, and happy.

It is not well for a woman to nurse her child soon after having suffered from fright, passion, etc. ; she should wait until she is perfectly composed, and perhaps it would be as well to draw off a portion of the milk before the child is again applied to the breast.

WEANING.

Perhaps it is a hobby of mine, but I certainly am of the most decided opinion, that nothing in this world is so productive of infantile diseases, as the early resort to artificial feeding. As I have already said, over and over again, the infant's stomach is not intended to receive or to digest anything except its mother's milk. And, with those who understand this fact, it is most certainly amusing to see with what avidity some nurses and most mothers hasten to have something prepared for the child, — a little cracker and water, or molasses and water, — immediately upon its arrival in this vale of tears. I presume they think, that as the child has been nine months within its mother's womb, without a mouthful to eat, it certainly must be hungry. Thus, at the very threshold of life, the seeds of disease are sown, and in the large majority of instances, they take root, and produce a plenteous crop of stomach, or intestinal diseases. Other children are more fortunate, and escape this early infliction ; few, however, pass the first six months of their existence, without suffering more than one attack of colic or diarrhœa, produced by the hand of the very mother who feels so much love for her darling babe. She early commences the process of weaning, even before the infant has cut a tooth, arguing, that by the time the child is able to do without the breast, it will have become so accustomed to other diet that the change will scarce be noticed. This argument always reminds me of the Irishman, who thought he could accustom his horse to live without eating, and so commenced gradually to deprive him of his food. He got him down to one straw a day, when the poor horse died ; and the failure simply convinced Pat that the ex-

periment had not been conducted with sufficient care.   Thus it is
with mothers ; one child is born, fed upon trash, sickens, suffers,
and dies ; another one appears, the experiment is repeated with a
like result, and so it goes on.   While, during the whole process,
physicians stand by, entering their protests, adducing facts and
statistics, offering arguments and illustrations, sufficient to convince any jury of men, "that until the child is at least eight
months old it should receive no food, except what it gets from the
breast, providing the mother or nurse has a sufficiency for it."
But this will not satisfy women, at least it has not as yet ; for they
still persist in giving the child a bone to suck, a little bread and
milk, or mashed potato to eat ; in fact, whatever the infant sees
that others eat, it wants, cries for, and usually gets.

This is the process, so they tell us, by which the child's digestive
apparatus is prepared for the great change of weaning.   But a
more erroneous idea never entered the head of any mortal, the
Irishman's, in regard to his horse, not excepted.

This, however, is exactly the process by which the digestive apparatus is ruined.   Some children suffer from dyspepsia before
they are six months old ; before they have cut a single tooth ;
and, as soon as the teeth do come through, they are blackened over
and decayed by the corroding influence of gassy eructations, arising from the fermenting stomach and deranged bowels.

It seems to me, that had our Creator intended that our diet
should consist of solid animal and vegetable food from the first of
our existence, we would have been provided, at birth, with teeth
for its proper division and mastication.

" The child comes into the world with toothless gums, and in-
stinctive powers, adapted, in the most perfect manner, for drawing
its nourishment from the maternal breast.   It is not furnished
with teeth, because neither the mode by which its appropriate
nourishment must be taken, nor the character of the nourishment
itself requires such organs." — TRACY.

As the office of the teeth is to divide and masticate the solid
portions of our food, one would very naturally suppose that their
appearance and growth might be taken as a fair index of the development of the child's digestive organs, and of the capabilities
and powers of the stomach, as well as of the demands of the general system, in regard to nutriment.

If we take the protrusion and growth of the teeth as a guide, by
which we are to regulate the diet of the infant, we shall find that

some children may be weaned far earlier than others; so that it is impossible to name any definite age at which all children may be entirely deprived of the breast.

We find that the eruption of the temporary teeth commences at or about the sixth or seventh month, and is complete about the end of the second year, — those of the lower jaw preceding the upper.

The temporary teeth appear in the following order: At about the seventh month the two inferior cutting or incisor teeth protrude through the gums; in the course of from four to six weeks, the two corresponding upper front teeth make their appearance; then, after a few weeks, the two lateral incisors of the lower jaw — one on either side of the two first — cut through the gums; and these are, after a few more weeks, followed by their corresponding lateral incisors of the upper jaw.

The first lower molars, or double teeth, are cut from the twelfth to the fourteenth month; the two lower canine, or stomach teeth, from the eighteenth to the thirtieth, and the upper molars and canine, or eye-teeth, soon after. See " TEETHING."

Now, until the first two teeth have made their appearance, or say till between the seventh and eighth month, the child's diet should consist solely of what it nurses from its mother, provided its mother has a sufficiency for it. But, soon after the first two teeth have cut themselves through, the use of other food of an appropriate nature may be advantageously commenced, and by the time the first eight or ten teeth have attained an equal length above the gum, that is to say, from the time the child is from twelve to eighteen months old, we can safely conclude that the digestive organs have acquired sufficient tone and activity to enable them to digest without difficulty an appropriate, artificial diet, — one of a more nourishing nature, and better adapted to the advanced state of the organization than the less substantial aliment derived from the mother.

This, then, should be the general rule for weaning children; namely, *soon after the evolution of the first ten teeth.* To this rule, as to all others, of course, there must necessarily be some exceptions, but of these we shall speak further on.

Some physicians have laid it down, as a general rule, that weaning should not take place till after the completion of first dentition. Few mothers, however, are able to nurse their children as long a time as this.

From a careful consideration of the whole subject, and after giving a fair hearing to all that has been said and written upon it, pro and con, we have concluded that the period above stated, namely, soon after the child has ten teeth, is as near correct as any that it is possible to arrive at. This rule is founded upon purely physiological principles, and is in accordance with the plain indications of nature. It is, therefore, necessary that we should pay strict obedience to it, unless we wish to compromise the best interest of the child, because, wherever nature's laws are broken, the penalty is sure to follow.

Having settled the question as to *when* the child should be entirely deprived of the breast, the next question that presents itself, is, *how* shall weaning be accomplished? Shall we accustom the child to artificial feeding gradually, or shall we deprive it of the breast without any such preparation?

We have already seen that the protrusion and growth of the teeth is a sure index of the development and capabilities of the digestive apparatus; therefore, soon after the first teeth have made their appearance, we conclude that a few articles of a bland and nourishing nature may be advantageously given, because the stomach now begins to digest without difficulty other food than the milk of its nurse, and it is as well to habituate it to the change which is soon to take place. For extended remarks upon the diet of young children, the reader is referred to page 136, article " SUP-PLEMENTARY DIET OF INFANTS."

We would here simply remark, that it is desirable to increase both the quantity and nutritious quality of the food, keeping pace with the gradual development of the teeth, yet being careful never to permit the child to swallow solid animal food until the process of first dentition is completed.

By this means you will gradually bring the infant up to the period where it is desirable to deprive it entirely of the breast; the change to the child will be insensible, and the trouble to yourself not worth mentioning.

We will now proceed to notice some of the reasons why it is not always possible to follow the rule which we have laid down. Various circumstances in connection with the mother may render it impossible for her to nurse her child the full period. Owing to fever, or some acute or chronic disease, the milk may spontaneously " dry up," in spite of the utmost care; or it may be, that, during the whole life of the mother there has been a latent tendency toward consumption, scrofula, or even cancer, which the excitement during pregnancy, or the nervous shock of confinement may have brought into activity, and either of which diseases would so contaminate her milk as to render it highly injurious to the child's health if she continue to nourish it at the breast.

Again, some mothers are unable to support this constant drain upon their system more than six months, without becoming pale, weak, and emaciated; their milk becomes thin and watery, and does not contain sufficient nutriment to support the child. In this case, as in the above, recourse should be had to a wet nurse, or the child must be weaned.

The return of the menses during the period of nursing, sometimes, but not always, produces a decidedly prejudicial effect upon the mother's milk; but, as a general rule, it does not render it necessary to wean the child, and never, so long as the milk agrees with it. The same is true, if pregnancy should occur while the child is too young to wean, especially if the mother is strong and healthy; but it is not well, perhaps, to continue the nursing longer than three or four, or at most five months, in any case after the commencement of pregnancy.

On the other hand, there are various reasons why it is advisable to protract the term of nursing beyond the ordinary period. In the first place, the child may be a delicate, weak little thing, with feeble digestive powers; or it may be suffering from some disease consequent, perhaps, upon teething, or any other temporary cause. You would, therefore, naturally wait till the sickness had passed off before you changed its food. Again, it would be hardly prudent to wean a child during the hot months of summer. The months of March, April, May, September, October, and November, may, all other things being equal, be regarded as the most favorable for weaning children. Some persons are very particular that weaning should take place during a certain phase of the moon; but this is all moonshine.

It would hardly be advisable to wean a child during the prevalence of an epidemic among children; because the morbific influence prevailing produces a strong disposition to diseases.

Caution upon the points which we have here glanced at, may be the means of preventing a severe fit of sickness, or even of saving the life of your infant.

SUPPLEMENTARY DIET OF INFANTS.

It is unnecessary for me to reiterate the importance of every mother's nursing her own infant. See article on "NURSING."

It is a well-ascertained fact, that the mortality among infants "brought up by hand," as it is termed, is far greater than among those who are not deprived of their natural food.

"I am convinced," remarks Dr. Merriman, "that the attempt to bring up children by hand proves fatal in London, to at least seven out of eight of these miserable sufferers; and this happens whether the child has never taken the breast, or, having been suckled for three or four weeks only, is then weaned. In the country the mortality among dry-nursed children is not quite so great as in London; but it is abundantly greater than is generally imagined."

However, it not unfrequently happens that mothers, owing to some constitutional defect, or from debility, ill-health, or from mental emotions, have not a sufficient secretion of milk to supply their children, and are thereby necessitated to employ a wet nurse, or resort to artificial feeding.

Artificial feeding we are opposed to in all cases, except where it is absolutely necessary for the welfare of the child. It will be understood, therefore, that in all cases where we recommend the administration of food to young children, it is because of the non-secretion of milk by the mother, or where the milk is of an unhealthy nature, or from some other extraneous cause, which makes it *impossible* for the child to nurse.

It often happens that a mother, though perfectly able and willing to nurse her child, fails to supply a sufficient quantity of milk for its nourishment, while the child does not thrive, but becomes lean and emaciated, is cross and fretful, simply from hunger. In such cases it becomes necessary to give it some additional nutriment. And it is desirable that the aliment which the child receives in addition to the mother's milk should approach the latter in quality as nearly as practicable; and from chemical analysis we find that, by adding a portion of loaf-sugar and water to cow's milk, we obtain a substitute nearly resembling breast-milk.

Dr. Tracy, in writing upon this subject says: "The food which I would recommend in most cases where it can be had, may be prepared by taking newly-raised cream from the milk of a cow that has a young calf, together with a little of the top of the milk.

" A young, healthy cow should be selected, that gives *rich* milk, — milk that will not look bluish after skimming it, and *her* milk alone should be used.

" At first, after the meconium has ceased to appear in the stools, you may take one table spoonful of this cream, and add to it twice that quantity of soft, warm water, and sweeten it with *loaf-sugar;* of this enough should be used to make it about as sweet as breast-milk.

" This preparation will do very well, and may be fed to your babe in such quantities as are necessary to satisfy its natural desire for food. As the age of the child advances, you may use a larger proportion of the cream, and may also take more of the top of the milk with the cream. This is an excellent food for babes, and many will thrive nicely upon it without any breast-milk at all."

This advice is all very fine for persons residing in the country, where cows are plenty, and milk " with cream on it," is cheap. But here in the city each individual cannot " select a young, healthy cow " whenever he chooses, and keep her for his own special use. No; we receive our milk from Orange county; always, of course, perfectly pure, but milked from several cows; some with old calves, and some with young ones; thoroughly mixed, however, by the railroad churning which it receives before it reaches our doors.

I have not the least doubt but that perfectly pure, fresh milk from one cow, sweetened with loaf-sugar, and diluted at first with two-thirds water, reducing the water after a week or two to one-half, and so on, until the child is four or five months old,— when it might be given pure, — would make the best possible diet for an infant. But the diabolical concoction obtained by filtering distillery swill through sick cows, which is peddled through our streets in wagons bearing the stereotyped lie of " Orange County Milk," and " Dry Feed Dairy," I unhesitatingly pronounce to be unfit for man or beast, — far more for tender infants and feeble children.

I am aware that physicians of observation, and writers upon diseases of children, to a man, agree that among the most pernicious kinds of nourishment for a young infant, may be named those miserable compounds of flour and milk, crackers, or bread and water, or oatmeal and water, which are fed to children under the names of pap, panada, and water-gruel.

" Let the child's stomach be once or twice filled during the

18

twenty-four hours with gruel, or any of the ordinary preparations prepared by nurses for this purpose, and the chances will probably be as ten to one, that acidity, vomiting, colic, griping, and jaundice will supervene."

True, Dr. Eberle; but of two evils choose the least. Now, which would you prefer, " swill milk," or water-gruel? Besides, I do not believe that it is the panada, the bread and cracker, or the water-gruel, that produces *all* the evil here spoken of; but rather, I take it, a combination of them all, and an over-feeding of the same. No doubt but that loading the infant's stomach with these, or a number of any other articles, would produce gastric derangement. I am thoroughly convinced, from close observation, that it is not so much the article given, as it is the *state* or *quantity*, in which it is given, that produces the trouble. For instance, you will find that the gruel prepared for children, is made from meal ground very coarse, and containing a great deal of feculent matter, as is also the case with panada, crackers, or bread and water, etc. Now, at best, this substance is unfit for the delicate stomach of a tender infant; but how much more so is it, when you come to feed it after it has been prepared two, three, or perhaps more hours; and, though not actually sour to your sense of perception, it has undergone some change which renders it unwholesome to the infant, occasioning the colic, and the gastric derangement, which writers attribute to the *kind*, instead of the *quality* of the food.

Experience also teaches us that we as frequently injure children by *over-feeding* them as we do by feeding them unwholesome food. We ourselves are not unfrequently reminded, by fits of indigestion, that we have indulged our appetite to too great an extent. Some mothers look upon every cry of their offspring as an indication of hunger, and every time the child worries or frets a little, it must be fed. By this means, the stomach is kept constantly distended with food, and the inevitable result of such a course, *indigestion*, will speedily follow.

" As a general rule, a healthy child from one to three weeks old, requires a *pint of breast-milk*, or other food equally nutritious, during the twenty-four hours. At the end of the first month, and in the course of the second, the quantity usually taken by the child increases gradually to about a pint and a half or a quart."

After thoroughly sifting the subject and coining it over, I have come to the conclusion that in cases where the mother does not furnish a sufficient supply of milk for the wants of her child, finely-ground rice or barley flour makes the best supplementary diet.

The flour, which comes in pound packages, though intended for, and no doubt is, when first put up, a superior article, is not as pure as that which comes loose, like ordinary meal; you will almost always find it impregnated with pepper, cloves, cinnamon, or some other spice, of course not by design, but simply, I presume, from contact with these articles while upon the grocers' shelves.

The following is the method by which these articles should be prepared for children's diet: For an infant take one table-spoonful of the flour, — more, of course, for an older child, — and moisten it with *cold water*, being careful to have it well stirred, so that it shall contain no lumps; then add a little salt, and a sufficient quantity of *hot water*, and boil it for *ten minutes*, during which time it should be constantly stirred to keep it from burning. After it has been removed from the fire, you should add a sufficient quantity of loaf-sugar to make it about as sweet as breast-milk.

The quantity of water which you should put to a spoonful of flour, will, of course, depend altogether upon the consistency you wish to give it. If it is to be fed through a nursing-bottle, it will have to be quite thin; if from a spoon, as will be advisable, when the child is old enough to take it thus, it can be made quite thick, — as thick nearly as an ordinary farina pudding.

For those children whose bowels are habitually inclined toward constipation, you will find the barley flour better adapted, as it has a slight loosening tendency. On the contrary, for those whose bowels are inclined to be lax, or tend in that direction, you will find the rice flour preferable.

You will observe, that I advise that the flour should be cooked with water, and not with milk. I do this, not specially on account of the impossibility to obtain pure milk here in the city, but because I have observed that when a child is taking breast-milk, other milk seldom agrees, the two having an antagonism to each other, and always managing to cause some disturbance.

When the mother does not supply any nourishment for the child from her breast, I would recommend you to add a portion of pure milk to the flour and water.

These two articles of diet, in addition to the milk furnished by the mother, are all the child will need or ought to have, and a

strict adherence to this simple diet, with as few variations as possible, except in case of sickness, until after the first teeth have
made their appearance, you will find more conducive to the general health, comfort, and happiness of the child than any other
you can adopt.

If it is perfectly true, that whatever is taken into the system
and digested is assimilated by the vital forces, and goes to make up
the tissues of which the body is composed, is it not important that
we, who have the selection of the warp and woof, should be particular as to the material from which the thread of life is spun?
Experience has taught observing mothers, as well as physicians
and nurses, after having made a proper selection of food for the
infant, the importance of adhering to one plain, simple course of
diet, and not to be constantly flying from one thing to another,
giving the child cracker and water to-day and panada or gruel to-
morrow.

I have chosen the rice-flour and barley because I have found it
to agree with the infant's digestive apparatus better than anything
else; and I recommend it as a constant diet, with the exceptions
which have already been mentioned, until after the period of first
dentition.

After the fifth or sixth month, the food may be made of a more
solid or substantial nature. At first it is but a simple gruel, and
should be fed from a nursing-bottle; but as the infant increases in
age, the food should be made thicker by the addition of a larger
proportion of flour; it will then become necessary to feed it from a
spoon. The nursing-bottle you will find to be a perfect pest, which,
in spite of the "complete and thorough scalding and washing"
which the nurse gives it, will get sour. You will, therefore, as
many a one before you has done, be resorting to all sorts of contrivances to induce the child to take its food from anything but a
nursing-bottle. It is a difficult matter to feed a young child from
a spoon; but, if you can induce the youngster to take a sufficient
quantity of food in this way, it will be quite as well for the child,
and a great deal easier for you than to make use of a bottle.

The better plan, however, I think, for feeding an infant is as
follows: Procure a *silver tube* about six inches in length, having a
flattened, oblong mouth-piece; then place your food in a cup and
let the young gent take it as his "pa" would a "sherry cobler."
But whenever a nursing-bottle is made use of, "particular care
should be taken to keep it perfectly clean and sweet. It should be

well washed, both inside and out, with *hot* water every morning and evening.

" After the child has satisfied his appetite, no new supply of nourishment should be added to what may have been left. Any that remains should be emptied out, and the bottle well rinsed, before more is put into it.

" The same food should not be allowed to remain in it more than three or four hours. When kept too long, even if not perceptibly changed in taste, it becomes injurious to the child's stomach and bowels.

" By these means, the food will always be sweet and free from offensive and irritating qualities. You will probably be able to obtain a bottle made for this express purpose at any druggists." — Dr. TRACY.

The mouth-piece, or artificial nipple, which is attached to these bottles, should immediately after the child is done nursing, be taken off and put in a cup of cold water, and there left until again wanted. These nipples will not last a great while, take as good care of them as you please ; they will get sour. You must therefore exchange one for another as often as it is necessary to keep it sweet.

Dr. Tracy, from whom I have quoted above, says : " When the child is taking its food, whether from the breast, the bottle, or the spoon, it should be supported in an easy, *semi-recumbent* position, upon the arm or lap of the person feeding it, and should be kept *quiet for at least thirty or forty minutes after having received its nourishment.*

" Rest is particularly favorable to digestion, because the digestive organs require a concentration of the vital energies upon themselves, to enable them to perform this important function with due rapidity and ease.

" Both experience and experiments upon the lower animals have shown that the process of digestion is particularly liable to be impeded by strong mental or corporeal exercise, or agitation, after a full meal. The practice, therefore, of dandling or jolting infants soon after they have taken nourishment is decidedly improper.

" You will notice that all lower animals, as well as your babe, manifest a disposition to this quietness and repose after eating."

In order, however, that you should have the advice of others, as well as myself, upon this important subject, I will, in addition to what I have already quoted, give you a few extracts from several eminent authors.

Dr. Meigs, in his work upon "Diseases of Children," recommends, especially for children of weak and irritable digestive organs, and those residing in large cities, where pure milk cannot be had, the following preparation : —

"It is made by dissolving a small quantity of prepared gelatin, or Russian isinglass, in water, to which is added milk, cream, and a little arrow-root, or any other farinaceous substance that may be preferred.

"The mode of preparation and the proportions are as follows: a scruple of gelatin — or a piece, two inches square, of the flat cake in which it is sold — is soaked for a short time in cold water, and then boiled in half a pint of water until it dissolves, — about ten or fifteen minutes. To this is added, with constant stirring, and just at the termination of the boiling, the milk and arrow-root, — the latter being previously mixed into paste with a little cold water. After the addition of the milk and arrow-root, and just before the removal from the fire, the cream is poured in, and a moderate quantity of loaf sugar added.

"The proportions of milk, cream, and arrow-root must depend upon the age and digestive powers of the child. For a healthy infant, within the month, from three to four ounces of milk, half an ounce of cream, and a teaspoonful of arrowroot, to a pint of water, is usually directed.

"For older children, the quantity of milk and cream should be gradually increased to a half or two-thirds milk, and from one to two ounces of cream. I seldom increase the quantity of gelatin.

"In cases of sick children, it ought sometimes to be made even weaker for a while than in the first proportion mentioned.

"It not unfrequently happens," remarks Dr. Ellis, "when the mother is not able to nurse her child, that it is impossible for the parents to obtain a wet-nurse, and there remains no resource but to bring it up by hand. If it is very important to select a proper wet-nurse, as it certainly is, it is even more important to select proper nourishment for the child when we are obliged to feed it. As neither goat's nor ass's milk, which is often used in Europe, is usually accessible in our country, cow's milk is generally used ; and it is true beyond question, perhaps, that this is the best food we can select until the child is at least six months' old. In cases of sickness, other articles may sometimes be required to take the place of milk for a temporary period ; but a physician who is acquainted with all the circumstances, in a given case, is alone qualified to judge when this is necessary, and what substitute should be chosen. But there are many points to be attended to in the selection and use of cow's milk, which it is very important, for the welfare of the child, should not be neglected. In the first place, it is very important that the milk should be taken from a single cow, and not a mixture of several. Then it is important, for a young infant, that the cow should not have been giving milk less than two or three weeks, or more than three or four months, if this can be well obtained. Cow's milk should be slightly alkaline ; but it sometimes occurs that it is slightly acid, in which case it is very apt to disagree with children. Hence, in selecting a cow from which to obtain milk for an infant, it is always well to test the milk by means of blue litmus-paper. Hold the end of a strip of this paper in fresh milk for a short time, and if it changes it to a red color, the milk is acid, and not suitable for a young child, but another cow should be selected. Good milk will change red litmus-paper to blue after some minutes' contact. Litmus-paper can be obtained at the druggists'. If milk which is being used disagree with a child, or cause disturbance of the stomach and bowels, it should be rejected, and the milk from another cow tried ; but test the milk, as above directed, before using it. For an infant, it is important to use the milk which is first drawn, as it is much weaker than the last which is obtained, and will not require diluting with water, which may impair its quality. The first-drawn milk need not be diluted, but should be sweetened a little with sugar of milk, or, in case that is not handy, a little white sugar. Milk which has been boiled is not as easily digested as unboiled milk, and it is generally better only to heat it to the right temperature for drinking ; and it is best that this should be done in a water-bath ; that is, by setting the dish containing the milk into a vessel of boiling-hot water." — DR. JOHN ELLIS's *Avoidable Causes of Diseases.*

WET NURSES.

It not unfrequently happens that a mother, either owing to some constitutional defect, or to some extraneous cause, is prevented from nursing her own offspring, and therefore it becomes necessary that some other means must be provided by which the child can receive its nourishment. We are left to choose between a *wet nurse* and *artificial feeding.*

That the nurse's milk is the best substitute for the mother's milk, we presume will not be questioned. Should any, however, be sceptical enough to doubt it, we have only to refer them to these children who have been "brought up by hand," in comparison with those who have had a nurse. The healthy appearance of the one beside the emaciated condition of the other offers proof stronger than any argument that we can adduce.

Inasmuch as the child will undoubtedly be influenced, to a greater or less extent, both by the moral and physical condition of the nurse, it is highly important that we should use great care and discrimination in selecting the person to whom we give the entire charge of the infant. It is true we are seldom left much margin for a choice; oftener we consider ourselves fortunate, indeed, if we are able to find a female with a breast of milk who is willing to give her whole time to the care and nursing of another's infant. But, in your eagerness to secure the object of your search, you should not accept the first that offers, irrespective of her general health or moral character, or else, in after years, when, perchance, your child developes a cross and sour disposition, or is afflicted with some ugly humor, you may have the unpleasant recollection that perhaps it took it from its nurse, and then forever blame yourself for what you can never, though you would fain, remove.

We have already seen, in a previous article, that errors in diet, moral and mental emotions, etc., have a decided and deleterious effect upon the milk, changing it from a source of nourishment to a substance which seems to act like poison on the infant. If, then, the delicate organism of the infant is so sensibly affected by these changes in the milk, — changes which the most delicate tests of the chemists are unable to detect, — perhaps our imaginations can catch an inkling of the manner in which the whole constitution of the infant might become radically changed; the whole moral and physical disposition, as inherited from the mother, becoming supplanted, or at least obscurated and superseded by the peculiarities of the moral and physical organization of the nurse in whose hands the infant has been placed.

Humanity, in the first flush of its tender existence, both in its moral and physical aspect, is not unlike the potter's clay; and, like the potter, he who has the handling of it can fashion it into *almost* any form he pleases.

I have watched a hen as she came forth with her first brood of ducklings, and wondered where they received their aquatic instructions, as the web-footed little rascals paddled off into the water to the amazement and consternation of the clucking hen that hatched them. The " swim " was born in them, and all the chicken argument in the world could not convince them that a green sward was to be preferred to a dirty puddle. "What is bred in the bone cannot come out in the flesh," and therefore I say that the child can be fashioned into *almost* any form of disposition. I do not contend that you can take a child of perverse and stubborn disposition, one who has had wickedness distilled into its veins through generations back, and can implant within him an obedient, kind, and loving disposition,—although to a great extent even that can be accomplished. He can be so brought up as to see the wickedness of wrong-doing, and, though ever sinning, be constantly endeavoring to do that which is right, and perseveringly fighting against " the wrong that is within him." However, it is a great deal easier to make a straight sapling grow crooked, than it is to make a crooked one grow straight. There is a natural tendency in us all toward evil, and you soon find, if you have not already found, that the evil traits of your child's character, like the weeds in your flower-garden, will grow quite as fast as is desirable, without any cultivation ; in fact, you will find that it will require your constant attention to keep the weeds from entirely choking out your flowers. You would not be content to trust a bed of choice flowers to a gardener, who, though he would not keep the ground mellow and support the tender stems against the rude storms, would solemnly swear not to sow a weed among them ? No, indeed !

Now, I contend, and I do not think you will oppose me, that a child of kind and loving disposition, one that is confiding, and easily persuaded to do another's bidding, is very apt to be led astray by an unprincipled or careless nurse, while a child who is perverse and shows a preternatural disposition to wrong, would, in such hands, be ruined beyond all hope of future redemption. At no period of life is a child so susceptible of being influenced by the unamiable qualities of a companion, as during the early months of infancy. It will not do simply to refrain from sowing cockle-seeds in your bed of flowers ; weeds will spring up of their own accord, and from whence came the seed it would be a useless waste of time to inquire ; your work, — which the sooner you begin, the better, — is to eradicate them, to see that they do not take root and multiply. Children take to wrong-doing just as naturally as ducks do to water. The seeds of wickedness are innate within them, and I tell you it will not suffice to have a nurse who simply refrains from setting them bad examples, or from cultivating their wayward inclinations; she must take every opportunity to impress upon them the difference between right and wrong, to cultivate by tender words of approbation and encouragement all these good qualities. "As the twig is bent, the tree's inclined."

These little " dew-drops in the breath of morn," as some one has beautifully called them, are but human twigs of tender growth; their susceptibilities are of the finest nature, and with a thread, as it were, we can lead them in any direction. But we should *constantly* add " line upon line, precept upon precept, here a little and there a little," of judicious care and culture, until they have passed through that period of life around which so many gilded lures of temptation, by the thoughtless and by the wicked ones of earth, are flung.

The impressions made upon an infant at this early period are not simply transient, as most persons are apt to think, but they sink deeply into the mind, and do seriously affect, either for good or evil, the whole future character of the subject of them. And, therefore, I would earnestly impress upon the minds of all parents the importance of early attention to the moral education of children. If there is anything in this world that a child does inherit from its parent or nurse, it is fretfulness, ill-humor, and the whole category of such like amiable qualities. Now, if the nurse, in whose society the child is constantly kept, possesses a genial disposition, the prominent points of which are cheerfulness, contentment, gratitude, hope, joy, and love, don't you suppose that as the child becomes developed, as each mental petal of that mind unfolds to the influence of surrounding objects, the impressions it receives are quite different from what they would have been had the nurse possessed all of those little satanic embellishments which we call moroseness, ill-humor, selfishness, envy, jealousy, hatred, revenge, and the like. I am pretty certain you will see in whose hands it is best to place the infant, especially when you come to remember, or if you do not already know, to learn, that " *the feelings constitute an ever-acting source of bodily health or disease,* and also a principal source of our enjoyments, as well as of our sufferings; and upon their proper regulation most of the happiness and true value of human life depends." — TRACY.

Plutarch, in his advice to those mothers who are unable to nurse their own children, remarks, " Such mothers should at least be cautious to choose carefully the nurse and attendants of their children, — not taking the first that offers, but selecting the best that can be had. These, in the first place, should be *Greeks in morals ;* for it is not more attention that the body of man requires, from the period of his birth, to insure the growth of his limbs in strength and symmetry, than does his mind, in order that to his moral qualities may be imparted the same firmness and perfection as to his physical. "

As yet I have said nothing, or at least next to nothing, in regard to the *physical* diseases that a child may inherit from its nurse. In fact, it seems to me that this part of the subject is so palpable to the perceptions of all, that it is unnecessary for me here to occupy the time or space with its elucidation.

The main object I have in view in this article is to impress upon the minds of parents the importance of using great care in the selection of a nurse to whom you are to entrust your child. I would advise you never to engage a wet-nurse, however favorably you may be impressed with her appearance, until your family physician, in whom you put implicit confidence, has first examined her. In fact, upon his decision should the question rest ; especially as far as her physical condition is concerned. If there is any disease about her, he will be able to detect it.

In conclusion, that you may know what the good points in a nurse are, permit me to call your attention to the following extracts : —

" The best nurses are those who possess all the evidence of good health ; the tongue clean ; teeth and gums sound, indicating healthy digestion ; the breath free from unpleasant odor ; the surface of the body free from eruptions, and the insensible perspiration inoffensive ; the breast smooth, firm, and prominent ; the nipples well developed, rosy-colored, and easily swelling when excited. The milk should flow easily, be thin, bland, of a bluish tint, and of a sweet taste, and, when allowed to remain in a cup or other vessel, be covered with considerable cream."

" It is said that women of a brownish complexion generally have an abundance of milk, and of an unusually rich quality ; and that those of a fair complexion have less substantial nourishment, which tends oftentimes to keep the bowels relaxed." — STEWART.

" Let the nurse, from whose breast the child is to derive its nourishment, be a healthy woman, free from any discoverable tendency to chronic diarrhœa, about the same age or younger than the mother, and delivered at least within a few months of the same time ; let her complexion be clear; skin smooth and healthy ;. eyes and eyelids free from any redness or swelling. She should be of an amiable disposition, not irritable, nor prone to anger or passion ; of regular habits, not indulging in any of the forms of dissipation ; naturally kind and fond of children." — SMALL.

" The nurse should be perfectly healthy ; hence she should not be afflicted with eruptions, ulcers, leucorrhœa, whites, syphilis, fetid sweat of the feet, foul breath, decayed teeth, bad digestion, glandular swellings, scrofula, epilepsy, or any other disease ; for it might entail years of suffering or a sickly constitution on the infant. The psoric dyscrania — ill-habit — which manifests itself in subsequent years, frequently dates from this period ; it is sucked with the mother's milk, and rankles in the recesses of the organism which it often prematurely destroys." — HARTMAN.

The nurse should make it her duty to guard the child as much as possible against diseases. This she will be best able to do by paying strict attention to her diet and her general mode of living. A nurse who loves children, will cheerfully deny herself the pleasure of eating or drinking any articles whatever which injuriously affects her milk. She should, by all means, avoid all heating or spirituous beverages, spices, flatulent food, or food that is very salt. In a word, her diet should be simple and easily digested, consisting of a proper proportion of animal and vegetable food. As little change as possible should be made from her former mode of living, lest the change should affect her health and thus disturb the child, causing flatulence, colic, diarrhœa, constipation, or some other of children's many ailments. For more extended remarks upon this subject, see " DIET DURING NURSING," page 124.

CRYING, WAKEFULNESS, AND RESTLESSNESS OF INFANTS.

It may be taken for granted, that infants do not cry, — that is, have frequent and long-continued fits of crying, — without there being some occasion for it. What that occasion is, can usually be ascertained upon careful examination. A fit of crying is not unfrequently caused by some mechanical irritation ; the child's dress may be wrinkled, or so adjusted as to be uncomfortable, or a pin may be displaced and pricking into the flesh.

Perhaps the most frequent cause of crying in infants is derangement of the stomach and intestines, such as cramps, colic, griping
pains, and so forth. These are indicated by writhing of the body,
drawing up of the legs, and diarrhœa.

Occasional crying of infants should cause no uneasiness in the
feelings of the mother, because this is the only method by which
the child can manifest its wants. It may cry or worry from hunger, or from lying too long in one position; but, when attention
to these and other particulars, which will suggest themselves to
every thoughtful parent, has been given, and the infant still refuses
to be pacified, the following remedies may be employed: —

Chamomilla. — Especially when there is colic, griping pains, the
child draws its feet up and contracts its body, and when there is
diarrhœa with green evacuations; also, when there is reason to
think the child has headache or earache. May be given alone or
in alternation with *Belladonna*, and especially with *Belladonna*
when the child starts suddenly out of sleep and begins to cry violently.

Rheum. — For violent griping pains and sour-smelling diarrhœic
stools.

Veratrum. — Colicky, abdominal spasms, with or without vomiting and diarrhœa; eructations of gas.

Nux-vomica. — When there is constipation and flatulent colic,
accompanied by sudden fits of crying.

Consult " DIARRHŒA," " COLIC," " TEETHING," or other diseases
which might occasion the disturbance.

Restlessness and wakefulness, like crying, is not a disease, but
simply a symptom of some derangement of the system. It is not
always possible to say just exactly what causes the child to worry
and prevents it from sleeping. We can often trace it to flatulency, and often to an overloaded stomach; but quite as often we
are in the dark as to its cause.

Chamomilla. — When flatulency and griping pains are the cause;
also when the child starts suddenly and jerks its limbs; great sensitiveness and irritability of the nervous system; feverish heat and
redness of one cheek.

Ipecac. or *Pulsatilla.* — For restlessness arising from an overloaded stomach.

Belladonna. — When the child is drowsy, but cannot sleep, or
when the child can sleep but a few minutes at a time. Should
Belladonna fail, try *Chamomilla* or *Pulsatilla*.

If, without any apparent cause, the child is fretful and cannot sleep, give *Coffea* and *Belladonna* in alternation. Should these prove insufficient, give *Opium.*

Should all the above remedies fail, try *Stramonium, Hyoscyamus, Nux-vomica, Phosphorus,* or *Lachesis.* See " COLIC " and " PIN-WORMS."

As the difficulty is sometimes occasioned by the condition of the mother's milk, it being in some way unwholesome, it will be occasionally necessary to prescribe for the mother, as well as to make some restrictions or regulations in her diet.

ADMINISTRATION OF REMEDIES. — Of the remedy selected, dissolve twelve globules in as many spoonfuls of water, and of the solution thus made, give one spoonful every half hour, hour, or two hours, according to the urgency of the case. Frequently there is some difficulty in getting the child to take the medicine in solution ; in such case, put two or three globules dry upon the tongue, and let them dissolve.

CHAPTER VII.

DISEASES OF THE AIR PASSAGES AND LUNGS.

GENERAL REMARKS. — When taking into consideration the alarming prevalence of diseases of the air passages and lungs, especially among young persons and children, half-grown maidens and tiny infants, together with the large percentage of deaths caused thereby, one would naturally suppose that, if those who had given this subject its due attention could devise any method by which these numerous affections could be warded off or prevented, their advice would be eagerly sought for, and implicitly followed. But no, it is not till grim disease, in the shape of some appalling epidemic, wrapped in a malarious robe, mounts his chariot and comes sweeping over fair sections of our country, spreading dismay and desolation on every side, snatching from a circle here upon the right, or there upon the left, a bud, a blossom, or perchance a full-blown rose, that the oft-repeated advice of the family physician, though listened to with marked attention, is actually heeded.

Every time a physician is called upon to prescribe for a patient, he is reminded of the necessity of administering a short lecture upon the general laws of health, including dress, diet, and the like. It is a noticeable fact that sick persons are very penitent, sorry for past transgressions, willing observants *now* of the decalogue, anxious beyond measure to obey implicitly every wish of the physician. But no sooner does the first glimmer of health irradiate their sickly forms than their self-reliance and independence returns:

> " God and the doctor they alike adore,
> But only when in danger; not before:
> The danger o'er, both are alike requited ;
> God is forgotten, and the doctor slighted."

The subject of dress, in connection with the class of diseases that we are about to consider, is a very important one. The

majority, nay, perhaps I should be justified in saying, all diseases
of the air-passages are caused by sudden chilling of the body.
Our climate, with its sudden vicissitudes of heat and cold, together
with the exquisite method of our American mothers of dressing,
or rather, I would say, of undressing, their children, the low neck
to show the beautiful contour of shoulders and of bust, the half-
pants, exposing the knee of small boys, — yet what beauty there is
in a *boy's* knee I never could ascertain, but I presume they must
be charming, or certainly they would not be left bare, — all these
add their quota toward the full development of throat and lung
affections.

The universal, deplorable ignorance or inattention, or both, in
regard to the subject of dress, is astonishing, and cannot be too
frequently brought before the minds of those who have the special
care of young children.

Prevention is in all cases better than cure, and certain it is that
by careful and wise attention to the physical education of young
children, you can ward off such diseases as croup, bronchitis,
laryngitis, pneumonia, and the like, even in those who have shown
a predisposition, or a liability to them. Undoubtedly one of the
most important means to be made use of is the adoption of a
proper dress, and this, in cold weather, should be one that will
cover the *whole* body.

You can see, at any time, ladies wearing warm and comfortable
dresses with high necks and long sleeves, sitting in the same room
with their children who are almost naked. The dear little crea-
tures, their arms and necks must not be covered up, they look
" so cunning" and " so sweet." Their dresses are made so low
and loose about the neck that the whole chest, down even to the
waist, is virtually exposed. Yet, mark you, as soon as the children grow older and therefore become stronger, and better able to
bear exposure, they are dressed warmer. What inconsistency!
Is it any wonder that children are more liable to diseases of the
air-passages and lungs than adults?

O Fashion! thy potent sway fills many an infant grave!

A distinguished physician, who died some years since in Paris,
declared: "I believe that during the twenty-six years I have practised medicine in this city, twenty thousand children have been
carried to the cemeteries a sacrifice to the absurd custom of exposing their arms and necks."

I would not wish to dictate to any parent how she should dress her children, at least any further than is necessary to preserve their health by protecting them against the evil effects of sudden transitions of temperature.

Children should never be dressed with low neck and short sleeves, except in the heat of summer. I am well aware that it is the custom so to dress them, even in mid-winter, but you yourself would be uncomfortable, to say the least, clothed in this manner, and how much more so must they be with their extreme sensibility of skin.

But, you may argue, the child, especially the infant, is never exposed ; the nursery is always warm and they seldom go out of it, why be so particular to cover the neck and arms ? That is true, the rooms are always warm and in the vast majority of cases they are almost too warm, but the doors are being continually opened and shut, subjecting the child to a constant fanning.

Now the nursery or room where children are kept should be large, airy, and well ventilated. Plenty of cool, fresh, and pure air should be constantly admitted for the purpose of respiration. The temperature, while the children are well, should never exceed seventy-two degrees, and generally speaking from sixty-seven degrees to seventy degrees will be found sufficient to be comfortable provided the children are properly clad.

I am aware that you will frequently be told, and that too by those who ought to know better, that early exposures harden the children and make them robust. Would you expect to harden a tender plant by exposing it to chilling winds, or to the cold and biting frost of a winter's night ? Would you expect your flowers to grow, your roses to bud and blossom without the genial warmth of a summer's sun ? No, indeed. Neither can you harden your children by allowing their little shoulders, arms, legs, and feet to be cold ; and you will often see them so cold that they are fairly blue.

It is cruel ; and you may rest assured, that, if these children do not suffer in infancy, they will, as they grow up, be more liable to diseases of the air-passages and of the lungs than those who have been properly cared for.

Croup is a rare disease among the Germans : they are very particular in regard to children's dresses, taking great care to have the throat and chest well protected.

Delicate children should *invariably* wear a flannel under-shirt, or a shirt made of some woollen material, next to the skin, made high up about the neck, and with sleeves to come below the elbows. Then put on the accustomed under-clothes, — and even these had better be made of woollen, not only on account of its warmth, but because it is lighter than other goods, — and over all a stout muslin, or a light woollen, dress.

The stockings should also be of woollen, and come high up, always above the knees. The old way of tying a garter around the leg, to hold the stocking up, is open to many objections. In the first place, it spoils the beauty of the leg, by preventing a full development of the calf, by cutting off, or at least retarding, the circulation. This alone would be sufficient reason to condemn it; but, what is of more consequence, it also produces cold feet, and causes congestion of the veins, making them knotty and uneven. An elastic strap, going from a button upon the outside of the top of the stocking to a button upon the waistband of the drawers, will answer every purpose, and be quite as convenient.

As I have before stated, in a previous article, all children should be accustomed to cold bathing. For puny, weak, and delicate children, subject to croup, catarrh, and cough, in fact, taking cold upon the slightest exposure, I have found bathing always, in conjunction with warm clothing, of valuable assistance in strengthening the child, — giving a good healthy tone to the system, and thus protecting it from many diseases to which it would otherwise have fallen a prey.

Our city houses are generally warmed —no, *heated*, that is the word — with furnaces, another prolific source of disease. The children are virtually parboiled, or, rather, baked, while in-doors, and, consequently, when they are taken out, the first draught of air that strikes the tender little hot-house plant produces a shock, drives the blood from the surface to the delicate membrane lining the throat or lungs, and thus produces some one of the innumerable diseases of the air-passages so prevalent in our midst.

CORYZA, SNUFFLES, COLD IN THE HEAD.

DEFINITION. — This disorder, which consists of an inflammation, and consequent thickening-up, of the mucous membrane lining the nasal passages, occurs as a distinct disease; but it is also frequently connected with inflammation of the lungs, with measles, but more particularly with scarlet fever.

It attacks all, indiscriminately, both old and young. In the older children, it is of but little account, never injuring the general health by its own action; but in the infant it is quite a different thing, and becomes a serious, even a dangerous, disease. In these little sufferers, who are unable or unwilling to breathe otherwise than through the nose, it is quite an impediment to respiration, especially after the first few days, when the head and nose become completely filled with a thick, tenacious secretion, which it is impossible to remove. Being prevented from breathing through the nose, the child, when nursing, is obliged to frequently relinquish the nipple in order to obtain breath, which makes it cross and fretful.

When coryza exists in connection with other diseases, it of course adds to their severity.

CAUSES. — As a general thing, cold is the exciting cause. Children, when put to sleep, should never lie with their head toward or near a window, or in any other position where there is the least liability of a draught of air, however slight, blowing upon them. A person takes cold much more readily while asleep than awake.

Nurses are apt to cover the child's face with a little blanket after it has been put to sleep. This, by confining the breath, invariably produces perspiration. Children covered in this way always waken with their head dripping with sweat, and, when taken up in this condition, are very liable to become chilled, and snuffles is the result. Do not cover the face.

SYMPTOMS. — All are acquainted with the symptoms of an ordinary cold in the head. It usually commences with shivering, some little fever, sneezing, obstruction, and dryness of the nose. This dryness is soon followed by a discharge, more or less profuse, with watering of the eyes, pain through forehead and temples, as well as about the root of the nose. Of course, the little infant does not complain of this pain; but the older children do: therefore we are led to infer that all suffer more or less from it.

The secretion from the nose interferes with respiration, and, when the passage from the head is completely filled, the patient is compelled to breathe through the mouth; and this soon causes dryness and stiffness of the tongue and throat.

TREATMENT. — For the premonitory symptoms of coryza, with shivering and headache, *Camphora* is the best remedy; and, if administered promptly, a few doses will, in the vast majority of cases, be sufficient to effect a cure. Three globules may be given every two hours; or dissolve twelve globules in twelve spoonfuls of water, and give of the solution one spoonful every two hours.

In case you have nothing but the ordinary spirits of camphor convenient, you may put one or two drops of that upon a lump of sugar, and then dissolve the sugar in a tumbler half full of water, and give according to the directions above.

Arsenicum. — This is a prominent remedy for the disorder. Therefore, if *Camphor* fails to arrest the first sysmptoms, and the difficulty increases, administer *Arsenicum*, and especially if the following symptoms are present: obstruction of the nose, with, at the same time, a discharge of watery, acrid mucus, and burning heat in the nose; the discharge from the nose producing excoriation and swelling of the adjacent parts; also when there is redness and watering of the eyes. When this remedy affords but partial relief, recourse may be had to *Ipecac.*

Nux-vomica. — This is another important remedy for coryza. I have cured a great many cases with it alone, especially in young infants at the breast. I always use the 200th potency of *Nux* and *Arsenicum* in this disease. Unlike *Arsenicum*, it should be given when there is obstruction, with little, if any, running from the nose, or, if there is running, it is in the morning, with dryness at night. It is also indicated when there is oppressive heaviness in the forehead, heat in the face, confusion of the head, constipation, and a sensation of weariness.

Chamomilla. — When the difficulty arises from checked perspiration; shivering, with heat and thirst; heaviness of the head; obstruction and watery discharge from the nose, producing excoriation and soreness; swelling of the face; redness of one cheek; redness and inflammation of the eyes; the eyelids often closed with mucus; the child cross and fretful, and wants to be carried in the arms all the time.

For all ordinary cases of coryza, one of the above remedies will be found sufficient; however, there are forms of the disease which at times prove obstinate, and call for other remedies than those enumerated. I will mention a few others, with the particular indications calling for their use.

Belladonna. — Swelling, redness, and burning of the nose; pain in the nose, aggravated by touch; throbbing pain in the head, aggravated by motion; coryza of one nostril; offensive putrid smell.

Sambucus. — When there is an accumulation of thick, tough mucus in the nose; wheezing and hurried breathing.

Mercurius. — Profuse discharge of acrid mucus, producing soreness of the parts with which it comes in contact; swelling and redness of the nose; tearing headache; pain even in the bones; scabs form in the nose; restlessness and feverish heat, with shivering; especially when the difficulty arises from suppressed perspiration. *Mercurius* may be given in alternation with *Nux-vomica;* especially in cases of alternate chills and fever.

Hepar-Sulph. — Particularly when but one nostril is implicated; boring headache, especially about the root of the nose, which is made worse by the slightest movement; swelling of the nose, with pain, as from a bruise, when touched; inflammation of the eyes and eyelids, with nightly agglutination. *Hepar* is especially indicated, when *Mercurius* affords but partial relief, or when *Mercurius*, though indicated, fails to afford relief.

Pulsatilla. — Thick, green, or yellowish discharge from the nose, which is very offensive; frequent sneezing; confusion of the head; pain in one-half of the head; loss of smell and taste; dry coryza, worse at night; relieved in the open air; painful pressure at the root of the nose; also, if there should be flying pains from place to place, or drawing pain, extending into the ears and side of the head; roaring in the ears.

Sulphur — may in some cases be of service, and particularly in obstinate cases, and where there is a profuse discharge of purulent mucus.

Euphrasia. — Especially when the eyes are red and watery.

Silicia. — For chronic coryza, with severe pains in the bones of the nose; also, to eradicate a disposition to colds in the head.

It is sometimes advisable, when the secretion becomes suppressed, or before it has commenced, when the nose is hot and dry, to apply, with a feather or camel-hair pencil, a little *almond-oil*, or cold cream to the interior of the nose, or let the vapor of hot water pass up the nostrils. Goose-grease rubbed upon the bridge of the nose in any quantity, is of no earthly use.

ADMINISTRATION OF REMEDIES. — Where the directions have not already been given, you may dissolve, of the remedy chosen, twelve pills in twelve spoonfuls of water; give one spoonful of the solution at a dose, every hour, second, or third hour, according to the severity of the case. In chronic cases a dose every evening, or every second evening will be sufficient.

COUGH, OR TUSSIS.

DEFINITION. — At the outset, let me state that cough is not a disease in itself; but rather a symptom denoting an abnormal condition of the lungs or throat. Cough " is a violent and sonorous expulsion of air from the lungs, preceded, rapidly followed by, or alternating with quick inspirations." — COPELAND'S MEDICAL DICTIONARY.

This, in fact, is but an effort on the part of nature to remove some obstruction, or to throw off some accumulation which disease has created. During the course of an inflammation of the lungs, there is always more or less mucus secreted; and, were it not for these forcible and violent expirations, the air-passages would become clogged up, and respiration materially interfered with.

This is but one of Nature's ways to rid herself of an offending substance; she has many. You will see an illustration of this parental care exhibited in the young infant; the child, not knowing how to eject air violently through the nose for the purpose of clearing that organ, has been provided with a " sneeze."

Cough is often combined with cold in the head, both originating from the same cause; namely, exposure. In the majority of cases, cough is but a slight inflammation or irritation of the throat or upper part of the wind-pipe, accompanied with more or less fever. A little *Bryonia* or *Nux-vomica* will be found sufficient to remove all the difficulty in a few days.

Sometimes, where cough originates from a high state of inflammation, the soreness in the throat, the fever, in fact all the acute inflammatory symptoms will have passed away, and the cough, though diminished, still remains. Such a cough should not be neglected, or it will become chronic, prove troublesome, and not easily to be gotten rid of.

CAUSES. — Like every other disease of the air-passages, cough, in the majority of instances, originates from exposure. But then there are a great many *indirect* causes which produce coughs; that is, it may be sympathetic, depending, as it not unfrequently does, upon some derangement of the digestive apparatus. A very troublesome kind, frequently met with, is one occasioned by an elongated palate; this keeps up a constant tickling, which is very provoking, and the cause being overlooked, it not unfrequently proves intractable.

It is unnecessary, in fact it would be impossible, for me, in the space which I have allotted to this topic, to enter into all the causes, direct and indirect, which give rise to the various forms of cough met with in practice.

TREATMENT. — In selecting a remedy for a particular case of cough, you should take into consideration all the circumstances attending it; for instance, from what does it originate; is there much fever, any chilliness or headache, sore throat or pain in the wind-pipe upon pressure; examine the throat to see if there is any inflammation, or if the palate is elongated, or the tonsils enlarged; ascertain if it be a nervous cough, worm-cough, or stomach cough.

Cough, simply as an isolated symptom, will be of no use to you whatever in selecting a remedy, because it is just as likely to depend upon an elongated palate as upon an affection of the air-passages. It may be occasioned by inflammation or congestion, irritation or the presence of a foreign substance, or it may be entirely sympathetic, originating from a derangement of some other important viscera besides the lungs.

Now, looking upon cough in this light, you will readily see the folly, the utter absurdity, of cough panaceas. I would, therefore, advise you never to have recourse to them. Their effect, to say the least, is uncertain, and not unfrequently they do a great deal of mischief.

The list from which you are to choose in treating the different varieties and forms of cough, comprises a great number of remedies. You should study each particular case carefully; ascertain, if possible, from whence comes the difficulty, and endeavor to select a remedy that will cover the most important and largest number of symptoms present. Then, if after waiting a reasonable length of time, it does not yield to your treatment, you had better consult an intelligent, homœopathic physician, and get him to point out the proper course to pursue.

In all ordinary cases, you will have no trouble whatever in making prompt and perfect cures, but occasionally you will meet with a chronic case — one of long standing — which will baffle all your skill. Such cases had better be at once turned over to a homœopathic physician, for here it is all desirable to ascertain to *a certainty* the *precise locality* and *nature* of the exciting cause. To illustrate this let me recite a case.

Not long ago a lady called at my office to consult me in regard to the health of her daughter, — a young lady of perhaps seventeen years of age. She had been suffering for the last eighteen months with a harassing cough, which was fast telling upon her general health. Her mother informed me that when her daughter was first taken, she thought it nothing but a common cold, for which she gave her hot teas, liquorice drops, hoarhound candy, and some few other domestic remedies. Since then, however, she had run the gauntlet of quack medicines from "Miss Susan Nipper's Cough Mixture" to "Ayer's Cherry Pectoral," of course including the abominable "Cod Liver Oil," all of which had proved of no avail; on the contrary, it had so deranged her stomach that she now, in addition to her cough, suffered intensely from dyspepsia.

After giving the case a critical examination, I turned to the mother and said : —

"Madam, your child needs careful attention. You have neglected her too long, already ; you have wasted valuable time in pottering with these wonderful cure-alls."

"I know, doctor, she has been sick a great while ; it is over a year and a half, but then I kept thinking she would get better."

"Your daughter's lungs are sound; there is nothing the matter with her throat ; this cough originates from *a disease of the spine.*"

"Why ! I never heard of such a thing before."

"Very well, you hear it now, and you would have heard it before, had you consulted a physician six months ago."

"Oh, dear ! why didn't I consult some one before ? what shall I do ? can she be cured ? "

"I think she can. I would recommend you to remove her from school ; take her into the country ; give her plenty of fresh, pure mountain air, and outdoor exercise, and put her under the care of an educated homœopathic physician."

Now you will readily observe that the prospect of curing the above case would have been slim, indeed, had not the cause of it been first ascertained. As it was, it yielded kindly, as I have since learned, to the treatment adopted, and the young lady now enjoys her former health.

The treatment of cough, where it appears in connection with other diseases, or where it is dependent upon functional derangement of some important viscera, will be given in connection with such affections under their appropriate headings. At present, we shall content ourselves with giving the indications for the important and prominent remedies, where cough is the principal symptom.

Aconite may be given when there is a violent, short, dry cough, especially at night, excited by a tickling in the larynx, or upper part of the windpipe, attended with constriction of the throat, or stinging pain in the chest, and difficulty of breathing; skin hot and dry. See " BRONCHITIS."

Belladonna. — Spasmodic cough, almost without intermission, day and night, caused by a tickling in the throat, or rather a sensation as though some foreign substance were in the windpipe; heat and redness in the face; fulness and pain in the head; pain in the nape of the neck; constriction of the throat; catarrhal cough, followed by sneezing; sharp, cutting, or griping pains in the abdomen. *Hyoscyamus* is frequently of service, when *Belladonna* affords but partial relief, particularly when the dry, tickling, nightly cough is relieved for the time by sitting up in bed.

Bryonia. — Dry, spasmodic, or catarrhal cough, coming on after eating, or immediately after entering a warm room, excited by an irritation in the throat, frequently accompanied by shivering, which is followed by fever and rheumatic or aching pains in the head and limbs. Also for a moist cough, with yellow expectoration, or expectoration streaked with blood. The fit of coughing frequently ends in vomiting. This remedy is particularly applicable for the coughs which come on in winter from taking cold, especially when the whole body feels lame and as though it had been pounded, with pain in all the limbs, through the chest, and under the shoulder-blades. Children do not like to be handled, are disposed to stretch and yawn, and are very drowsy in the daytime.

Chamomilla. — Catarrh and hoarseness after taking cold; cough arising from a tickling in the pit of the throat, and extending down under the breastbone, especially when talking; cough in children, excited by crying or by anger; dry cough; worse at night and continuing during sleep; wheezing; suffocative stoppage of breath, or a sensation of something rising in the throat and taking away the breath; cough after midnight; fever with nightly exacerbations; paleness of the face, or paleness of one cheek and redness of the other; great irritability of the nervous system; child frets and worries all the time; is only quiet when being carried about.

Calcarea-carb. — Tedious cough through the day, as from dust

21

in the throat; violent, dry cough at night, with palpitation of the heart, and beating of the arteries; shooting pain in the head from within outward; also pain in the chest and side. For cough with expectoration of thick, yellow, or lumpy and very offensive mucus; chills in the evening, followed by fever and perspiration. This remedy is particularly applicable to fat children with fair hair, pale, clear skin, disposed to eruptions, and taking cold upon the slightest exposure.

Cina. — Dry, spasmodic cough, with difficulty of breathing; anxiety, paleness of the face, and moaning after the cough; cough at night, with sudden starting up, and loss of consciousness, particularly in children suffering from worms; fluent coryza, with burning in the nose, picking of the nose; nausea early in the morning, and while eating; cough, with expectoration of mucus.

Drosera. — Morning cough, with bitter and mucous expectoration; dry, spasmodic cough, worse on lying down at night, or toward evening, aggravated by laughing or talking, frequently followed by vomiting, or bleeding at the nose. See "HOOPING-COUGH."

Dulcamara. — Moist cough, with copious secretion of mucus; mucus streaked with light-red blood; barking cough, excited by deep breathing. For coughs arising from taking cold by getting wet; tightness about the chest; anxiety and fulness in the fore part of the chest; cough is worse when in a warm room, or when lying still; better when moving about.

Euphrasia. — Cough, with violent running and soreness of the nose; eyes inflamed and watery; cough worse during the day; morning cough, with copious expectoration, and oppressed breathing.

Hepar-sulph. — Paroxysms of dry, hoarse cough, ending in a fit of crying; worse at night; hollow, suffocative cough, aggravated by exposing any part of the body to the cold air, especially during sleep; wheezing respiration, with danger of suffocation when lying down; also a dry, deep cough, excited by a feeling of tightness in the chest, or by talking, stooping, ascending stairs, or much exertion of any kind; sneezing after coughing.

Hyoscyamus. — Nightly, dry, spasmodic cough, worse when lying down; cough after meals; attacks resembling hooping-cough; greenish expectoration. Compare *Belladonna.*

Ignatia. — Dry, short, hacking cough, as from a feather in the throat, which becomes aggravated the longer the paroxysms of coughing continue ; nightly cough, which becomes worse after eating, on lying down at night, or on rising in the morning ; spasmodic cough, with fluent coryza ; constriction of the throat and chest.

Ipecacuanha. — Catarrhal, nervous, or spasmodic cough, especially at night, with painful shocks in the head ; also with nausea, gagging, and vomiting, attended with pain in the abdomen ; violent fits of coughing, which continue until the child becomes fairly purple in the face, and the limbs grow quite stiff ; oppressed breathing, as though the lungs were filled with mucus, which, on coughing, almost suffocates the patient. This remedy is also serviceable when there is a tickling sensation in the throat, dry cough, or cough with expectoration of offensive mucus. See " HOOPING-COUGH."

Lachesis. — Dry, racking cough, with tickling in the throat or chest, and dryness of the throat ; cough from a sensation as of a dry spot on one side of the throat ; soreness of the larynx ; constriction, with a sensation of swelling in the throat ; the patient is unable to bear anything tight about the neck ; cough from ulcers in the throat ; sore throat, with a feeling as though something had lodged there, or as though something liquid had gotten into the wind-pipe ; spitting up of blood ; hawking up of mucus ; difficult expectoration ; pain in the throat, chest, and ears ; cough soon after lying down, or when sleeping.

Mercurius. — Dry, convulsive cough, particularly increased by talking ; pain in the head and chest when coughing ; hoarse, catarrhal cough, with watery secretion from the mouth or nose, or with watery diarrhœa ; fatiguing cough, with tickling and dryness in the chest, sometimes attended with retching, with expectoration of blood, or bleeding from the nose ; catarrh, with coryza, cough, and sore-throat.

Nux-vomica. — Particularly for a dry cough, or, if dry through the day and evening, with expectoration towards morning ; oppression of the chest on lying down, with a feeling of heat and dryness in the mouth ; catarrhal cough, with hoarseness ; cough excited by tickling or scraping, or a rough, acrid sensation and itching in the throat, or a feeling of roughness or soreness, and followed by stinging pains, expectoration of tenacious mucus, which is detached with difficulty, sometimes streaked with blood ; cough accompanied with rending pain through the head, and a bruised sensation at the pit of the stomach ; cough after eating, and occasionally followed by vomiting; cough aggravated by movement. Nux is specially applicable to persons who are of a constipated habit.

Phosphorus. — Dry cough, from irritation of the throat, or with stinging pain in the chest, worse when lying upon the left side, and during motion ; hoarseness and pain in the chest, as from excoriation ; cough, with expectoration of blood or tenacious mucus; heaviness, fulness, and tightness of the chest ; stitches in the chest.　See " PNEUMONIA."

Pulsatilla. — Cough, with easy expectoration of mucus; also severe, shaking cough, mostly in the morning, with retching and inclination to vomit ; loose cough, with salt, bitter, disgusting expectoration, sometimes streaked with blood.

Sulphur. — Cough, particularly during the night, with painful stitches through the chest, or under the ribs, in the back and loins ; cough, with expectoration of yellowish or thick mucus, of a salt or sweetish taste ; cough worse when lying down.　Especially applicable to obstinate cases.

Tartar Emetic. — Hollow, rattling cough; cough with nausea and vomiting of food ; cough, with rattling of mucus in the chest; rapid and difficult breathing; fits of coughing and hoarseness. This remedy answers well for the first stages of influenza.

ADMINISTRATION OF REMEDIES. — After carefully selecting the remedy, dissolve twelve globules in as many spoonfuls of water, and give of the solution one spoonful for a dose.　Or, if you prefer it, and it is sometimes more convenient, you can give three or four — to older children six or eight — globules, dry, upon the tongue.　In acute cases, from a recent cold, you can repeat the remedy every hour, hour and a half, or two hours, according to. the urgency of the case.　In coughs of long standing, one dose, night and morning, or at most every four hours, will be sufficient.

DIET. — Patients suffering from cough, particularly if it is a chronic cough, should live upon a good, plain, substantial diet, avoiding all articles of food which are found to disagree.　Avoid all rich, high-seasoned food, fat meats, new bread, and all articles of a stimulating nature, or having a strong pungent taste or smell, strong drinks, acids, beer, and so forth ; also spices of every description.

REGIMEN. — Free exercise in the open air is highly beneficial ; a morning walk ; exercise with the dumb bells ; drawing large quantities of air into the lungs, then beating upon the chest with the hand; all this will expand and strengthen the lungs not only, but the whole bodily frame. Children should be encouraged in lively out-door play; it makes them active; let them run, skip, and jump; give them a hoop to trundle, a top to whip, a rope to jump, a kite to fly, or a ball to throw. Supply them with corncobs and blocks; let them erect and destroy house, church, or castle.

During some part of the day children should be allowed the most perfect liberty.

The daily bathing with cold water is the best means of overcoming a predisposition to coughs and colds.

A sponge or shower-bath should be taken every morning, the skin afterwards should be rapidly dried and rubbed to a glow, either with the hand or a coarse towel, after which the child should be warmly dressed.

BRONCHITIS.

DEFINITION. — This disease has several appellations; by some it is called Catarrhal Fever, or Catarrh on the Chest; by others Cold on the Chest, etc. It is simply an inflammation of the mucous membrane lining the bronchial tubes. The bronchial tubes, you will remember, are formed by the bifurcation or division of the windpipe, — trachea, — and they lead directly to the lungs. The one intended for the left lung is two inches in length, while the one for the right is but about one. Upon entering the lungs, they divide into two branches, and each branch divides and subdivides, and ultimately terminates in small sacks, or cells of various sizes, from the twentieth to the one-hundredth of an inch in diameter. The office of these tubes is to convey air into the lungs.

Now, in mild cases, — ordinary bronchitis, or as we commonly call it, cold on the chest, — the inflammation, which is only slight, is confined to the larger tubes; there is little or no difficulty of breathing, moderate cough, and slight fever, while in the severer forms the inflammation extends down into the most minute bronchial ramifications, and all the symptoms from the onset are of a severe nature.

CAUSES. — The reader is referred to the GENERAL REMARKS at the commencement of this section. The most frequent, and perhaps the only exciting, cause of the disease is sudden transitions from a warm to a cold atmosphere. " I believe myself from what I have seen in this city (Philadelphia), during the last eleven years, that the most fruitful cause of bronchitis and also of pneumonia, croup, and angina, in early life, is the style of dress almost universally used for children." — MEIGS, " *Diseases of Children*," 2d edit. p. 180.   Now, that is the opinion of one of the most celebrated writers upon diseases of children, and I presume there is not one physician in a thousand but that agrees with him.   " The dress is insufficient.   It consists usually of a small flannel shirt, cut *very* low in the neck, scarcely covering the shoulders, and without sleeves; of a flannel petticoat; a muslin petticoat; and an outer dress made, in nearly every case, of cotton.   The dress, like the flannel shirt, is cut low in the neck, is without sleeves, and fits very loosely about the chest, so that not only are the whole neck, the shoulders, and the arms exposed to the air, but, in consequence of the looseness of the dress about the neck, it is fair to say, that the upper half of the thorax is also without covering." — *Ibid.*

Now, children are dressed in about this style the year round, both winter and summer, and physicians go their daily rounds, year in and year out, constantly warning parents of the danger of such exposure, especially during the fall and winter months, and parents heed them not.   There is many a long row of little white stones in Greenwood Cemetery that would not have been erected had it not been for Fashion's dictum.

Every physician, as soon as he commences to treat a case of bronchitis, orders the child to be warmly dressed about the chest and arms, and to be kept from the cold air; he knows that without this precaution, in the large majority of cases, his remedies would be prescribed in vain.

SYMPTOMS. — For convenience sake, we divide this disease into three forms, — 1st, Simple Acute Bronchitis; 2d, Acute Suffocative or Capillary Bronchitis; 3d, Chronic Bronchitis.

The first form, Simple Acute Bronchitis, — cold on the chest, or as it is usually called, catarrh on the chest, — is a very frequent disease among children of all ages.   It seldoms sets in suddenly, as an inflammatory affection, but gradually developes itself from an ordinary catarrh, or cold in the head.

The breathing becomes somewhat accelerated; there is more or less cough, stuffing of the chest, some fever, skin a little hotter than natural.   On applying your ear to the chest you will hear a wheezing sound, or a rattling of mucus in the air-tubes;

sometimes, after a severe coughing spell, vomiting will take place. As a general thing, toward night, the patient is more restless and uneasy, fever higher, and cough more troublesome.

Remarkable remissions at times take place in the course of this disease, the child appearing quite well for hours at a time, or it may wake up quite bright in the morning, but, as the day wears on, the fever rises ; the skin again becomes hot and dry ; respiration hurried and anxious ; cough frequent, with a sensation or an appearance of tightness across the chest, so that during the day and forepart of the night he appears to be quite ill, but, as morn approaches, the fever diminishes ; the skin becomes moist, the cough less frequent, and the child gets a quiet nap, which so much refreshes him, that, during the next forenoon, he appears quite like himself. These symptoms may run along for four or five days, when the difficulty of breathing, with the fever and the restlessness, disappears ; the cough grows less, gradually diminishes, and the child soon regains its accustomed health.

In cases rather more severe than this, the cough is a prominent symptom from the beginning ; at first, dry and violent, very frequent and harassing as well as painful, the paroxysms of coughing sometimes lasting a quarter of an hour, during which the child cries, throws his arms up, or his head back, thus evincing his anxiety and pain. The cough is excited by crying and sucking.

As the disease progresses, the cough becomes loose ; small children vomit up quantities of phlegm, while larger children expectorate quite freely. The mucous rattle may now be heard over almost every part of the lung ; the fever is high ; breathing quick and oppressed ; skin hot and dry ; pulse frequent ; child fretful and restless. Older children complain of pain when coughing, and the infant evinces it by its wincing, as well as by its endeavor to suppress the cough.

The expectoration, at first scanty and viscid, later becomes copious and streaked with blood. There is an entire loss of appetite, foul tongue, great weakness, paleness of the lips, countenance anxious or dull, and the child drowsy.

Symptoms of improvement, which generally take place in three or four days, are diminution of the fever ; the skin, instead of continuing hot and dry, becomes moist, and feels more natural to the touch ; respiration becomes less frequent ; soreness and pain diminished ; the cough becomes loose and less frequent ; the appetite returns, and the child rests better.

Ordinary bronchitis is a very frequent disease among children, and often follows in the wake of hooping-cough, scarlet-fever, or measles. This form of the disease is rarely fatal.

TREATMENT. — *Aconitum* is generally the first remedy called for, particularly should there be much fever, with skin hot and dry; pulse hard and rapid; respiration quick, anxious, or difficult; great thirst; a short, dry, and frequent cough, seemingly excited by a tickling sensation in the throat or chest; great anxiety and restlessness; constant desire to be taken up or laid down; not contented long with any one position; more or less pain in the chest, particularly at night.

Pulsatilla. — When there is moderate fever and heat of skin; cough shrill and racking; huskiness of the chest, or hoarseness when crying; obstructed respiration, especially when lying upon the back; no anxiety; little or no thirst; tongue coated yellow; vomiting or expectoration of thick yellowish or greenish phlegm; coryza, with discharge from the head. Child appears better, even quite bright, through the day, but, as evening approaches, the symptoms increase, and the patient passes a very uncomfortable night.

During convalescence there is profuse secretion of mucus, which can be heard rattling in the chest; contractive tightness of the chest.

Phosphorus. — This remedy is indicated when, after the inflammatory symptoms have subsided, there is still great oppression of breathing, hoarseness, and roughness of the larynx, or upper part of the windpipe; great anxiety and heat in the chest; dry, hacking cough, which is disposed to get worse after having once abated. The cough is excited by a tickling in the throat, by talking, or by laughing.

This remedy is particularly valuable when there is danger of a complication with inflammation of the lungs.

Tartar Emetic. — Especially if the disease is marked from the commencement with a rattling of mucus in the chest, which can be heard or even *felt* on placing the ear to the chest. *Ipecac*, is also a good remedy for these symptoms, especially should the child be threatened with suffocation, when coughing, from the excessive secretion of mucus. *Tartar Emetic*, also, when the paroxysms end in the expectoration of a quantity of mucus, or when a fit of coughing ends in vomiting, especially in young children. Coryza, with profuse discharge from the head; the child sleeps with its eyes half open; cries when touched.

Chamomilla. — When, after the fever is subdued with *Aconite*, there is a dry cough, worse at night, or even during sleep; child cross and fretful. Other remedies than those enumerated are often serviceable, but we believe the above will be found quite sufficient to control all ordinary cases. We might, however, add *Spongia*, and *Hepar-Snphur*, which may be given in alternation, when there is hoarseness, a dry, hollow cough day and night, but more particularly at night, with scanty, ropy, and sticky expectoration; skin hot and dry; respiration anxious and laborious, with burning in the chest.

ADMINISTRATION. — The remedies may be given either dry or in solution. When given in solution, dissolve twelve globules of the chosen remedy in twelve teaspoonfuls of water; give one spoonful every two or four hours, according to the severity of the symptoms. When given dry, three to six pills make a dose; put them upon the tongue and let them dissolve.

CAPILLARY BRONCHITIS.

DEFINITION. — Capillary Bronchitis is so named from the fact that the inflammation extends down into the capillaries or small sub-divisions of the bronchial tubes. It may appear as an idiopathic, or primary affection, but, as a general thing, it succeeds the form just described, particularly when that form has been neglected or improperly treated. We shall not enter into a description of it here, for two reasons: first, it is but seldom met with; second, when met with, it should be treated only by an experienced physician.

CHRONIC BRONCHITIS.

DEFINITION AND SYMPTOMS. — This form of the disease usually follows an acute attack, either on account of improper treatment, or the presence of some hereditary taint, predisposing the child to scrofula or consumption.

The cough from the acute form never entirely ceases; it becomes loose, and the expectoration may be considerable; the difficulty of breathing, though diminished, never entirely disappears; every night, or perhaps only every other night, fever arises, and is followed by more or less perspiration; the lips crack and become ulcerated; sores break out around the nostrils; the skin looks

22

blanched; eyes are sunken; appetite lost; the strength diminished; thirst is excessive. The neighbors and friends remark, that the child is going into a " decline."

These symptoms may last for weeks, or months, or even years; but at any time, a colliquative, or watery diarrhœa may set in; and this will soon put out the last ray of its glimmering existence, and the little sufferer dies of marasmus.

It needs a good deal of nice discrimination in the selection of remedies to combat successfully this form of the disease. A great many things in regard to the general health and constitution of the patient, must be taken into consideration. I would, therefore, advise you to consult an educated Homœopathic physician.

PLEURISY, OR PLEURITIS.

DEFINITION. — The lungs are enclosed and their structure maintained by a serous membrane called the *pleura*. This membrane forms a shut sack, as, in fact, do all serous membranes; and the lungs fit into it as does a boy's head into his tippet, when it is inverted or folded partially within itself. You will observe, therefore, that the lungs, though enclosed by this membrane, are still upon the outside of it. After covering the lungs as far as their roots, the pleura is reflected over the inner surface of the chest.

This may be illustrated as follows: — Take a glass tumbler, the inner surface of which may represent the inner surface of the chest; place within the tumbler a hollow India-rubber ball; call this the pleura; now, with your finger, indentate the ball; this forms a little *cul de sac*, within which the finger rests. Thus you have a shut sack; the pleura — ball — enclosing the lung — finger — and reflected upon the inner surface of the chest — tumbler.

Very well; now pleurisy, or, as physicians call it, pleuritis, consists of an inflammation of this membrane. You will observe, that at every act of respiration, or in other words, every time the lungs expand and contract, the opposing surfaces of this membrane must glide upon each other; and, when in a healthy state, they do this freely, for the parts are well lubricated with serum, just as a piece of machinery is with oil, and for the same purpose; but when inflamed, the pleura becomes hot and dry; the supply of serum is diminished or entirely suppressed, and the *friction* thus inevitably produced, causes the pain or stitches in the side and chest.

Pleurisy may either terminate by an adhesion or a gluing together of the opposed surfaces of the empty sack, or its walls may be widely separated by a pouring forth of serum; this latter effect constitutes Dropsy of the Chest.

This disease seldom attacks infants and young children; it is not as frequent, neither is it as dangerous a disease, as inflammation of the lungs, with which, however, it is often connected.

CAUSES. — The exciting cause, as a general thing, is exposure to cold or damp. It may also arise from severe injuries to the chest, as from a blow or a fall. See " GENERAL REMARKS," at the head of this chapter.

SYMPTOMS. — Pleurisy, from the onset, is marked by a sharp, stabbing pain, on a level with or just beneath one or the other of the breasts, preceded or accompanied by chilliness, or shivering; a dry ineffectual cough is usually present, with no expectoration, or, if any, very little, and of a frothy, whitish look; some difficulty of respiration; high fever; pulse quick and hard; great thirst; hot dry skin; loss of appetite; headache; and sometimes bilious vomiting.

The pain beneath the breast may diffuse itself throughout the chest, but usually it is confined to a small space, and is of a sharp, stabbing nature, seemingly as though a knife were thrust into the side, which prevents the patient from taking a long breath, and produces great suffering; when coughing or sneezing, the child endeavors to suppress the cough. The pain is always aggravated by deep inspirations, change of position, or by pressing upon the parts; it usually lasts three or four days, and then subsides. In some cases, — but these are few indeed, — there is little or no pain.

The patient cannot lie upon the affected side, at least during the first stages of the disease; that position increases the pain; however, as the pain subsides, and effusion takes place, he is unable to lie on either side, on account of the pressure made upon the sound lungs by the effused serum, which produces great difficulty of breathing. The patient is, therefore, compelled to lie upon his back, or nearly so. This effusion into the pleura sack, sometimes amounting to several pints, causes the affected side to bulge out, and become evidently larger than the other.

TREATMENT. — *Aconite* and *Bryonia* are the two principal remedies for this disease, and, in a great majority of cases, will be all that is necessary to effect a cure. In severe cases, or when the attack is sudden, they may be given in alternation, every half-hour, increasing the intervals between the doses as the severity of the symptoms diminishes.

Mercurius. — This remedy may be given when there are copious night-sweats, more or less difficulty and shortness of breathing, *after* the fever has been subdued with other remedies.

Arnica, of course, should be administered where pleurisy arises from external violence.

Arsenicum. — If extensive effusion has taken place, and there is considerable prostration.

ADMINISTRATION OF REMEDIES. — Of the chosen remedy, dissolve twelve globules in twelve teaspoonfuls of water, and give one teaspoonful at a dose every hour, or every two hours, according to the severity of the symptoms.

DIET AND REGIMEN. — As in pneumonia, the application of cold bandages is often of great service.

When pleurisy appears in connection with pneumonia, other remedies than those enumerated may be called for. Consult " PNEUMONIA."

PNEUMONIA, OR INFLAMMATION OF THE LUNGS.

DEFINITION. — Pneumonia is an inflammation of the *substance* of the lungs ; but the majority of the cases of pneumonia are attended with more or less inflammation of the serous membrane lining the interior of the chest, and inverting over the lungs ; that is, there is some pleurisy. Bronchitis is also a frequent accompaniment.

Pneumonia may be either single or double : one lung may be affected, or both. It is more common upon the right side than upon the left, and generally commences in the lower lobes. Why it does so, is not known ; but such is the fact.

CAUSES. — Inflammation of the lungs, or lung-fever as you will hear some call it, is a very important, because frequent, disease of childhood. As a general thing, it does not occur as a primary affection, but supervenes as a complication, either in scarlet-fever, measles, hooping-cough, inflammation of the bowels, or bilious remittent fever.

As cold is an active, exciting cause, you will find pneumonia much more frequent during the winter than during the summer months. A severe blow or fall upon the chest, the inhalation of noxious or irritating gases may, and often does, produce it. I have known children to inhale hot steam from the spout of a coffee-pot, or from the spout of a tea-kettle, and thereby excite an inflammation of the lungs.

Children of all ages are liable to its invasion; but, from statistical reports, we are forced to believe that it is more frequent from the third to the fourth year; nursing infants, and children under two years of age, being less liable to it than those older.

SYMPTOMS. — Pneumonia, in the majority of cases, commences, as do all inflammatory or febrile diseases, with a chill or shivering, followed by heat, and an increased frequency of the pulse. Cough is always present, at first dry and deep, or quick and spontaneous. The respiration is accelerated, the breathing from 50 to 60, sometimes even 60 to 80, in a minute. Pain, or, more properly speaking, a stitch in the side, usually the right, on taking a long breath or deep inspiration. If you will now in this, the *first*, stage of the disease, place your ear to the patient's chest, you will hear a peculiar crackling sound, similar to that produced by throwing salt upon hot coals, or like the sound produced by rubbing between your finger and thumb, a lock of one's own hair, close to the ear. This is an important symptom: it gives an early and sure intimation that engorgement, or congestion, the forerunner of inflammation, has taken place.

The expectoration, which, however, is seldom present in children under four or five years of age, is at first tough and sticky, but soon changes to a bloody mucus; sometimes, especially in older children, the sputa is of a rusty color. The face is flushed, and wears an anxious look: it is, in severe cases, blanched, and the features pinched. The skin is hot and dry, and of a shiny or glazed appearance; thirst is excessive; the pulse ranges from 130 to 140; in young children, it may run as high as 160, or even 180. The tongue may be hot and parched; but, as a general thing, you will find it moist, and covered with a yellowish or whitish fur. The patient does not wish to be disturbed, would much rather be let alone, usually lies upon his back, and desires nothing but plenty of cold water.

Now, the train of symptoms presented in a young infant — a babe at the breast — differs in some respects from the preceding. Of course, the child cannot tell you that it has a pain in its side: it cannot express its suffering in words. How, then, are you going to ascertain what is going on within that little chest? In fact, *how* are you to know what is the difficulty, and where it is located? Children are not deceitful; and if you are attentive, and at all discriminating, you will have but little trouble in interpreting their look of anguish or their cry of pain.

The child will be peevish, restless, and uneasy; cries and frets all the time, does not care to nurse; skin is hot and dry; respiration is short and hurried. You will observe that the chest does not rise and fall regularly with each inspiration, but the movements are short, uneven, or jerking.

Respiration is carried on chiefly through the action of the abdominal muscles.

From the onset, cough is present, at first dry, short, and hacking, but it soon becomes loose; vomiting is frequently present; sometimes a spell of coughing will end in vomiting, and thereby the expulsion of a quantity of glutinous mucus, or mucus tinged with blood. That the child suffers from pain when coughing is evident from the expression of its face; the grimaces and twistings of the features are always marked, and then, as you will observe, when the cough comes on, the little sufferer attempts to smother it; instead of taking a full inspiration, as it would if its chest were not sore, it tries to make it short and sudden; it tries to suppress it.

Each spell of coughing is accompanied, or *instantly* followed by a screech of pain, or a fit of crying. The cry, also, is peculiar; it is not a healthy cry, but a kind of a suppressed cry, more of a sobbing nature, but still sharp and shrill, indicative of real suffering.

When the inflammation has reached its height, which it does generally by the fifth or sixth day, the symptoms, not invariably but usually, remain stationary for one or two days, and then begin to subside. The fever diminishes; the skin loses its hot and harsh feel, becomes soft and moist; the cough becomes quite loose, less frequent, and ceases to be painful; the child can take a deep inspiration, or even cry aloud, without suffering pain. The flushing of the cheeks passes away; the expression of the face becomes more natural; the child looks around and notices all that is going on.

At this period of the disease, children are apt to be quite cross and fretful, wanting everything, and throwing all away as soon as gotten. Mothers say, that this is a good symptom.

When pneumonia ends unfavorably, the patient lingers along for a great while; the disease runs the same course as above described, but, instead of taking a favorable turn, the fever continues; the breathing becomes less frequent but more laborious and irregular; the child gradually fails; the strength diminishes; the face looks blanched and sunken; low muttering delirium may be present, but usually intelligence is retained to the last.

TREATMENT. — *Aconite* is the prominent remedy in the first stages of pneumonia; that is, during the chilliness and fever.

The following symptoms call loudly for its employment: high fever; pulse full and bounding; cough short, dry, rough, and racking; excessive thirst; redness of the face or cheeks; great nervous irritation; respiration hurried and anxious, or sobbing; violent pain in the chest and side when drawing a long breath, or when crying. The pains about the chest are particularly troublesome at night, causing the little sufferer great uneasiness, not allowing it to lie down; and, as it is not free from pain when sitting up, it constantly wishes to change its position, to be passed from mother to nurse, from nurse to the cradle, from the cradle to the bed, and so on. Now, as the disease increases, this restlessness subsides, and the child is willing to lie perfectly quiet upon its back.

Aconite should be continued as long as the fever remains high, and this nervous, restless disposition continues, and the cough remains as above described; but if, regardless of aconite, the fever should continue to increase, and there should be considerable congestion about the head, with violent throbbing of the arteries, which pass up on either side of the neck, *Belladonna* should be given, or *Aconite* and *Belladonna* may be given, in alternation, one hour apart.

Bryonia. — Particularly when the fever has somewhat diminished under the use of *Aconite.* When the breathing is short, quick and anxious; there is pain and oppression in the chest, which is aggravated by movement; the pain appears to be in the upper part of the chest; the cough, which is spasmodic, and almost constant, is generally loose, and sometimes ends in vomiting; the expectoration consists of white, slimy mucus streaked with blood. When the disease arises from exposure, from taking cold, especially if there should be rheumatic pain about the extremities, or pain in the small of the back, or, if in young children, there should be constipation, *Bryonia* is frequently indicated, in alternation with *Aconite.*

Phosphorus. — This is a valuable remedy and, in *severe* cases, may precede *Bryonia* or *Belladonna*, either alone or in alternation with *Aconite.* It is indicated by the following symptoms: decided increase of the short, hacking cough, especially in the evening, with suffocative sensation in the chest; little or no expectoration. The child cries, when coughing, evidently from the cutting pain in the chest; it also endeavors to restrain the cough, or make it short and sudden. There is also great difficulty of breathing, which interferes with nursing; the child nurses but little, or refrains from the breast altogether; heaviness, fulness, and tightness, as though a band were drawn around the chest; stitches in the side, especially the left; depression of the mental faculties; great prostration; paleness of the face; dimness of the eyes; muttering delirium; picking at the bed-clothes; extremely laborious respiration; small, quick pulse.

Tartar-emetic — may be given in alternation with *Phosphorus,* especially when *Aconite* and *Bryonia* afford no relief; also, for paroxysms of cough, with great oppression of breathing; loose cough, with profuse expectoration; rattling, hollow cough, or rattling of mucus in the chest; little or no pain; nausea and vomiting, especially after coughing; feeble and accelerated pulse.

Pulsatilla. — Difficulty of breathing, especially when lying upon the back; young children wish to be held in an upright position; hoarseness and roughness of the chest; racking, shrill cough, or cough with greenish or bloody expectoration.

Arnica. — Should be exhibited in cases arising from mechanical injuries.

Mercurius. — Where Aconite has diminished the fever, but still there is some difficulty of breathing, and the patient is very much exhausted from copious night-sweats.

Arsenicum. — Great restlessness and excessive thirst; small, weak pulse; cold extremities; rapid prostration of strength.

ADMINISTRATION OF REMEDIES. — Of the selected remedy dissolve ten globules in as many teaspoonfuls of water, and give one teaspoonful at a dose every hour, or every two, three, or four hours, according to the severity of the symptoms.

DIET AND REGIMEN. — The diet should be plain, consisting of light, easily-digested substances, panadas, gruel, etc. Cocoa makes an excellent drink. Cold water may be allowed, when desired. The breast, of course, is the diet for infants.

While suffering under a sharp attack of pneumonia, or more

particularly when recovering from it, great care should be taken that the child is not exposed by taking it from one room to another, or, if it should have been confined up-stairs, by carrying it down through cold halls, or into a damp basement.

HOARSENESS, OR RAUCITUS.

DEFINITION. — This affection, like that of cough, does not, in itself, constitute a disease. It is rather dependent upon some morbid condition of the larynx or throat, such as irritation, inflammation, or a congested condition of the parts.

CAUSES. — The causes are the same as those which are productive of almost all chest difficulties. It often occurs in connection with those diseases, which produce great disturbance about the chest and throat, and is frequently an accompaniment or sequence of a common cold. It is a suspicious symptom, under any circumstances, and ought never to be tampered with or neglected. You will remember that this is one of the premonitory symptoms of membranous croup.

TREATMENT. — Where hoarseness appears in connection with other diseases, such diseases must be treated according to the directions given in the articles under their respective heads; and, in all probability, as the original difficulty disappears, the hoarseness, which is a mere attendant, will pass away. But should it still remain, or recur by itself, the individual peculiarities of the patient, his general habit and constitution, together with any complaint to which he may be subject, or from which of late he has been suffering, should be taken into consideration and carefully studied. Then, if you can select a remedy which will cover the totality of the symptoms, you can rest pretty well satisfied of making a perfect cure.

Arsenicum. — If, with hoarseness, there should be an excessive discharge of acrid water from the nose.

Causticum. — For obstinate cases, combined with influenza; catarrh, with cough and smarting, as from excoriation in the chest; chronic hoarseness, worse morning and evening.

Capsicum. — Hoarseness, attended with a dry obstruction and tickling or crawling in the nose; violent cough, worse toward evening, with pain in the head and other parts of the body; smarting in the throat extending up into the ears.

Chamomilla. — Hoarseness after a cold, with an accumulation

23

of tough mucus in the throat; hoarseness, with pain in the throat after expectoration; cough morning and evening, with tickling in the pit of the throat; fever and great irritability, especially toward night.

Carbo-veg. — Chronic hoarseness, worse morning and evening, aggravated by talking or crying; dry cough, with hoarseness and roughness of the chest; tickling in the throat; for hoarseness and cough after measles.

Hepar-sulphur. — Hoarseness, with low, hollow voice; deep-seated cough; hoarseness, either acute or chronic, with a dry, evening cough, which is accompanied with a sensation of soreness in the throat and chest, stinging in the throat as from splinters; especially for persons who have suffered from large doses of calomel.

Kali-carb. — Hoarseness and roughness of the throat, with constant sneezing, or a choking sensation, as if there were a plug in the throat; dry cough, with tickling; for young girls suffering from leucorrhœa.

Mercurius. — Hoarseness, attended with a burning and tickling sensation in the throat; a thin, watery discharge from the head; also when there is profuse perspiration, especially at night, *Mercurius* may be given in alternation with *Nux-vomica*, or it may follow *Pulsatilla.*

Nux-vomica. — Catarrhal hoarseness, worse during the morning, with dry obstruction of the nose; rough, dry, fatiguing cough; the patient feels alternately chilly and hot, is impatient and morose.

Phosphorus. — For chronic hoarseness, with roughness and dryness of the throat; catarrh, with cough and fever; cough, with stinging in the throat; voice almost extinct.

Pulsatilla. — Hoarseness, with almost total loss of voice; pain when swallowing; loose cough, with yellow or greenish and offensive discharge from the nose; loose cough; pain in the chest; especially if there be suppression of the catamenia, or when hoarseness appears in connection with leucorrhœa. If *Pulsatilla* does not suffice, follow it with *Sulphur.*

Sulphur. — Especially for chronic cases, or obstinate ones, attended with roughness and scraping in the throat; also for hoarseness, coming on during damp, cold weather; deep, rough voice, especially at night. Follows well after *Mercury* or *Pulsatilla.*

ADMINISTRATION OF REMEDIES. — Either one of the above medicines may be given dry or, in water. When given in water, dissolve twelve pills in as many spoonfuls of water. One spoonful of this solution may be given for a dose, and repeated every two or four hours, according to the nature and severity of the case. If, in the course of forty-eight hours, no relief is obtained, another remedy should be selected.

In chronic cases, the remedy should not be so frequently repeated : a dose night and morning will be sufficient. Change the remedy in six or seven days, if no improvement is perceptible.

DIET AND REGIMEN. — The same as in " COUGH."

CROUP.

Every mother, of any experience, is more or less acquainted with this dreaded disease ; and few subjects possess greater interest for her. It is divided into two separate and distinct forms ; namely, spasmodic and membranous. Spasmodic croup — the form which I shall describe first — is that which is usually called by the laity in general, simply croup ; by authors, it is described as false, or pseudo-croup, in contradistinction to the true, or membranous.

SPASMODIC CROUP.

DEFINITION. — This is one of the most frequent diseases to which childhood is exposed. It is almost peculiar to children, and occurs, as a general thing, during the period of first dentition ; that is, about the second year. However, children from one to ten or twelve years of age are liable to it. It is said to be more common among boys than girls : whether this is the case or not, I am unable to say. A popular belief, too, is, that fat, lymphatic children are more liable to croup than those of a spare habit ; but, as far as my observation goes, it occurs indifferently in the weak and strong.

Though not contagious, as some erroneously suppose, there is strong argument in favor of its being hereditary. I can call to mind several families that can trace back, for three or four generations, a predisposition to this disease.

Spasmodic croup, from whatever cause it originates, consists in a simple, ordinary inflammation of the upper part of the windpipe, — the larynx, — with a violent spasmodic action of that organ.

It is very apt to manifest itself abruptly, without any premonitory
signs whatever, — the child going to bed as well as common, or,
at most, suffering from a slight cold, with huskiness of voice, but
before midnight, usually at about eleven o'clock, is aroused from
a quiet slumber with a spasmodic fit of coughing, which, once
heard, will never be forgotten.

CAUSES. — Croup is more common in cold, damp climates than
in warm, dry ones. Rapid and frequent changes of season, weather,
and temperature have considerable influence in producing it. I
think there is but little doubt that certain states of the weather or
season predispose to it in a greater degree than usual, and thus
occasion a larger number of children to be attacked upon the
slightest exposure ; and this has undoubtedly given rise to the
supposition that croup is a contagious disease, and that it frequently appears as an epidemic.

Children, hereditarily predisposed to croup, are liable to an
attack at any time, upon the least exposure ; and I have not unfrequently known it to be produced or excited in such by a long
and severe fit of crying.

No doubt, the most frequent *exciting* cause is exposure to cold,
— carrying the child through cold halls, or from a warm atmosphere into a cold one, or sitting in a draft of air. I have also
known it to be produced by cutting the hair in rough, cold
weather.

SYMPTOMS. — The symptoms of croup are well marked, and
need never be mistaken for those of any other disease. In the
evening, or before midnight, the child will be aroused by a paroxysm of spasmodic coughing. The cough is rough, barking, and is
accompanied by a shrill, sharp sound ; during the paroxysms of
cough, the breathing is spasmodically oppressed, at times seemingly
almost to suffocation. The face and neck are at first highly
flushed, but, as the paroxysm becomes more violent, assume a
dark, livid red, which afterward passes into a deadly paleness, if
the fit is of long duration. The veins swell, and beads of perspiration stand out upon the forehead : sometimes the whole head is
wet with sweat.

The voice, during a paroxysm, becomes almost extinct. The
child, almost frightened to death, not only by his own condition,
but by the excitement of those around him, throws himself from
side to side, grappling, as it were, the disease which seems to
threaten immediate suffocation, his countenance presenting a pic-

ture of the utmost anxiety. The patient may remain in this condition from fifteen to twenty minutes, or from half an hour to even an hour. As soon as the violent symptoms abate, the child falls asleep, when afterwards, on awaking, all that remains is a little hoarseness, a loose cough, and some fever. The patient may have but one of these attacks during the night ; but more frequently there are several in succession, perhaps not so severe as we have described. The attack, however, is very apt to recur toward morning, and, if not then, the following night. This is about the course of the disease when left to itself; but if cut short, as it easily can be, at the very commencement of the first paroxysm, the child is as well the next morning as though nothing had happened, and remains so. On the contrary, if improperly treated, or allowed to continue, the second night will usher in a scene equally frightful as the one preceding it ; and, before the third day has passed, the inflammation will have extended down through the trachea, or lower part of the windpipe, toward, and sometimes even into, the bronchial tubes ; and then, with all its attending horrors, you will have a true case of membranous croup.

TREATMENT. — An ordinary case of croup no mother need fear, if she but have the proper homœopathic remedies at hand. *Aconite*, *Hepar-sulph.*, *Spongia*, and *Tartar-emetic* are the remedies, and with them you will be able speedily to dissipate the crowing, croupal cough ; and the little sufferer, who but a moment before, was in imminent danger of suffocation, will drop into a sweet and peaceful slumber, and that, too, without the nauseating dose of Syrup of Ipecac. ; without the onion-draughts to the feet, and goose-grease to the chest, to which ignorant people trust their all.

If premonitory symptoms are present, such as a hoarse cough with fever, *Aconite* and *Spongia* should be given in alternation, every hour. If, during the evening, the symptom should assume a distinct form of croup, or should the child be startled from sleep with a suffocative, crowing, barking cough, the *Aconite* must be discontinued, and *Tartar-emetic* substituted for it, and given, in alternation with *Spongia*, every ten or fifteen minutes.

A warm bath is a valuable auxiliary, and should be resorted to in all severe cases, as it lessens the agitation, and makes the symptoms yield more easily. The temperature of the water, when the child is immersed, should be about 96°, and gradually raised by the addition of hot water.

The child may remain in this bath from ten to fifteen minutes, or until the choking cough ceases. When the child is taken from the water, it should be *quickly* wiped dry, and then well wrapped up, to prevent it from taking cold. Cloths wrung out in cold water and applied to the throat, are of great service ; these cloths should be covered with dry flannels.

This treatment will usually break up the most violent attacks of croup in a very short time ; in fact, the prompt administration of *Aconite* in the first stage, that is, when the inflammation first sets in, will, in the majority of cases, cut short the disease without any other assistance.

For the cough, which is ringing, but more moist and loose, that remains after the violent paroxysm has subsided, *Spongia* and *Hepar-sulph.* may be given in alternation every hour, lengthening the interval as the severity of the symptom subsides.

If a recurrence of the severe paroxysm should occur, the same treatment should be repeated.

Hepar-sulph. — Should be given when the cough is loose, with rattling of mucus in the air-passages, or when the expectoration consists of thick, tenacious phlegm. *Tartar-emetic* may be given in alternation with *Hepar*, from half an hour to one or two hours apart.

For the hoarseness remaining after an attack of croup, and to obviate a tendency to a relapse, *Phosphorus* or *Hepar-sulph.* may be given once in three or four hours.

Kali-bichrom. — This is a valuable remedy for a hoarse, dry, or whistling cough, or when there is expectoration of ropy mucus ; but of more service in severe cases of membranous croup.

ADMINISTRATION OF REMEDIES. — When Tartar-emetic is given, dissolve about as much of the first trituration as can be put upon a five-cent piece, in twelve spoonfuls of water, and of this solution give one spoonful at a dose, as above directed. When any of the other remedies are chosen, dissolve twelve globules in twelve spoonfuls of water, and give one spoonful at a dose.

For DIET AND REGIMEN, see " COUGH," and " MEMBRANOUS CROUP."

MEMBRANOUS CROUP.

DEFINITION. — This disease is closely allied to, and often connected with, the one we have just had under consideration, — spasmodic croup. " It consists in inflammation, generally of a highly acute character, of the larynx —upper part of the wind-

pipe — or the trachea, windpipe proper, or of both, which terminates, in the majority of cases, in the exudation of false membrane
more or less abundantly upon the affected surface."

The inflammation usually begins high up, near that part which
contains the vocal cords, or what physicians call the larynx. Perhaps you would better understand me, if I should say that it commences in the region of that projecting cartilage, by some called
Adam's Apple, and extends down into the bronchial tubes.

This form of croup differs from the preceding one in this particular, — the formation of a false membrane upon the surface inflamed; this of course obstructs the air-passages, and, in severe
cases, it completely closes them up, so that the patient dies from
actual suffocation. This membrane when coughed up, or when
taken from the dead body, looks about like a stick of boiled mac-
aroni; is commonly of a yellowish color, and from a sixteenth to
a twelfth of an inch in thickness.

Why one form of croup should be marked by violent spasmodic
action of the larynx, and the other by the formation or exudation
of a false membrane, when they both consist in an inflammation
of the same tissue, no cause can be assigned.

CAUSES. — What has been said in regard to the causes of spasmodic croup applies equally well to this disease. The reader
is also referred to the general remarks at the commencement
of chapter VII., preceding " Diseases of the Air-Passages and
Lungs."

SYMPTOMS. — Croup usually commences as a common cold, with
sneezing, cough, and more or less fever; seldom presenting any
symptoms by which we can distinguish it from ordinary catarrh.
Like catarrh, it is attended with slight fever, drowsiness, watering
of the eye, and running from the nose.

The child complains for two or three days of a croupy cough
and hoarseness. Now hoarseness, under any circumstance is a
suspicious symptom, and ought never to be neglected; children
suffering from ordinary cold seldom exhibit this symptom. After
being simply hoarse for two or three days, the voice becomes weak;
the child speaks or cries in a whisper. In the last stage, and in
severe cases, the child is wholly unable to speak, even in a whisper,
or cry; the only noise it is at all able to make is the peculiar violent, short, shrill, barking cough. This shrill, harsh cough, peculiar
to membranous croup, has been compared to the sound produced
by the attempt of a young cock at crowing.

Between the paroxysms of cough, which are excited by speaking, drinking, etc., the wheezing in the air-passages is heard at every inspiration; occasionally something is expelled from the windpipe by the cough. As the disease progresses, the voice becomes more and more hoarse, and less distinct, the windpipe becoming clogged up with mucus or with false membrane.

From day to day and hour to hour, we can trace the steady, onward march of this disease toward its fatal termination. Unlike spasmodic croup, after a severe paroxysm of cough, or attack of strangulation, it does not retreat, and leave the way open for a free return of respiration, but maintains the ground which it has gained, and throws out barricades to obstruct returning health. These barricades of false membrane may consist of patches here and there thrown out upon the surface of the windpipe, or the whole tube may be completely lined or even entirely filled up, so as to preclude the possibility of respiration. Death is the inevitable result.

Dr. Watson, an eminent writer, says, " As the obstruction to the passage of air increases, the blood ceases to receive its proper quantity of oxygen; the skin grows dusky; the pulse feeble and irregular; the feet and legs cold; the cough also ceases to be loud and clanging; it becomes husky and inaudible at a short distance; and the voice sinks to a whisper; the head is thrown back; the nostrils dilate widely, and are in perpetual motion; the face pale, and sometimes livid; the pupils often dilated."

The duration of the disease may be stated to be from three to twelve or fourteen days. Death, however, has been known to occur as early as the first day.

TREATMENT. — " This dreadful disease," says Dr. Hering, of Philadelphia, " may, in the majority of cases, be easily and promptly cured with homœopathic remedies. We hardly lose one-fifth of the number who die under the old treatment."

What that old treatment consists in is of no earthly use to us; but as a curiosity in medicine I present it here. I shall quote only from standard allopathic works. Dr. West (*Dis. of Child.*, p. 223), after recommending an emetic of ipecac and antimony, to be followed by nauseating doses of antimonial wine, says: " The abstraction of blood, and the administration of tartar-emetic, are the two measures on which your main reliance must be placed; and you must bleed largely, and give tartar-emetic freely, remembering, that, if relief do not come soon, it will not come at all. . . .

I have never met an exception to the rule which prescribes the free abstraction of blood in every case of severe idiopathic croup."

Dr. Eberle says (*Dis. of Child.*, p. 273), " Without doubt, however, the remedy upon which our principal reliance should be placed, for the removal of tracheal inflammation (croup), is bloodletting." Underwood says (*Bell's* ed., p. 273), " Bleeding is always necessary, if the physician be called at the commencement of the disease or stridulous noise. If the patient be visited too late to endure this evacuation, I believe no hope can remain of his being benefited without it, unless the infant be very young, which, however, in another view, cannot but add to the danger." Dr. Condie says (*Dis. of Child.*, 2d ed., p. 305), " The practitioner who, in violent cases, neglects this important measure (bloodletting), and places his hopes on any other remedy, or combination of remedies, will have but little reason to flatter himself upon his success in the management of the disease."

Yet M. Guersent (p. 373) asserts that " bleeding has not the power of arresting this specific inflammation." M. Bretonneau (*Meigs, Dis. Child.*, 3d ed. p. 96) " is of the opinion that it has no effect in preventing the formation of the false membrane." M. Valleix (p. 353) says, " From the examination of a large number of cases, I am convinced that blood-letting, whether general or local, is not a powerful curative agent ; and it does not obviously arrest the progress of the disease." Dr. Wood says (*Prac. of Med.*, vol. i. p. 788), " Blood-letting, in this variety of croup, is much less efficient than in the catarrhal."

Dr. Douglas, of Boston, was the first to use calomel in this disease. In regard to mercurials, Dr. Samuel Bard says : " The more freely I have used them, the better effects I have seen from them." " The remedy principally relied on in the present day, and which in many cases has acted like a charm, is large and repeated doses of calomel." — (*Gardner's Med. Dic.*) M. Valleix, as quoted by Meigs (*Dis. of Child.*, 3d ed., p. 100), on the contrary, doubts " whether there are any true cases of croup on record cured by calomel alone."

Dr. Meigs says (*Dis. of Child.*, 3d. ed., p. 101), " The largest quantity (of calomel) exhibited in any one of my cases, was between forty and fifty grains ; the smallest, six. In most of the cases, the quantity varied between fifteen and thirty grains."

Dr. Bard, whom we mentioned above, says he gave calomel, in

24

the quantity of *thirty or forty grains, in five or six days, to children three or four years old.*

Dr. Meigs, after recommending calomel, quaintly remarks, that it (calomel) " has been known to produce gangrene (mortification) of the mouth, and necrosis (death) of the cheek-bones."

Burning the throat with caustic has been recommended; but, as M. Bouchut remarks, it has this slight disadvantage : " Immediate suffocation may be the consequence, should the sponge be left too long upon the glottis, and should too large a quantity of the liquid enter the glottis."

Tracheotomy (opening the windpipe) has its adherents and opponents.

I think I mentioned somewhere, in a previous article, that the old practice of treating diseases was, to say the least, unsatisfactory. Does not the above series of quotations savor of something of the kind ?

Four — and I could quote forty, were it necessary — eminent men recommend blood-letting; an equal number, equally eminent, condemn it. One asserts that calomel is the remedy, *par excellence ;* another proves it to be so ; but a third denies it all, and says that he does not believe there is a case on record cured with it. Besides that, a fourth asserts that its effects are bad, — it destroys the mouth and cheek-bones.

" Burning the throat with caustic might be of service ; " but " then you suffocate the patient." Thus they wrangle.

In fact, there is but one class of remedies that any two allopathic physicians can agree upon recommending, — emetics; and the good effect of these is erroneously attributed to their nauseating properties. But this is a mistake ; for Ipecac. and Tartaremetic are *homœopathic* to the disease in question ; their action is direct and specific. Every allopathic physician knows, or, at least, he ought to know, that large doses of these drugs will produce symptoms upon a healthy person *similar* to those of croup. And, had he a mind, he could soon prove it to his entire satisfaction, by experimentation, that minute doses of these remedies will produce results more prompt and satisfactory than large ones.

I am aware that I have already taken up too much room in quoting and commenting upon this irrational mode of practice. Let us hasten to an intelligent and rational method of treatment, leaving the old for those " who, having eyes, see not," etc.

Aconitum. — Is the first remedy in all cases. It is peculiarly adapted to this specific form of inflammation, and may be given, in alternation with *Spongia*, every fifteen or twenty minutes, from that to an hour or an hour and a half, according to the severity of the symptoms. If no decided improvement follows the administration of these two remedies, but, on the contrary, the disease becomes visibly worse, and the danger increases, give *Hepar-sulph.* in alternation with the *Spongia.*

I have never yet seen a case of croup prove fatal, unless it were complicated with some other disease, where Aconite and Spongia had been properly administered at the outset. Just as soon as you perceive that the child has taken cold, and exhibits any of the premonitory symptoms of croup, such as a *dry, hoarse cough, huskiness of voice,* more or less fever, etc., give a dose of Aconite, and repeat it once an hour. Two or three doses will put an end to all the difficulty, and thus save your child a severe fit of sickness, and yourself a vast amount of anxiety.

Should the case have become very violent, either from neglect, or, worse yet, from the tampering of some ignoramus, with plain indications of the formation of false membrane, and be threatened with suffocation, you must resort immediately to *Kali Bichromicum.* A small powder of the first trituration should be given every few minutes. Should no good effect follow this, give *Arsenicum.* Hot applications to the throat are now of service.

Bromine. — Is a valuable remedy in this stage of the disease, particularly should there be a great deal of mucus rattling in the throat, with labored respiration, wheezing, and rough cough, with danger of suffocation, gasping for air, expectoration of thick mucus. Should the expectoration be of this *ropy mucus,* give *Kali Bichromaticum.* When *Bromine* is used, a low dilution is preferable.

Phosphorus. — This remedy is also of service in some of the worst forms of croup. It may be given, in alternation, with *Lachesis.*

Any hoarseness that may remain, after the more dangerous symptoms of this disease have passed away, can often be removed by *Phosphorus* or *Hepar;* should these not answer, you can give *Carbo-v., Belladonna,* or *Arnica.*

Warm baths are always of valuable assistance in croup; they relax the system, lessen the agitation, and make the symptoms yield more readily to the remedies given. You can either make them general, that is, immerse the child all over, or partial; bathe the arms and legs; keep the feet warm.

No application need be made to the head, unless it be hot; then a cloth wrung from cold water, and applied to the head, would be advisable.

Severe forms of the disease should always be treated by a physician, when one can possibly be had.

ADMINISTRATION OF REMEDIES. — The directions for Bichromate of Potash accompany its indications. Of the other remedies, whichever is chosen, dissolve twelve globules in twelve teaspoonfuls of water, and give one teaspoonful at a dose; as above directed.

DIET AND REGIMEN. — The diet of a patient, while suffering from this disease, should be plain and of a non-stimulating nature. A mucilaginous diet is the best, as oat-meal gruel, barley-water, etc. As a drink, cold water or toast-water may be allowed, or perhaps milk. Oat-meal gruel, sweetened with raspberry or strawberry syrup, if palatable, will answer both for nourishment and drink.

Children subject to croup should be carefully guarded against taking cold, and a predisposition to it will be best eradicated by an occasional dose of *Phosphorus* or *Hepar*. *Lycopodium* is also recommended for this purpose.

Dr. Von Boenninghausen, a justly celebrated German physician, makes use of the following simple treatment in *all* cases of croup, and, according to report, with a success far exceeding that of any other practice. He takes five powders and numbers them respectively, as follows: No. 1, Aconite; 2, Hepar-sulph.; 3, Spongia; 4, Hepar-sulph.; 5, Spongia. All of the two hundredth potency. These he directs to be given, in the order of their numbers, thirty minutes apart, until relief is manifest, when their administration is to be suspended. Of three hundred cases treated in this manner by Dr. Boenninghausen, and a majority of them being what are termed *membranous*, not one was lost, two hundred and ninety out of the three hundred were cured in less than two hours. Better success than this certainly could not be wished for.

Boenninghausen's treatment, as it is called, was introduced into this country in the year 1860, by Dr. Carroll Dunham of New York, to whom homœopathia is indebted for many valuable contributions: since then, it has been tested by, and received the sanction of, all the most eminent homœopathists of the United States.

HOOPING-COUGH, OR PERTUSSIS.

DEFINITION. — This is one of that peculiar class of diseases that seldom, if ever, attacks the same individual but once during a lifetime. It is essentially a disease of childhood. Not but that adults would be just as liable to its invasion as children, but for the fact that they are protected by previously having had the disease.

CAUSES. — Some authors pretend that girls are more liable to it than boys, but I never could observe any difference. One thing is evident: let it once enter a family of children, and the whole group is pretty certain to have an attack, let them be boys or girls. It is undoubtedly contagious, as has been proved beyond dispute. It appears at times in the form of an epidemic, usually in the spring and fall. When appearing in the fall, it is apt to be more severe, from the fact that it is more liable to become complicated with lung difficulties, catarrhs, etc., and, of course, when connected with other diseases, they add to its severity.

Particular care should be taken that feeble children, as well as young and delicate infants, are not exposed to hooping-cough during the fall months. Such children might pass through an uncomplicated attack during the warm and genial months of summer unharmed; whereas in winter, during cold, inclement weather, they might succumb, even if they escaped the usual crop of autumnal coughs and colds.

Its duration, according to allopathic authority, is from six weeks to six months, and from their statements one is led to conclude that they never yet have had the good fortune to mitigate its severity a particle, or to abridge its duration a moment. Some, looking upon it as an inflammatory affection, have applied their antiphlogistic treatment with heroic severity. Others, taking into consideration the essential features of the disease, — the absence of fever, the little or no signs of inflammation upon inspection of the parts involved, — consider it purely a spasmodic affection of the air-passages, " calling for *anti-spasmodics*," cry they, and, as a matter of course assafœtida, leads the van.

Another clique, whose opinions are brought to a focus by Dr. Dixon, of the " New York Scalpel," assert that it is sheer nonsense, worse than folly, to talk about " cutting short the disease ; " " it must have its regular period of duration," say they. But did you ever hear of one of them refusing to treat a case, and honestly

confessing that over this disease his remedies had no control ?
No, indeed.   As Dr. Dixon asserts, " it is all a mere matter of
money-getting."

SYMPTOMS. — Physicians, as a general thing, divide hooping-
cough into three stages, as follows : 1st, catarrhal ; 2d, spasmodic ; and 3d, the stage of decline.

The catarrhal, or stage of invasion, commences with the ordinary symptoms of a common cold.   For several days, more or
less, — seldom, however, more than ten or twelve, — the child will
present all the symptoms of catarrh ; sneezing ; watering from
the eyes and nose ; irritation and tickling of the throat ; loss of
appetite ; restless uneasiness ; often chilliness, with flushes of heat ;
indisposition to do anything but worry and complain.   Sometimes
there is considerable fever, especially toward night, with a hollow
cough, which, at first, is quite dry, but afterward with copious expectoration of thick, tough mucus.   The cough at this stage has
no peculiarity by which it can be distinguished from an ordinary
cold.   Sometimes this first or catarrhal stage is entirely wanting.

The second — the spasmodic or convulsive stage — consists of
violent spasmodic paroxysms, or fits of coughing.   These paroxysms occur at longer or shorter intervals, and last from a quarter
to a half or three quarters of a minute.   A rapid succession of
them may occur so close together as to make almost one continual
paroxysm, lasting from ten to fifteen minutes.

These paroxysms are made up of a succession of quick, forced
expirations, without any intervening *inspiration*, until the little
sufferer gets fairly blue, yes, at times, even black in the face, and
appears upon the very point of suffocation.   This is followed by one
long-drawn act of inspiration, which produces that peculiar, shrill
sound, or hoop, as it is called, from which the disease derives its
name.   Immediately after this deep inspiration, the same series of
short coughs or *expiratory* movements take place, until all the air
is expelled from the lungs ; then is repeated the long, deep-drawn
inspiration with its accompanying hoop.   This alternation of
several short coughs, expelling all the air from the lungs, followed
by one long inspiration, again filling them, usually ends in the
expulsion of a quantity of thick, ropy mucus, or else in vomiting.

When one of these attacks is approaching, if the child is lying
down, he will suddenly jump up ; or, if playing, he will drop his
playthings, and run to the nearest fixed object for support ; catch-

ing hold of his mother's dress, a chair, table, anything to hold on to, while the fury of the paroxysm passes over. For a few moments, after it has passed, he stands quite exhausted ; but soon returns to his playthings, as unconcerned as though nothing had happened.

In some very severe cases, during a fit of coughing, blood will fly from the nose and mouth, and occasionally from the eyes and ears. The eyes, blood-shot and sunken, will fairly start from their sockets, presenting a horrid spectacle of suffering.

The third, or stage of decline, consists in an amelioration of the severe symptoms ; the paroxysms become less frequent, and of shorter duration ; the child's appetite returns, and he again resumes his natural habits and disposition.

In this stage of improvement, when all is going on smoothly, a slight cold may reproduce all the distinct characteristics of this peculiar cough.

COMPLICATIONS. — Simple hooping-cough, when unconnected with any other disease, is seldom or never attended with much danger. But its complications are many, and of various forms ; therefore it is highly important that all the accidents that are apt to occur in the course of this disease should receive a careful consideration.

BRONCHITIS. — This is a frequent complication of hooping-cough. It may be recognized by a greater amount of fever than usual ; an incessant cough during the first stage, with difficult breathing. The expectoration will be more difficult, and less profuse, and have a frothy or a yellowish look. You will also notice, when the child coughs, a marked expression of pain pass over his face. Sometimes the spasmodic cough of pertussis will give way entirely to that of bronchitis ; and here let me remark, in passing, *Pulsatilla* would be the remedy.

CONVULSIONS. — This is by no means a rare complication, but is by far the most dangerous of any that we meet with. Parents whose children are subject, or show a predisposition, to convulsions, would do well to keep a close watch over them, while passing through a siege of hooping-cough.

This complication, with that of head difficulties, — congestion of the brain, and the like, — is more apt to occur at about the second year, or rather at any time during dentition. And we augur ill from their approach, when the cough is of great severity, and the face remains livid for a considerable length of time after the paroxysm has passed ; or, perhaps you will perceive slight twitchings of the extremities, — the fingers and toes ; or of the muscles of the face, particularly around the eyes and mouth. The child is apt to be languid and sleepy after a fit of coughing is over.

Hooping-cough complicated with convulsions, is apt to prove a serious affair. I would, therefore, advise you not to trust to your own judgment, but to procure the attendance of a good homœopathic physiscian. Should this be impossible, do not, by any means, allow delusive hope to persuade you into the belief that an allopathist would do better than yourself ; as, my word for it, he will not. Consult article on " CONVULSIONS."

PNEUMONIA. — Inflammation of the lungs is another complication often met with ; for the particular symptoms indicative of its presence, see article on " INFLAMMATION OF THE LUNGS."

There are other accidents than those enumerated, such as emphysema, tuberculization, etc., which at times occur in connection with hooping-cough ; but to the unprofessional reader, their etiology and diagnosis would be like so much Greek ; and, as they appear but seldom, we will therefore pass on to the consideration of the

TREATMENT OF SIMPLE HOOPING-COUGH. — Gratifying it is to be able to assure the reader that over all the alarming symptoms of pertussis, as well as its milder forms, homœopathy holds specific control, and often " cuts it short," Dr. Dixon to the contrary notwithstanding.

As it is sometimes doubtful whether your catarrhal symptoms are the precursors of hooping-cough, or the accompaniments of an ordinary cold, you should treat them as simple catarrh. But the moment you hear a spasmodic bark, or the moment you become satisfied that you have a genuine case of pertussis to deal with, no matter in what stage you find it, give either *Mephitis putorius* or *Corallia*, one dose every four hours, and continue it for four or five days.

Drosera. — This is another invaluable remedy, especially during the second stage of the disease. It is indicated by the following symptoms : dry, spasmodic cough, with retching, worse at night, or upon repose ; pain in the side, just under the short ribs ; when coughing, the child presses with its hand upon the pit of the stomach ; severe fits of coughing following each other in quick succession, with hemorrhage from mouth and nose ; expectoration of thick, tough phlegm ; the cough is excited by talking or laughing.

I would recommend a high attenuation of this remedy to be used, say the two hundredth, of which one dose may be given every six hours.

In all probability, one or the other of the above remedies will break up the severe, spasmodic cough; when, for the remaining symptoms, you can give an occasional dose of *Causticum*, and, especially, if there be present a rough, dry cough, worse at night, with hoarseness, and pain in the chest when coughing.

Now, as hooping-cough does not always appear dressed in the same set of symptoms, other remedies than those enumerated may be called for. I will mention a few of them.

Aconitum. — When there is much fever, with short, dry cough, and pain in the chest. This remedy may be called for at any stage of the disease, and may be given in alternation with *Bryonia* or *Phosphorus*, particularly when hooping-cough threatens to become complicated with inflammation of the lungs.

Coccionella. — For a violent, spasmodic cough, with expectoration of white or ropy, tenacious mucus.

Stibium. — Tartar-emetic. — When, at the commencement of the disease, there is a hard, suffocative cough, or when there is a quantity of mucus which can be heard rattling in the chest; paroxysms of cough, with apparent danger of immediate suffocation; watery discharge from the eyes and nose; pain in the eyes. If, in addition to the above symptoms, the cough is excited by a tickling in the larynx, short fits of coughing, following each other in quick succession, each inspiration seemingly producing a fresh fit of coughing, Ipecacuanha would demand the preference.

Chamomilla. — When there is a wheezing, rattling at each inspiration; cough excited by an irritation in the windpipe; the child is cross and fretful; crying irritates the throat which causes it to cough, consequently the more it cries, the worse it coughs.

Cuprum. — Frequent fits of coughing, with rigidity of the whole body; rattling of mucus in the bronchial tubes; entire prostration after a fit of coughing; the paroxysms end in vomiting, when complicated with convulsions.

Hepar-sulph. — This remedy is particularly adapted to the period of convalesence when the cough is subsiding. Also, for a hollow-sounding cough; oppression of breathing; paleness of the face; hands hot and dry.

This remedy will tend to prevent a recurrence of the cough, from taking cold, by removing the sensitiveness, and the suseeptibility of the mucous membrane from atmospheric changes.

ADMINISTRATION OF REMEDIES. — Of the remedy ehosen dissolve twelve pills in twelve teaspoonfuls of water ; of this solution, give one teaspoonful for a dose, and repeat it every two, three, or six hours.

When giving Drosera, do not alternate it with any other remedy.

DIET AND REGIMEN. — The diet and regimen of a child, while passing through a siege of hooping-eough, is an item of no little eonsideration. A stimulating diet will eause an inordinate aetivity, and therefore an inereased susceptibility of the whole system to slight exposures. The diet should be plain and nutritious. Light and easily digested food is the best. All spiees and hot, stimulating drinks should be strietly avoided. For a drink, we may allow eold water, oatmeal-gruel, barley-water, riee-water, toast-water, ete.

Exposure to cold will very mueh aggravate the eough, and even reproduee all the severe symptoms, when the ehild is in a fair way of reeovery. The dress should be so regulated as to guard against all sudden atmospheric changes, so as to keep the body at about an even temperature.

ASTHMA OF MILLER.

DEFINITION. — Great differenee of belief has existed as to the nature and exeiting eause of this disease. Many eonsidered it synonymous with spasmodie eroup, and as sueh treated it ; others regarded it as dependent upon an enlargement of the thymus gland, calling it thymic asthma. But the generally received opinion now is, that it eonsists in spasms of the glottis, or opening at the top of the windpipe, eaused by an affeetion of the spinal system of nerves. It oeeurs during the first, second, or third year of life, and appears to be frequently eonneeted with dentition, and a deranged state of the digestive system. Asthma of young ehildren somewhat resembles, and is sometimes mistaken for, spasmodie croup ; but instead of being, like eroup, a speeifie inflammation, it is purely a nervous affeetion, and manifests itself in children of a strumous or serofulous habit, in feeble, delieate, and exeitable subjeets.

SYMPTOMS. — The entranee of air into the lungs is impeded or entirely prevented by the spasmodic contraction of the opening into the windpipe. Difficult breathing is the first symptom; the inspirations are prolonged and arduous; the air, as it passes through the narrowed opening (rima glottidis), produces a wheezing sound, or the breathing may be momentarily arrested, after which the child catches its breath with a shrill cry. In severe cases, when the closure is complete, the child gasps for breath; the body is thrown violently backward; the face turns pale, sometimes blue; the forehead is bathed in sweat; the nostrils dilate; the eyes are fixed and staring. If the paroxysms continue many seconds, the extremities become cold, the fingers and toes contracted, or the whole body may become convulsed. After a time, varying from a few seconds to a few moments, but which appears a much longer time, the spasm of the glottis ceases, and a loud, full inspiration takes place, followed by a fit of crying. The child looks around somewhat frightened, sobs for a little while, but soon regains its accustomed spirits. An attack of this kind, occurring suddenly, may befall a child every week, — yet, a fortnight or a month may intervene between the attacks; and, during the intervals, the child is as well as ever, respiration perfectly free, no fever, appetite good.

The paroxysms at first manifest themselves in a mild form; but if allowed, from neglect or improper treatment, to continue, they increase in frequency, as well as severity, and finally become associated with other spasmodic affections. At first, they come on only at night, or after the child has been asleep; gradually, as they become frequent, they occur indifferently in the daytime or at night, after a crying-spell or a fit of anger.

This disease, which is, as I have already observed, sometimes mistaken for spasmodic croup, may be distinguished from that affection, if you will remember that asthma presents no premonitory symptoms, occurs in the daytime as well as at night, is not accompanied with fever, and leaves no cough or hoarseness behind; while, in croup, these symptoms are always present.

TREATMENT. — The principal remedies for this disease are, *Sambucus*, *Ipecac.*, and *Arsenicum*.

ADMINISTRATION OF REMEDIES. — Of whichever remedy chosen, — and I would advise you to commence with *Sambucus* and *Ipecac.*, in alternation, — dissolve twelve globules in twelve teaspoonfuls of water, and give one teaspoonful of this solution at a dose, and repeat it every ten or fifteen minutes. If this should afford no relief, administer *Arsenicum* in the same way. Should these remedies fail, *Mosch.*, *Phosphorus*, or *Belladonna* may be indicated.

LARYNGITIS, OR INFLAMMATION OF THE LARYNX.

DEFINITION. — The larynx is a cartilaginous cavity or tube, forming the upper part of the windpipe. It consists of several movable pieces, forming that complex and beautiful instrument which produces every variety of tone, from the harsh, unmelodious noise of a midnight brawler, to the soft, sweet, flute-like sound, that flows from the warbling throat of a Jenny Lind.

Now, laryngitis is simply an inflammation of the mucous membrane lining the larynx, just as bronchitis is an inflammation of the mucous membrane lining the bronchial tubes.

It occurs at all ages of life ; but, as a general thing, is confined to children of from three to six years of age. It very much resembles an ordinary case of croup, but differs from that disease by being devoid of that peculiar, spasmodic, stridulous cough.

An acute attack of laryngitis sometimes runs its course with alarming rapidity, producing death by suffocation before you are hardly aware that your child is in danger. It is, therefore, highly important that you should early recognize its first or premonitory symptoms, that you may combat it at the outset with prompt and energetic treatment.

This disease it was that laid low the best, the purest man whom earth ever saw, George Washington.

CAUSES. — Exposure to cold is the most frequent cause of this disease. The reader is referred to what has been said upon diseases of the air passages in general at the head of this chapter; also to the article on " CROUP."

SYMPTOMS. — This disease is usually preceded by the ordinary symptoms of catarrh, — sneezing, with mucous discharge from the head. But frequently the first thing amiss that you observe, is an alteration in the voice or the cry of the child ; he speaks quite hoarse, or the voice is deep, or entirely lost. The cough is hoarse, and at first dry, but it soon becomes loose ; it is usually quite moderate, and, as a general thing, more frequent during the evening, or early in the morning. On looking into the throat, you will observe more or less inflammation about the tonsils and palate ; this inflammation may be diffused over the whole surface, or may be in patches, and varies in color from a mere blush, as in mild cases, to a deep rose, or even a violet-red.

Now the first symptom which should excite your alarm, is the *difficulty* of *swallowing*, for which you can find no *adequate* cause; certainly the slight inflammation observable upon inspection of the throat, is not sufficient. No, it is lower down than you can see; you ask the child to put his finger upon the sore spot, or seat of his distress, and he will point to that projecting cartilage called *Pomum Adami*, or Adam's apple.

To this difficulty of swallowing you will soon find added a difficulty of breathing. The respiration is peculiar; it is attended with a throttling noise; each inspiration produces a wheezing sound, just as though the air were drawn through a narrow reed. In fact, this sound is produced by the air being drawn through the narrowed opening at the top of the windpipe, and this narrowing is produced by the thickened or puffed-up state of the lining of membrane, which results from the inflammation therein existing.

You will find the larynx painful, upon pressure being made externally. The face is flushed; the skin hot and dry; the pulse more frequent than in health, rising to 120° or 130° to the minute; the child is thirsty, restless, and uneasy.

As the disease advances, the general distress increases, the face loses its flushed appearance, and assumes a livid, anxious, ghastly aspect; the eyes protrude; intense suffering is depicted upon every lineament of the countenance; the little sufferer throws up his arms and declares or makes signs that he wants air, that he must have it. And now, if relief is not soon obtained, death closes the frightful scene; he perishes, he dies from actual strangulation.

I am happy to state that this disease is not a very frequent one, at least the severer form of it. I have never seen a case that proved fatal.

It was quite prevalent during the winter of 1859, but yielded promptly to the usual treatment.

When improvement sets in, the cough becomes less frequent, looser, and easier; the fever, soreness of the throat, and difficulty of swallowing gradually disappears; the voice loses its harshness; the appetite returns, and, in a few days, all traces of the disease has vanished.

You will have no trouble in distinguishing this disease from a common sore throat, by the peculiarity and difficulty of the breathing. It is true, extreme enlargement of the tonsils obstructs respiration, but then, on inspection, this swelling will be *visible*.

In laryngitis the inflammation and swelling is slight, at least all that you can *see* of it.

You will recognize it as not croup, from the fact that in that disease *swallowing is not interfered with*, besides the cough is quite different.

TREATMENT. — As soon as the inflammatory process is lighted up within the larynx, Aconitum should be administered in doses frequently repeated, and often it will arrest the severest attack at its very commencement.

Aconite. — This remedy is especially indicated, when the following symptoms are present: skin hot and dry; short breathing; quick pulse; great thirst; face flushed; short, dry cough, and irritability of the nervous system. *Aconite* should be continued until there is an evident abatement of the febrile symptoms, or until the pain and sensibility in the upper part of the windpipe become more decided; the breathing and cough shrill, with an increase of hoarseness, and difficulty of articulation; then *Spongia* must be substituted for, or given in alternation with, *Aconite.*

Hepar-sulph. — This remedy should be given in preference to *Spongia*, when the febrile symptoms remain unabated after the use of *Aconite.* *Hepar* and *Spongia* may be given in alternation from one to two hours apart.

Belladonna. — Highly sensitive and inflamed state of the throat, inability to swallow liquids, every attempt produces a spasmodic choking; the tongue hot and dry; dry, short cough, worse at night.

Tartar-emetic. — When there is hoarseness from the first, cough hard and ringing, or paroxysmal fits of coughing, with suffocative arrests of breathing.

Phosphorus. — For the remaining hoarseness, with more or less pain, and a feeling of fulness or tightness about the chest; expectoration of viscid mucus.

Lachesis. — Hoarseness, with a sensation as though something had to be hawked up; great sensitiveness to external pressure.

ADMINISTRATION. — Dissolve six globules in twelve teaspoonfuls of water, and give of the solution one teaspoonful at a dose, from one to three hours apart. In extreme cases a dose may be given every ten or fifteen minutes.

DIET AND REGIMEN. — A child suffering from this disease should be confined to a warm room, and not allowed to roam all over the house, through cold rooms and in drafts of air.

A slight reduction of diet is advisable, forbidding all condiments, or anything of a stimulating nature. A farinaceous diet is the best.

The application of cold water is always advisable, often affording great relief.

Rubbing the throat with camphorated oil, goose-grease, ready-reliefs, — in fact, the application of stimulating lotions or embrocations of any kind is highly objectionable.

COLDS.

The term cold is a relative one, and is used in this connection to express a certain condition or sensation produced by the abstraction of heat from the system by any substance of a lower temperature than that of the body. This condition or sensation, it will therefore be observed, may not always be occasioned by the same degree of temperature. For instance, a temperature that, to a healthy, active, vigorous man, would seem warm, or at least comfortable, would to one enfeebled by disease, appear right the reverse. Or, the temperature of a room to a properly-clad female, might seem warm, while to her delicate child, with arms and chest entirely naked, it would be cold. What would be cold weather to a Southerner, would be warm to an Esquimaux. Again, a man or child, though perhaps not of the strongest physical constitution, but full of *vim*, with high, moral and physical courage and excitement, will resist a greater amount of cold, than one who is faint-hearted, nervously depressed or despondent.

Children are more susceptible to atmospheric depressions than adults, and simply because the power of generating heat within themselves is weak, undeveloped.

In considering the effect of cold upon the system, due reference should always be had to the capabilities of those functions whose duty it is to supply heat, and whose capabilities, it should ever be remembered, are in direct proportion to the healthfulness and vigor of the constitutional powers. When the vital energies are weak, a less degree of cold will depress them than when they are vigorous and energetic.

A robust, healthy boy, full of life and energy, with free and full respiration, one who is accustomed to much out-door exercise, and can trundle his hoop till the blood goes dancing through his veins, distributing vital warmth throughout his system, is vastly more

capable ot withstanding the injurious effect of cold than the child
who has been housed up in warm rooms, whose respiration is
feeble, and circulation sluggish. The latter, instead of being able
to spare some heat, must be so clad as not to allow a particle of
that which he evolves to escape.

Some persons, owing to an exhausted or inactive condition of
the digestive, nervous, circulating or respiratory functions, are
peculiarly sensitive to the least atmospheric depression ; they are
unable to generate vital heat, sufficient to supply the loss which is
extracted from the exposed surfaces of the body. This is especially the condition with infants, aged persons, and convalescents.

There are many conditions which favor the injurious action of
cold upon the body, among which may be mentioned the undeveloped and the exhausted state of the " heating apparatus,". as
in infancy and old age ; debility ; exhaustion from previous illness ; fatigue ; dissipation ; over-heating of the body ; too long
fasting ; over-eating, or excesses of any kind which depress or
diminish the vital forces of the nervous system. The power of
resisting cold is also diminished in a wonderful degree by sleep.

Other things being equal, cold is more likely to prove injurious,
when it is applied by a wind, or a current of air, and especially
when but a part of the body is exposed to its influence, as when
sitting by an open window, or in such a position that a current is
allowed to impress any definite part of the body for a length of
time. One may walk or stand, out-doors, in the wind and feel no
bad results from it, but let him be exposed to a current of air of
the same temperature, passing through a chink in the door,
directly upon his back, or upon his head from a partially opened
window, and it will soon produce results injurious to a greater or
less extent, according to the individual's constitutional powers of
resistance.   Cold does not always cause disease in the exact part
to which it has been applied, that is to say, because a person sits
through a tedious concert with a draft of air continuously playing
a disease-producing tattoo upon his back, he must not necessarily
have rheumatic pains or some other trouble in the part, although
this may be the case. As a general rule, the cold, by diminishing
vital action in the parts on which it acts, so determines and increases the same in distant parts, as to give rise to congestions and
inflammations, or to a train of diseased action, more or less definite, which, by common consent, is usually termed a cold, such as
chills, general soreness and lameness, pains and aches in the head

and limbs, followed, as soon as reaction comes on, by accelerated respiration and circulation, as well as other symptoms which constitute fever.

Cold does not affect all persons alike. Two ladies, exposed to the same current of air, may, as the result, suffer from diseases quite dissimilar. This depends upon peculiarities in temperament, predisposition, and habit of the individual. As a general rule, however, those organs or parts of the system are first affected which are the weakest. If the lungs are predisposed to disease, cold will develop some difficulty in these organs. Should a person be subject to catarrh, cold will act as an exciting cause to bring it into action. Children subject to croup, to glandular enlargements, or to gatherings in the head, need but a cold to set the disease in motion. The same principle is true with other organs and structures of the system. The extent and severity of the disease thus excited will depend upon the amount of exposure, and the delicacy of the part affected.

The most common results of taking cold are catarrh and cough, sometimes fever, colic, dysenteria, diarrhœa, neuralgia, sore-throat, pains in the teeth, ears, or general pain and soreness throughout the whole system.

TREATMENT. — In the selection of remedies for the evil effects of cold, it will frequently be necessary to refer to those chapters where the diseases are treated of more at large. In this place, we shall only treat of those medicines which are applicable to the more common cases.

In many instances, one is aware of having taken cold long in advance of any definite symptoms manifesting themselves. Or, in other words, the individual is aware of having taken cold before its injurious effects have settled upon any particular organ or part. There may be a general feeling of lassitude, some chilliness, and an inclination to yawn and stretch. In most instances, these symptoms may be dissipated, and all further evil results warded off, by a few doses of *Aconite*.

Some persons are in the habit, for removing these first symptoms of a cold, of taking a hot sling on retiring at night, and thus induce free perspiration. This frequently answers every purpose, affording prompt relief. A few globules of *Aconite* and a glass of cold water, however, will produce like results, and are certainly much more agreeable.

Should the cold have been occasioned by the patient getting wet, give *Dulcamara*, *Bryonia*, or *Rhus*.

For cold in the head, from wet feet, give *Cepa* or *Dulcamara ;* for pain in the limbs, *Rhus* or *Mercurius ;* for general soreness and lameness, *Bryonia*.

COLIC. — When colic results from a cold, give *China*, *Chamomilla*, *Nux-mos.*, *Mercurius*, or *Nux-vomica*. See " COLIC."

COUGH. — For coughs provoked by a cold, we give *Aconite*. When there is a sensation of dryness and roughness in the larynx, or even throughout the whole chest, occasioning an incessant, short, dry, hollow cough, accompanied with more or less fever, if the cough excites vomiting, *Nux-vom.* or *Carbo-v.* If accompanied with a tough expectoration, which children cannot get up, *Chamomilla ;* if loose, *Dulcamara* or *Pulsatilla*. *Bryonia* may follow, or be given in alternation with *Aconite*, especially if there should be soreness of the chest, pain under the ribs, stitches in the side, cough dry and convulsive, expectoration streaked with blood, or cough from a tickling in the throat.

Hepar-sulph. — For a loose cough, attended with mucous rattling in the chest, pain in the throat when coughing, and a feeling in the chest which renders talking oppressive. If the cough is always worse after retiring, *Belladonna* or *Nux-mos.* Consult article on " COUGH."

COLD IN THE HEAD. — For catarrh, with great heat in the eyes and head, and soreness of the nose, give *Belladonna*. If the nose is entirely stopped, *Nux-vom.*

Mercurius. — When the lining membrane of the eyes and nose is highly irritated, giving rise to copious discharge ; a feeling of fulness in the head ; pains in the limbs, accompanied with profuse perspiration ; constant sneezing ; profuse, excoriating coryza.

Cepa. — When there is running from the nose and eyes ; great heat and thirst, accompanied with headache ; worse at night and while in-doors.

Arsenicum. — Excessive discharge of an acrid, burning water from the nose, with hoarseness and sleeplessness.

Pulsatilla. — Stoppage of the nose in the evening, with a discharge of thick, yellow mucus in the morning ; catarrh, with loss of taste and smell ; may be followed by *Nux-vom.* If a catarrh, becomes checked by a fresh cold, give *Pulsatilla*.

Consult article on " CATARRH and CORYZA."

HEADACHE, TOOTHACHE, EARACHE. — *Belladonna* for headache,

when there is a fulness, as though the head would burst, especially when going up-stairs, and worse in the open air. For pain, or pressure in single spots, with buzzing in the ears, and hardness of hearing, give *Dulcamara*. When headache is accompanied with giddiness, give *Nux-mos*. If accompanied with nausea and vomiting, *Nux-vom.*, *Coculus* or *Antimonium-c.*

For earache, arising from cold, give *Rhus*, *Chamomilla*, *Nux-vomica*, *Mercurius*, *Dulcamara*, *Bryonia*, or *Sulphur*. See " EAR-ACHE."

For toothache, arising from cold, give *Aconite*, *Bryonia*, *Chamomilla*, *Rhus*, *Nux-mos.*, or *Mercurius*. See " TOOTHACHE."

SORE THROAT. — For sore throat, arising from cold, *Belladonna* or *Mercurius* will usually suffice. See " SORE THROAT."

DIARRHŒA. — Diarrhœa, occasioned by cold drinks, generally yields to *Arsenicum* or *Bryonia*. For diarrhœa immediately after taking cold, give *Opium* or *Dulcamara*. For diarrhœa from checked perspiration, especially in the summer time, or during warm weather, *Bryonia*. From getting wet, *Dulcamara*, followed by *Bryonia*. When caused by eating *ice-cream*, give *Opium*, *Glonoine*, or *Bryonia*. Consult article on " DIARRHŒA."

ADMINISTRATION OF REMEDIES. — Dissolve of the remedy selected, twelve globules in six spoonfuls of water; and of this solution give one spoonful at a dose, and repeat it as often as the exigency of the case seems to demand.

CHAPTER VII.

DISEASES OF STOMACH AND INTESTINES.

GENERAL REMARKS. — As a large number of the various stomach and intestinal disorders originate from the same disturbing influences, I have deemed it expedient to make here a few observations embraced under the heading of General Remarks, upon their frequency, cause, and prevention. By this means we shall save numerous repetitions, when we come to speak of specific diseases.

Of the importance of a just understanding, and a due consideration of this class of diseases, it is but necessary for me to assert, that they are the occasion of about one-fourth of all the deaths under fifteen years of age. They are by far the most frequent and fatal diseases to which childhood is exposed.

The particular age at which children are most liable to these affections is from birth to the termination of first dentition; this of course includes the second summer. From this period onward, as the child increases in years, it becomes less liable to their invasions.

CAUSES. — By far the most frequent exciting cause of all gastric diseases during infancy, is an improper, or an unwholesome diet. They are not unfrequently, in nursing infants, dependent upon an unhealthy condition of the mother's milk; but it seems to me that the chief source of difficulty is the too early resort to an artificial diet, or an artificial diet badly chosen. Of course, the natural aliment of an infant is its mother's milk, which, during the first few months is very thin, and possesses properties peculiar to itself. Now contrast this with the various articles of food prepared for children; the latter consist of pap, or thick bread and milk, or crackers moistened with milk and water, to which a little sugar is added; gruels of all kinds; coarse preparations of rice, barley, etc., etc. As before intimated, the stomach of an infant is only intended to receive the milk provided by its parent; and it is entirely incapable of digesting the thick or coarse food, as well as often too rich, which is so frequently substituted for that which nature has provided.

However, it is not always the quality of the food only that is at fault, but, often, I imagine, the quantity as well. Physicians and discriminating nurses have long recognized the over-feeding of children as a frequent source of mischief. Children fed upon an artificial diet scarcely ever escape these intestinal derangements. Every mother of any experience, knows that diarrhœa is very apt to set in immediately after weaning a child; and any one who has given it a moment's thought must certainly have inferred that this is but the result of an irritation of the mucous membrane lining the intestinal canal, produced by the change of food made at the time.

An indigestion, or a loss of digestive power, and the consequent enervation and wasting-away of the system, from imperfect assimilation, is but the direct effect of an improper diet, or the over-taxation of the digestive apparatus from excessive feeding.

The heats of summer and sudden atmospheric changes are undoubtedly powerful predisposing causes to infantile bowel complaints; in fact, we seldom have these diseases to any great extent, presenting all their characteristic features, except during the hot months of summer. To the heats of summer, we have usually to add, impure air and badly ventilated houses. As you pass through some of the streets of our cities, and inhale the effluvia from the dirty gutters, you wonder, not that so many are taken sick, but that all do not die; and then, when you come to enter the damp basements, and find huddled together whole families of ten or a dozen persons, occupying one room, in which they cook, eat, and sleep, you are actually bewildered and in amazement, wondering how any mortal can draw the breath of *life* from such a contaminated atmosphere.

To people living thus, and all those who reside in narrow, crowded streets and alleys, these diseases are as scourges. But these disorders are not confined exclusively to the poor, and to those living beneath the ground, and away from the light and air which God has given us. No, they are only too common among all classes of the inhabitants of our large cities.

Dentition, being a natural, physiological process, we should not expect it to be productive of any evil results; nevertheless, it is a well-established fact, that the cutting of teeth is a powerful pre-disposing cause to intestinal irritation, and it frequently impairs or diminishes the tone of the digestive function, so that a child is often unable, during the period of cutting teeth, to digest food, which at other times agreed with it perfectly well. In considera-tion of these facts, it would seem advisable that a child should not be weaned until after the period of first dentition is completed. The protrusion of the teeth appears to be the first indication on the part of nature, that the digestive organs are sufficiently de-veloped to receive and digest other food than milk. It can now masticate for itself. This is a fact which I have always endeavored to impress upon the minds of my patients, and, I am sorry to say, with but indifferent success. They complain that nursing a child is so fatiguing ; it is such a drain upon their system, that they are really unable to bear it ; or, if they are willing, some *other* obsta-cle presents itself, which makes it advisable that the child should be weaned.

The preceding remarks apply exclusively to infants. Gastric derangements of children, from the completion of the first denti-tion to the age of eight or ten years, may, in the majority of cases, be traced directly to the persistent inattention on the part of mothers and nurses to the general laws of health. It is the strangest thing in all the world to me, that poor human nature, who is so plain in all her requirements, should be so wholly disre-garded. I often wonder, whether it is from ignorance or from thoughtlessness, that people give so little attention to what they know is necessary for their well-being, if they would preserve good health. It appears to be a studied endeavor on the part of some people, to see how far the laws of nature can be perverted, or wholly ignored. Sometimes I think it is the evil one himself who is leading us astray, and that by these invasions upon a rational mode of living, he is endeavoring to enfeeble the race, both physically and mentally, and thus render us an easier prey to his infernal machinations. Scripture tells us, "Whatsoever a man soweth, that shall he also reap." This applies as well to the phys-ical as to the spiritual man, and nowhere, perhaps, would a ser-mon on this text be more appropriate than just here. As, if ye sow tares, ye expect not to reap wheat, so neither ought ye to expect a harvest of good health after sowing the seeds of dys-pepsia.

Since the first contemplation of this work, I have made extensive inquiries in regard to the diet of children. These investigations have amply confirmed my own observations, that it is quite a common thing, in fact the general custom, to allow children of from two to three years to sit at the table and partake of the same food that is prepared for the adult members of the family. Now, any one who will take the trouble to pause a moment, and consider the number and variety of dishes concocted to suit the delicate palates of this fastidious people, and at the same time remember that the children have an indiscriminate and free choice from amid such a profusion, certainly cannot be surprised at the thin, pale, puny specimens of humanity that meet his gaze on every hand, continual subjects of intestinal and gastric derangements, whose systems, like a wilted plant, droop with the slightest exposure, and are continually harassed by some one of the mighty host of ills following in the train of an enfeebled digestion.

Dr. Meigs, the most celebrated writer upon diseases of children, and from whom we have already made extensive quotations, when speaking of diet in relation to intestinal diseases, makes use of the following language : —

" The chief causes of the disease, after the first dentition, are, according to my experience, the habitual use of improper food. . . . . I am acquainted with some families in this city, the children of which, from the age of two years, are allowed habitually, to breakfast upon hot rolls and butter, hot buckwheat cakes, hot Indian cake, rice cakes, sausages, salt fish, ham or dried beef, and coffee or tea ; and to dine upon a choice of various meats, and a great variety of vegetables, which latter they often prefer to the exclusion of meat, and then make a rich dessert of pies, puddings, preserves, or fruits ; and lastly, to make an evening meal of tea and bread and butter, almost always relished, as the term is, with preserves, stewed fruits, hot cakes of some kind, or with radishes, cucumbers, or some similar dish. Add to such meals the eating, between whiles, of all kinds of candies and comfits, which many children here regularly expect in larger or smaller quantity, cakes, both rich and plain, fruits to excess, and at all hours, from soon after breakfast to just before going to bed, raisins and almonds and nuts of various kinds; and the wonder is, not that we are a pale, thin, dyspeptic, and anxious-looking race of people, compared with Europeans, but that we have any health at all, when our children are allowed to make use of the indiscriminate and unwholesome diet just described."

Now one would think this alone enough to degenerate the whole race in a short time; but to all this, is yet to be added that vile, pernicious habit of drugging children with medicines. Most mothers and nurses have, each, their little collection of remedies; some choose the simple, — those that are usually called domestic remedies, that are chiefly concocted from roots and herbs; while others make free use of all the patent medicines which the shops afford; and are ever ready to descant learnedly upon the relative merits of all the vermifuges, blood-purifiers, and human regulators generally, that quackery ever yet imposed upon an innocent public. For every little ailment that may overtake a child, brought about, as we have already said, in most cases, by some error in diet, a dose of medicine must be given, — usually a cathartic. Now this, in the first place, is all unnecessary, even if it were ever so harmless. The seeming disorder from which the child is suffering is but an effort on the part of nature to rid herself of some offending substance, as, in the great majority of cases, she would readily do, if meddlesome hands would but let her alone. But no. The child is sick, perhaps "bilious," so a dose of medicine must be given, to work it off. And what is the result? Why, a slight indisposition, which a judicious restriction of diet, and a little care would have speedily removed, is, by officious hands, otherwise changed to some serious disorder. The medicine given is so repulsive to nature, that the whole system is thrown into commotion, in the effort to reject it; the child is vomited, physicked, in fact, "thoroughly cleaned out," as the saying is; and this is looked upon as salutary; but a greater mistake was never made. If the child recovers, it does so *in spite* of the treatment; but in the majority of cases, the extra irritation and exhaustion thus produced, if not the direct cause of some immediate mischief, is surely laying the foundation for future disease, by enfeebling the whole digestive apparatus.

And thus the nervous system becomes shattered also; the child grows up irritable, cross, morose, and a constant subject for all sorts of nervous affections.

These little innocent domestic remedies that mothers are so fond of giving are not as innocent as they have been wont to think. I am of the opinion, — and you can have this opinion endorsed by questioning any intelligent physician, — that their effects are far deeper and far more lasting than people generally suppose.

I do not know how the nervous system can be more speedily

affected and permanently injured than by this eternal drugging. As for the patent medicines of our day, their ill effects are incalculable. There is scarce one of them that does not contain some rank poison, and that, too, in no "infinitesimal" quantities. It is singular with what audacity these nostrums are placed before the public. The proprietress of one advertises herself as "The Florence Nightingale of the Nursery!" "Angels and ministers of" health defend the children! say we, from her somniferous hand. There is no doubt but that any of their syrups or cordials would put a fretting child to sleep, even while "teething," and you might attend a party or a ball without a fear of its waking; opium would produce the same result; either, however, does it only at the hazard of the child's future health. If these domestic remedies were but given occasionally, so much need not be thought of it; but just think what slight ailments call for their use, and with what a free and generous hand they are given. It is astonishing how slight a recommendation they need; a mere hearsay is all that is required.

If you had a valuable gold watch, and from some cause, to you unknown, it should stop running, and a friend should come along and tell you that all it needed was a little oiling, would you open the case and pour in at random a few drops of oil? No; you would take it to the best jeweller that you could find, and charge even him to be careful with it. But your child, whose organization is far more intricate, far more delicate, and far more susceptible of permanent injury, when tampered with by ignorant hands, than any watch ever manufactured, — if its system becomes deranged, or deviates in the least from its normal condition, and some old lady comes along and says, "Give it this, or give it that," why, down it goes. You simply take her recommendation for it. If the lock upon your front-door refuses to bolt, you know not why, or your gas refuses to burn, from some obstruction in the meter, you do not commence tinkering at the one, or overhauling the other, but immediately send for a man skilled in the business, and place the article in his charge. How different with your child! If he is sick, you know just exactly what to do; if he picks his nose, he has got worms; and, if he has got worms, why, of course, he must take Dr. Taenia's Vermifuge, — it is a certain cure, for you read so in a medical almanac. I saw a lady, this very day, who was mourning because she did not find out sooner that her child had the scarlet fever, for old Mrs. Blixen had said,

27

" that if a sheep's melt were bound upon the child's feet, at the outset, it was a certain cure ! " How often will some people take up a newspaper, read over a string of symptoms attached to an advertisement of some popular quack nostrum, and immediately recognize the very disease from which their child is suffering, send for a bottle, and force it down, whipping the child, perhaps, because its nature revolts against the nauseous stuff. Besides, all children will get bilious, — at least, parents are apt to think so, — and therefore they must occasionally take an antibilious pill, and, by way of variety, perhaps an emetic.

In addition to all this, every spring and fall, the children, like an old house, must have a " thorough cleaning out," perhaps be fumigated, under the absurd notion of *purifying* the blood. Now, who, when in his right mind, reasoning rationally, can help arriving at the fact, that this eternal drugging is one great source of stomach and intestinal derangements. I do not believe there is another nation upon the face of the earth where drugging is so universally practised as it is here in America.

In passing through our cemeteries, one cannot help being struck with the number of little white stones, all nestled among the grass, and upon many of them you will find this beautiful little couplet, beautiful, when true : —

> " Sleep on, sweet babe, and take thy rest,
> God called thee home; he thought it best."

I never read it but I sigh, as 1 think how often it might be changed, and truly, too, to read as follows : —

> Sleep on, sweet child, thy trouble's past;
> Physic has freed thy soul at last.

THRUSH, OR APHTHÆ.

DEFINITION. — The term thrush or aphthæ is applied to an ulcerative sore mouth, peculiar to infants, which makes its appearance during the first year; as a general thing, within the first fortnight.

Nurses and women of experience generally anticipate the arrival of this unwelcome visitor, and, as a preventive, are very careful, after the child has nursed to wash its mouth with a soft linen rag dipped in cold water, and to remove all particles of milk, which, if allowed to remain there and ferment, would tend to induce an irritation of the delicate membrane lining the mouth. This general expectation, on the part of nurses, shows that the disease in question is one of no uncommon occurrence.

CAUSES. — I do not think there is any great reason to be surprised, that parts so tender and delicate in their structure should become abraded and inflamed, when we take into consideration the fact, that of all the thousands of children born annually within the United States, few, yes, very few indeed, escape having their stomachs crammed with some pernicious mixture of sugar or molasses, or some one of the forty thousand other outlandish things, which grandmothers and old-fogy nurses invariably have ready at hand upon their arrival. Why, the irritation of the spoon, even, in punching such trash into an infant's mouth is enough itself to excite an inflammation.

I know it is common to attribute the sore mouth of children to some derangement of the digestive organs, but seldom, indeed, do you find children suffering from thrush, or, in fact, many of the diseases of the mouth and intestinal canal, which are so common, unless where the stomach has been made a vat in which to ferment some nauseous mixture of honey or molasses.

SYMPTOMS. — Thrush or aphthæ is characterized by the eruption of vesicles, capped with small, white spots, which break, and are followed by small, round ulcers, with edges more or less thickened, and surrounded by a red circle of inflammation ; the bottom of the ulcer is of a grayish color, and secretes a whitish, cheesy humor, which adheres more or less to the surface.

When these ulcers or aphthæous patches are isolated, they usually occupy the internal surface of the under lip and cheeks, the edges of the tongue and gums ; when numerous or confluent, the inner surface of the mouth is quite covered with them, while the matter secreted extends or spreads from one to another, forming a complete coating, of greater or less thickness. When this layer becomes detached, upon close inspection the ulcerated points are visible beneath.

As a general thing, there is little or no fever, neither is the child as restless or as fretful as one would expect. The mouth is hot, and the saliva is secreted in larger quantities than is natural. These ulcers, even when in small numbers, cause severe pain, and if situated far back in the mouth, interfere with swallowing.

Now, aphthæ may be confined exclusively to the mouth, or it may penetrate into the windpipe, the œsophagus, or the stomach ; it may occupy spots upon different parts of the alimentary canal, or the whole mucous membrane, from the mouth to the rectum, may be involved.

When aphthæ extends over a large surface, affecting the stomach and bowels, the child grows pale and thin, diarrhœa sets in, and the affection, which at first seemed insignificant, may assume quite a serious character.   See " CANKER OF THE MOUTH."

TREATMENT. — Borax is an excellent remedy for aphthæ, and may be given in the form of pills, dry, upon the tongue, or twelve globules may be dissolved in as many teaspoonfuls of water, and given, one teaspoonful of the solution every three hours.

A weak solution of Borax — a few grains to a teacupful of water — is frequently used as a gargle or wash for the mouth; when this is done, no other internal administration of this remedy will be necessary, as the child will swallow quite a sufficiency for a dose.

Mercurius. — When there is profuse salivation, and a tendency toward ulceration. If *Mercurius* fails to effect a cure, follow it with *Sulphur.*

Arsenicum. — For bad cases, when the ulcers assume a livid hue, or if the mouth and throat become covered with ulcerations, attended with diarrhœa and great prostration of strength.   When *Arsenicum* does not prove sufficient, give *Nitric-acid.*

Chamomilla is sometimes of service, especially when the mouth is hot, considerable fever, and great restlessness.

Nux-vomica and *Bryonia* may also, in some cases, be of service.

ADMINISTRATION OF REMEDIES. — The directions for Borax have already been given.   The other remedies may be given in the same manner.

CANKER OF THE MOUTH.

DEFINITION. — This form of sore mouth is usually found in children of from five to ten years of age : by many it is considered contagious, but, upon this point, physicians are divided; all agree, however, in considering it epidemical.   It is an inflammation of the mucous membrane, with an exudation upon the surface, of a yellowish, plastic lymph, with erosion, or ulceration, which occasionally, particularly if improperly treated, assumes a very destructive character, running into deep, dark, sloughing sores.

This affection is also known as cancrum oris, scurvy of the mouth, or canker-sores.

SYMPTOMS. — The peculiar characteristics of this disease are, first, pain and uneasy sensations in the gums, which soon become hot, red, and very sensitive; they also swell, become spongy, and bleed when touched.

The gums, and internal surface of the cheeks are covered, or rather, spotted over with patches of false membrane which adheres, with considerable force, to the tissue beneath; under this layer of exudation, small ulcers make their appearance on the gums, the inside of the lips, and cheeks, on the soft palate, and edges of the tongue. Sometimes this false membrane is entirely wanting, when the ulcers are plainly visible, and present a grayish or livid appearance, with swollen, softened, or bleeding edges.

These ulcerated spots may be but few in number, either upon the inner surface of the lips, and cheeks, or edges of the gums, or they may be studded over the whole cavity of the cheek.

The breath is always more or less fetid, and not unfrequently putrid, or almost gangrenous, and sometimes, especially in severe cases, there is a copious discharge of offensive bloody serum from the mouth.

The glands about the throat and neck are swollen and painful; the movements of the under-jaw are stiff; this, together with the looseness of the teeth, makes mastication very difficult, while swallowing is interfered with, from soreness of the tongue and throat.

There is generally more or less of a low grade of fever; the patient loses his strength, and sometimes becomes very much prostrated.

The course of this disease is short, if under judicious treatment; but, not unfrequently, sudden, severe, and destructive salivation is set up by the intemperate — allopathic — administration of calomel, which, if not ending in gangrene of the mouth, prolongs the difficulty to an indefinite length of time.

TREATMENT. — *Mercurius.* — This remedy is indicated in almost every case, and may always be given at the commencement of the disease; unless it was brought about by calomel or mercury in some form, in which case *Carbo-v.* should be administered, to be followed if necessary, by *Hepar-s.* or *Nitric-acid.*

Natrum Muriaticum. — Particularly when the gums are swollen, and bleed when touched; when blisters and small ulcers appear upon the tongue, which smart and burn, rendering talking painful.

Nux-vomica. — For putrid and painful ulcers; swellings of the gums; fetid ulcers over the whole inside of the mouth; emaciation, constipation, and irritability.

Sulphur — At the end of the cure, or when other remedies fail; also when there is swelling of the gums, with pulsative pains.

Arsenicum, Carbo-v., Dulcamara, and *Capsicum* are sometimes serviceable.

ADMINISTRATION OF REMEDIES. — Of the chosen remedy, dissolve twelve globules in twelve teaspoonfuls of water; one teaspoonful of this solution may be given for a dose. At the commencement of a case, or in severe cases, the remedy may be repeated every two hours, until amelioration or change. In cases less urgent, a dose may be given every four hours.

DIET AND REGIMEN. — The diet should be plain, and of either a farinaceous or vegetable form; animal food, either solid or in soups or broth, had better be dispensed with.

It is desirable that the mouth should be frequently gargled or rinsed out, and especially after eating, that no offensive matter, or particles of food may remain to irritate the parts. A weak solution of brandy and water makes the best wash, lemon-juice and water is frequently used, so is a decoction of sage. I advise, however, the brandy and water.

Decayed teeth, or stumps of teeth, remaining in the mouth, are often the source of irritation : when such is the case, they should be speedily removed.

GANGRENE OF THE MOUTH.

DEFINITION. — The term gangrene, you will please bear in mind, is synonymous with mortification.

Gangrene of the mouth is, justly, the terror of all those who have it to contend with. It generally commences with ulceration of the mucous membrane lining the cheeks and covering the gums, which, if not soon arrested, runs into gangrene. The mucous tissue of the mouth, the gums, the lips, and the substance of the cheeks, are destroyed, turn black, and slough away, leaving the teeth loosened, the jaw-bone denuded and exposed, while from the mouth there stream quantities of offensive, thick, black mucus. Such is gangrene of the mouth.

CAUSES. — This affection is seldom met with in private practice. It is almost exclusively confined to public institutions, such as nurseries, almshouses, hospitals, and the like, where large numbers of children are gathered promiscuously together. It almost always follows upon some previous acute or chronic disease, such as long-continued fevers, measles, or other acute exanthema, and where, I take it, the patient suffered more from the treatment than from the actual disease.

Unfavorable hygienic conditions, debilitated constitutions, a scrofulous habit, etc., are conceded on all sides to constitute the predisposing cause of this affection; but the *exciting cause* has been, and still is, a bone of contention among those physicians encountering this disease. It is perfectly plain, however, to those who are disposed to see, that gangrene of the mouth is nothing more nor less than poisoning by mercury.

This disease, as has been observed, only occurs in public institutions, — those hot-beds of routine practice and experimentation. Private practice sees but little of it, because here *heroic* treatment is under some restraint.

SYMPTOMS. — Having never treated a case of gangrene of the mouth, and seen but few, I shall borrow from Rillet and Barthez a description of the disease, which Dr. Teste assures us is as true as it is striking.

" Gangrene of the mouth begins during the course of convalescence from another acute or chronic disease, by ulceration, aphthæ, or more rarely by œdema, — swelling of the part where gangrene is about to be developed. At this time, the face is pale, the breath fetid; the fever not very intense, unless there also exists a febrile disease, and then the pulse may be considerably accelerated; the child becomes more sad; ordinarily complains little, or none at all, of the mouth; sometimes, though rarely, he suffers severe pain.

" The ulceration, slight at first, and with a grayish base, situated upon the middle of the internal surface of the mouth, or in the folds between the gums and the lips, is soon covered with a grayish, putrilaginous excretion, of a fetid and peculiar odor; at the same time, an infiltration of the diseased cheek or lip takes place; the œdema is soft, rather regularly circumscribed; it soon becomes increased, and there is formed, deep in its centre, a hard, regular, round nucleus. Then the cheek becomes tense, shining, and pale, or marked with a violet-colored marbling, more decided upon the prominent parts of the tumor. In the interior of the mouth, the eschar has taken a brown color; it has spread considerably, has reached the gums; it is sometimes surrounded by a violet-colored circle.

The child is seated in his bed, and occupies himself with the

objects around him; sometimes without strength, he lies in a state
of indifference, his face, puffed and without expression on one
side, is sad and depressed on the other; a bloody, or already
blackish saliva flows from his half-open lips. He asks, however,
for food, and takes with avidity what is offered him, and swallows,
together, his food and the putrid matter detached from the gangrenous parts.

"His skin is cool, and his pulse but little developed, and of moderate frequency; unless there exists some febrile complication, his
mind is clear, but sometimes, during the night, he has more or
less intense delirium.

"From the third to the sixth day of the disease, the scene
changes, an eschar is formed upon the most purple and prominent
part of the tumor, either upon the cheek, or upon the under lip,
small, black, and dry, this eschar extends itself from day to day,
and sometimes attains considerable dimensions, invading almost
the whole side of the face, or even descending upon the neck; at
the same time, that of the mucous membrane is increased in the
interior. The aspect of the child is as sad as it is hideous to the
sight: sometimes, in a sitting posture, and availing himself of all
his strength, he tears from the interior of his mouth the gangrenous fragments; sometimes lying dejected, depressed, he allows
to flow out and cover him a blackish and fetid sanies.

"This appearance, however, may become still more repulsive,
when the slough is partially detached, and the mass is seen hang-
ing from the cheeks, or, even worse, when, falling off, it leaves a
perforation, through which the bare and loosened teeth, and the
blackened maxillary bones are visible. The odor is then of the
most offensive character; the child still retains some strength, and
asks for food, or, in the last state of prostration, he refuses all
nourishment; there is always great thirst, and the patient drinks
with avidity; he does not vomit, but there is great relaxation of
the bowels; he becomes rapidly more and more emaciated, his
skin is dry, but not very warm, his pulse, very small, becomes insensible, and death arrives, without other phenomena."

TREATMENT.—Hartman recommends Secale-cornutum, and
Arsenicum, to which Teste adds Ipecac., Muriatic-acd., and Kreosotum.

PTYALISM, OR SALIVATION.

DEFINITION. — This disease consists in an irritation, inflammation, and swelling of the salivary glands of the mouth and throat, with a profuse discharge of saliva, or spittle.

CAUSES. — Most persons, when hearing of a patient suffering from salivation, are very apt to attribute it to the injudicious use of mercury, and lay all the blame upon the head of the physician who happens to be in attendance. Now, because one of your children's parotid or submaxillary glands takes a notion to swell up, and secrete an unnatural amount of saliva, which keeps the poor patient constantly spitting, do not blame some sapient son of allopathy, who has happened lately, or perhaps years ago, to prescribe for your child, when it had the measles or scarlet fever; because salivation is not necessarily the effect of mercury.

We often see patients recovering from smart attacks of fever, with all the symptoms of salivation, where there has not been one particle of mercury given. It is well to know and understand these things; because justice is due to all, and poor, fading allopathy has enough to bear, without fathering the whims of every fickle gland.

Several other substances, besides mercury, are known to have the occasional effect of producing an increased, and even a profuse, flow of saliva; for example, preparations of gold, copper, antimony, arsenic, and potassium; and it is asserted, upon good authority, that castor-oil, digitalis, and opium have occasionally the same consequences.

Salivation sometimes occurs spontaneously, that is, without any obvious cause: occasionally it results from some local irritation within the mouth, from decayed teeth, etc.; sometimes it owes its origin to colds, and the various forms of fevers, particularly the cutaneous variety; it sometimes occurs as a critical discharge, by the action of nature, and is then beneficial; and *occasionally* it is induced by mercury.

SYMPTOMS. — As salivation is a disorder of the salivary glands and mucous membrane of the mouth and throat, it is always advisable to examine the parts closely, and on so doing, in the majority of cases you will find them red and swollen, sometimes considerably inflamed; the glands beneath the under jaw are usually enlarged and very tender. These glands, when in health, secrete only the necessary amount of saliva; but, when diseased, they discharge it in large quantities, and not unfrequently you will

28

find it very much changed in its character and appearance. Instead of being thin, watery, colorless, inodorous, and tasteless, as it is in health, it may become dark, thick, stringy, fetid, and very offensive.

TREATMENT. — If salivation has been produced by calomel, or any mercurial preparations, the remedies will be *Hepar-s., Lachesis, Belladonna, Nitric-ac.,* and *Sulphur.* When caused by cold, *Mercurius* will be the appropriate remedy; also when there is painful swelling of the salivary glands, fetor of the mouth, ulcers on the inner cheek, profuse discharge of fetid saliva.

All astringent washes or gargles, which directly diminish the salivary discharge, are injurious. Mild washes or gargles, such as milk and water, may be used, and are sometimes attended with considerable benefit.

ADMINISTRATION OF REMEDIES. — When Mercurius is given, dissolve twelve globules in twelve teaspoonfuls of cold water; and one teaspoonful of this solution may be given once in two or four hours. Other remedies the same.

DIET. — The diet must be of the mildest kind, — gruels, milk and water, crackers soaked in water, plain puddings, and the like. For a drink, cold water may be used, or cocoa, if the patient likes it.

RANULA, SWELLING UNDER THE TONGUE.

DEFINITION. — CAUSE. — Ranula is a swelling of the salivary glands under the tongue, or, as an anatomist would say, of the *sub-lingual* glands, caused by some obstruction of the salivary duct, — which is the little canal that carries the saliva from the gland to the mouth, — from cold, inflammation, or some irritating cause.

Tumors of this kind are not generally painful; but when they are of any considerable size, they interfere with the free motion of the tongue, and thus materially interfere with speaking.

TREATMENT. — *Mercurius, Calcarea-c., Thuja, Sulphur* are the principal remedies.

Mercurius. — When the tumor is of an inflammatory nature. Should the swelling burst, and leave a troublesome ulcer, *Mercurius* and *Calcarea-c.* may be given, in alternation, every night and morning.

ADMINISTRATION OF REMEDIES. — Of the selected remedy, give three or four globules, dry, upon the tongue, night and morning, until the difficulty is removed. If the first remedy fails to produce any favorable result, proceed to select another, which you can administer in the same manner.

GUM-BOILS. ABSCESS IN THE GUMS.

DEFINITION. — CAUSES. — Almost every form of swelling with inflammation, that affects the gums passes under the head of gumboils ; even abscesses and inflammations are thus generally designated. These troublesome affections, — for they are sometimes very annoying, — arise from various causes; not unfrequently they are the primary disease, depending upon an inflammation from some common or accidental cause; generally, however, they are but the result of some irritation or disease going on within the gum. For instance, a decayed tooth may be the primary trouble, or the cutting of a tooth ; the wisdom teeth are almost always preceded by considerable inflammation and swelling.

TREATMENT. — When gum-boils are caused by decayed teeth, extraction is the only remedy. Almost always before an abscess, or a gum-boil is formed, there is considerable inflammation and swelling, with heat and pain, for which you should give *Aconitum* and *Belladonna*, in alternation, every two hours, until the heat and tension is relieved.

Mercurius, — when there is considerable throbbing or pulsative pain, may be given in alternation with *Hepar-s.* *Mercurius* may also be given when *Aconite* and *Belladonna* fail to afford relief.

In swelling of the jaw with suppuration, whether in consequence of decayed teeth, or the unskilful abstraction of a tooth, *Silicea* will be the appropriate remedy. *Silicea* should also be given where the preceding remedies have failed to arrest the progress of the boil, and suppuration has already taken place. · *Calcarea* is also another excellent remedy, under the same circumstances.

In gum-boils from irritation arising from the cutting of the wisdom teeth, *Aconite* and *Chamomilla* will be the appropriate remedies. Sometimes it will be found necessary to make a slight incision with the lancet or a knife.

Staphysa. — Against bleeding from the gums.

Hyoscyamus. — For throbbing pain in the bone, attended with fever.

ADMINISTRATION OF REMEDIES. — During the inflammatory stage, the remedies may be given as often as every hour; from that to two or three hours.   When giving Silicea or Calcaria, one dose night and morning.   Dose, three pills.

MUMPS, OR PAROTITIS.

DEFINITION. — CAUSES. — The salivary glands are six in number, three upon either side of the throat; and are named, respectively, the *parotid,* — so called from being situated below and in front of the ear; the *submaxillary,* — because situated beneath the sub-maxillary or under jaw bone; and the *sublingual,* — that is, under the tongue.

The office of these glands is to furnish saliva or spittle, with which the food, during mastication, is moistened; so that when carried into the throat, it passes with ease through the œsophagus into the stomach.

Now mumps is an inflammation of the largest and most important gland in this group, the *parotid ;* hence the name parotitis. It often prevails as an epidemic ; when it attacks one child in a family, or a school, several others are pretty sure to be affected also, either simultaneously or in succession.   It is undoubtedly contagious.   It chiefly attacks children and young persons; and what is rather curious, it seldom, I might almost say never, attacks them the second time.

SYMPTOMS. — At the commencement of the disease, there are no marked symptoms, except the tumefaction and swelling, which you will find just below the ear.   The swelling generally extends from the parotid, where it commences, to the submaxillary, and even to the sublingual glands.   Sometimes only one side is affected; sometimes both at once ; but, I presume most frequently first one side is affected, and then the other.   The swelling is hot, dry, and painful ; very tender to the touch.   There is usually some fever ; the motion of the under jaw is interfered with from the swelling in the vicinity of the joint.   The inflammation reaches its height in about four days, and then begins to decline ; its whole duration may be stated, on an average, at eight or ten days.

Mumps is not considered dangerous, unless from imprudent exposure the patient takes cold, or from any other cause the disease " strikes in," that is, becomes thrown back upon the system, so as to involve some of the vital organs.   In many cases, under these circumstances, the swelling about the neck and throat subsides quickly on the fifth or seventh day, and shows itself upon the testicles in the male sex, and upon the breast in the female, and these parts become hot, swollen, and painful. Another dangerous transfer of this disease, but particularly rare, is from the testicles to the brain.

TREATMENT. — *Mercurius* is the principal remedy, and often the only one required; two or three doses in most cases will effect a cure; one dose every night until four doses are taken. Dose, four pills.

Belladonna. — When the swelling gets hot and dry, or when it is very red, having an erysipelatous appearance; also when it recedes and affects the brain, producing delirium and other head symptoms, — give a dose, three globules every hour. If *Belladonna* does not afford relief, follow it with *Hyoscyamus*. When the swelling suddenly disappears and affects the testicles, give *Pulsatilla*, a dose every two or three hours.

ADMINISTRATION OF REMEDIES. — The globules may be given dry, upon the tongue, about three at a dose, or you can dissolve twelve globules in twelve teaspoonfuls of water, and of the solution give one teaspoonful at a dose.

DIET AND REGIMEN. — The diet must be light. Toast and black tea, cocoa, custards without spice, bread puddings, baked apples, and stewed prunes may be allowed. If it is during cold weather, the patient should be kept in a moderately warm room; if there is much fever, he had better lie in bed. No external application need be made, unless it be simply a handkerchief tied around the neck. Should the neck get very tense, hot, and dry, it will be advisable to apply hot flannel cloths. Great care must be taken to prevent the patient from taking cold. Never apply cold water, or any of the many lotions; follow simply the directions above given.

INFLAMMATION AND SWELLING OF THE TONGUE. GLOSSITIS.

DEFINITION. — Glossitis is an inflammation of the substance of the tongue, characterized by pain, redness, hardness, and swelling, either with dryness of the mouth, or a profuse discharge of saliva, and accompanied with the usual symptoms of inflammatory fevers. The inflammation may be confined to one side of the tongue, or the whole organ may be implicated.

CAUSES. — It usually arises from mechanical injuries, or from the contact with chemical agents or acrid substances, which may excite an irritation. In many cases, however, the attack is very sudden; a severe inflammatory action setting in, without any apparent cause. "This affection is sometimes induced," says Dr. Copland (vide Copland's Med. Dict.) "by exposure to cold, or to currents of cold air about the head after the use of mercurials, or from the suppression of the salivary discharge by these causes."

SYMPTOMS. — The first symptom complained of is usually an acrid, stinging sense of heat, or burning pain in the tongue; the inflammation, as a general thing, sets in suddenly and proceeds rapidly; the pain and swelling is very great; the tongue presents a livid or dark red appearance. The inflammation may commence upon one side, or be restricted to a very small portion, but gradually it may extend until the whole organ becomes involved. During the progress of the disease, the pain becomes more acute and of a burning and lancinating character, which is aggravated by the slightest movement; the attempt to talk or swallow causes great suffering.

In severe cases, the tongue becomes enormously swollen, filling the entire mouth, speaking and swallowing being entirely prevented, while respiration is obstructed, even to threatened suffocation. In other cases, the swollen and inflamed organ is protruded from the mouth, presenting a horrid picture of suffering. The tongue is usually furred over with a thick coating, and a profuse secretion of saliva flows from the mouth.

TREATMENT. — *Aconite* should be administered at the commencement, when the fever and inflammation are severe, and accompanied with acute, tense, cutting pain. If the inflammation should have arisen from mechanical injuries, as it often does, particularly in children subject to convulsions, where the teeth at times lacerate the tongue severely, *Aconite* and *Arnica* should be given in alternation every hour or two hours, according to the severity of the case. Four globules may be given for a dose, dry, upon the tongue, or dissolve six in twelve teaspoonfuls of water, and of this solution give one teaspoonful at a dose.

Mercurius and *Belladonna* are the principal remedies to be relied upon in the majority of cases. *Mercurius* may be given first, when, at the commencement, there is violent pain, swelling, hardness and salivation; also, when the tongue becomes involved with ulceration of the throat. *Belladonna* should have the preference, where the inflammation assumes the character of erysipelas, as well as where *Mercurius* has proved insufficient, and the inflammation has extended to the neighboring parts; also, when numerous little ulcers make their appearance upon the tongue and gums.

Where the two remedies are apparently indicated, and you are undecided which to choose, the two may be given in alternation, from one to four hours apart, according to the severity of the case. Dose, same as Aconite.

Should you meet with a case where the swelling had become so enormous as to threaten suffocation before a physician could arrive, do not hesitate to take your knife, or any sharp instrument, and make a free, longitudinal incision in the tongue. This gives exit to the blood, which removes the congestion and relieves the patient.

CASE. — I was called last January to see a young lady whose tongue suddenly inflamed without any cause, — at least as far as we could ascertain, — and became so enormously swollen as to threaten suffocation, the whole buccal cavity was entirely filled up, and a large portion of the inflamed organ protruded from the mouth. The slightest motion of the tongue caused great suffering; swallowing or talking, even the lisping of a word, was impossible. The tongue was covered with mucus, and a fetid discharge of saliva flowed constantly from the mouth.

I prescribed for this case Mercurius, 30th, and in less than six hours after taking the second dose, the tongue had resumed nearly its ordinary size. The improvement continued, and in a few days the patient was as well as ever.

DENTITION, OR TEETHING.

The cutting or eruption of the teeth being a perfectly natural physiological process, we should scarcely expect it to occasion disease or suffering of any kind, and, perhaps, were all children in a perfectly healthy condition at the time of its commencement, they would suffer but little, if any, during this period. However, in the first place, all children are not born healthy; and, secondly, those few that are so born, soon — by mismanagement in dress, diet, and exercise — have all their functions so blunted and despoiled as to be in no better condition than those who, at first, possessed unhealthy constitutions. Under these conditions, or from these causes, dentition not unfrequently becomes complicated, difficult, and even dangerous. Its most common complications are derangements of the digestive organs and the nervous system. Being so frequently and intimately allied with disorders of various kinds, difficult dentition has been rightly classed among the diseases of infancy.

The first, milk, or temporary teeth, as they are indifferently called, are twenty in number. Their eruption should commence at the sixth month, and be completed about the end of the second year, those of the lower jaw preceding the upper.

As a general thing, they make their due appearance in the following order: at about the sixth month, the two middle lower incisors, or cutting teeth, as they are called, come through; in from three to four weeks these are followed by the corresponding ones in the upper jaw; from the seventh to the tenth month, the lower lateral incisors appear, soon after these the two upper ones; from the twelfth to the fourteenth month, the anterior molars, or first jaw-teeth, two below and two above, are cut, and shortly after these the stomach and eye teeth, and, finally, at about the end of the second year, the four back jaw-teeth, or posterior molars, two below and two above, make their appearance, completing the set.

This regular order and time of teething, however, is not always observed; there are considerable variations. Some children get their teeth two or three weeks after birth, or indeed are born with them, while others again do not cut any teeth until they are ten or twelve months old. The order of succession, mentioned above, is also frequently violated; the upper incisors making their appearance before the lower, the molars before the stomach and eye teeth; frequently, also, they do not appear in pairs, there being a difference of weeks or even months, between the appearance of the first ones.

Teething, in the most favorable cases, is preceded by slight salivation, or, as it is commonly called, drooling; by heat and swelling of the gums, increased thirst, restlessness, or fretfulness, and frequent desire to thrust things into the mouth, evidently to allay irritation and itching.

In some cases the irritation, swelling, and inflammation of the gums become severe; the mouth hot and dry; the gums extremely sensitive and intolerant to the slightest pressure; the child starts in its sleep, or, on waking, the head is hot, fever high, great thirst for cold water; there are also frequent spasmodic twitchings of the hands and feet while sleeping.

As a general thing, these symptoms are all occasioned by the pressure and irritation of the young tooth or teeth, and as they push forward, the gum wastes from absorption, and is at last cut through, the tooth making its appearance, and the symptoms of complaint gradually vanish, leaving the child bright and happy.

Occasionally we find, instead of a hot and dry skin, with thirst and other fever-symptoms, the very opposite of these, namely, a profuse perspiration, great flow of saliva, and general relaxation of the whole system. These symptoms, however, are as a general thing, but temporary, and are afterward followed by dryness of the skin and mouth, fever and thirst. At times the patient is restless, fretful, and irritable; again, he is heavy, dull, and drowsy; sometimes there is a rash upon the skin, which is called "Red Gum," or "Tooth-rash."

Connected with teething, there are often many sympathetic affections, such as determination of blood to the head, convulsions, constipation, swelling, and suppuration of glands, eruptions of various kinds, both upon the head and body, gatherings, and discharges from the ears, cough, and always, I believe, great irritability of the nervous system.

During the process of teething, the whole system is in a peculiarly excitable condition, so that trifling causes, such as at other times would make no impression whatever upon the child, may, during this period, excite a train of acute and serious symptoms, which only prompt and judicious treatment can successfully combat. In children of deficient vital power, a cold, an error in diet, or some undiscoverable cause, may excite a slight derangement which is at first perhaps scarcely noticeable, or at least considered of no account, but which, by neglect, or improper treatment, eventually leads to a permanent state of bad health, ending in tubercular degeneration of the lungs, or of the digestive apparatus.

The necessity, therefore, of jealously guarding the children from every source of disease, at this time, to which they might otherwise be exposed, will be obvious to all.

Unfortunately, however, for the children, most young mothers have an aunt or a kind female acquaintance, who has brought up, or at least seen some one else bring up, a large family of children, and, therefore, "knows all about these little complaints of teething children, and can treat them just as well as any doctor." Every friend, that calls and observes, or is informed, that the child is ill, at once suggests a remedy, "which has never been known to fail in such cases," and, as soon as they get home, their kind hearts prompt them "to send one of the children over with a bottle."

29

Of course the patient must take a little from each of these contributions, no matter how numerous they are. The inexperienced mother remarks that, "certainly from so many infallibles one must be found, that would just hit her child's case," and, indeed, I have frequently seen the case hit, or rather seen it, after it was hit, but I have usually found it hit upon the wrong side.

I have been frequently not only amused, but utterly confounded, on observing the paraphernalia of some nurseries, and especially those not previously under the charge of a homœopathic physician. The mantel-piece is graced with lotions, pills, and powders; bottles stand arrayed in warlike order; ipecac, and squills, Godfrey's cordial, paregoric, hive-syrup, castor-oil, and the like, fill up the front rank, flanked by the redoubtable, never-failing, ever-to-be-exalted "soothing syrup for children teething;" while sulphur and molasses, peppermint, goose-grease, and catnip bring up the rear; mustard and onion-draughts being held as a reserve.

Now, the frequent and persistent administration of these choice remedial agents from the domestic armamentarum is not, in my humble judgment, exactly the best way to preserve health, or even to restore it, when once lost. I argue that the delicate organization of a child is unable to withstand the rude shocks which such treatment must inevitably produce. Many of the little divergencies from health, which, taking place at this period, if left to nature, or treated rationally, would amount to nothing, are often, I am satisfied, "doctored" into some serious disease.

A diarrhœa, which in itself is not unfrequently salutary, or a slight cold, no matter how trifling, must, in most people's opinion, needs have something; and, not knowing exactly what is best, they give whatever they happen to have in the house, or whatever a neighbor may suggest. If the first prescription does not afford relief, or, perhaps, more properly speaking, if the child does not rally, in spite of the prescription, another dose is concocted, and forced down its throat. This may be continued for several days, more or less, or until all the domestic and patent medicines have been tried, when the mother finally makes up her mind to send for a physician. The doctor arrives; but alas! often too late to be of any service, and the child dies, either from the disease or the treatment, or both combined; seldom from the disease alone.

This giving such quantities of such barbarous stuff to delicate children, most certainly exhibits a degree of recklessness and ignorance, pitiable to behold. If grown persons have a mind to scour themselves out weekly, with the most drastic purgatives; if they have a mind to take stimulants, tonics, and correctives; if they have a mind to cover the whole surface of their bodies with blisters, ointments, and plasters; if they have a mind to be bled annually, and take purifying medicines every spring, I have certainly not the least objection; but I do protest, in the name of humanity, that little, delicate children should be spared such inhuman treatment.

Listen to what Dr. Trall, in an article in his " Water Cure Journal " says, in commenting upon the death of President Lincoln's son.

" Little Willie, a healthy, robust, playful, genial, and happy child as ever was seen, had a slight cold; it was doctored into a continued fever; this was drugged into the typhoid; and then the typhoid was dosed into *death;* and the sprightly, joyous boy of a few days ago, now lies pale and mouldering in the cold and silent grave. As Willie Lincoln died, so do thousands every year."

Perhaps at no period of a child's existence, is it so often and so thoroughly drugged as during that of teething. Of late years, we have had thrust upon the public, with an impudence only equalled by the barbarity of the treatment, a host of " simple " remedies, discovered by experienced nurses, especially, for " children teething; " nine-tenths of which are concocted and manufactured by brazen-faced men. I have more than once in these pages pointed out the pernicious effects of all such *narcotics.* I use the term narcotic advisedly, though well aware that these preparations are all guaranteed not to contain opium or morphine. It is not necessary to analyze a drug to ascertain its medicinal properties; its effect upon the human organism is sufficient to demonstrate to what class it belongs. If it stupefies and puts to sleep, it is a narcotic, no matter of what it is composed.

Opium, when given in any form, first excites or exalts the brain and nervous system, afterward calms, or " soothes," which appears to be a favorite word, the child into a quiet sleep. But its early and persistent use arrests the growth and activity of the brain, the bodily and intellectual faculties are blunted or dwarfed into insignificance, and the subject, if he lives to grow up, which, indeed, is doubtful, presents a physiognomy painful to behold. Various diseases of the nervous system, as paralysis, convulsions, etc., can frequently be traced directly to the abuse of this drug.

I am well aware how pleasant and happy a thing it is, for a

mother to possess a magic wand by which, magician-like, in the
twinkling of an eye, she can " soothe " her fretful child, and gently
put him away in the arms of Morpheus, where he will quietly lie,
oblivious to all earthly cares or pains, while she dresses for a ball,
or flirts away a few hours in innocent amusement. But I assert,
without fear of contradiction, and would to heaven every mother
in the land might hear and heed the assertion, that such practice
cannot be indulged in except at the expense of the future intelligence, health, and happiness of the child.

TREATMENT. — As the local irritation, attending the eruption of
the teeth, is generally the exciting cause of most, if not all, the
diseases and disturbances connected with dentition, our first endeavor should be to moderate or remove the irritation as speedily
as possible ; and this, in many instances, can readily be accomplished by making a free incision through the gum down upon the
offending tooth. The lancet should always be resorted to when
the gums are found hot and swollen, or when you can see or feel
the tooth through the tissues, and especially if there should be a
great determination of blood to the head, accompanied with
twitching of the muscles, — symptoms indicating a tendency to
convulsions.

This lancing the gums, I am aware, is looked upon with great
horror by many mothers, but the pain which it causes is really
insignificant ; in fact, in most cases instead of causing pain, it
affords instant relief. It is true, children always cry when it is
done, but you will notice they commence crying before the lance
touches the gum ; certainly it is not the pain of the lance that
makes them cry thus early. Besides, some people have the notion,
which undoubtedly they obtained from physicians, that if the
gums are lanced too soon, the cut will heal up, forming a scar
through which it will be more difficult for the tooth to break.
This is an antediluvian notion : the idea that the tooth pushes its
way by main force through the opposing tissues, has been exploded
long ago, if, indeed, it were ever entertained by thinking men.

If, for any reason, it is not deemed expedient to lance the gums,
and they are swollen and sensitive, and the child wants to press or
bite something, to relieve the intolerable itching and irritation underneath the gum, give it an ivory ring or something of the kind
to bite upon ; or, what is frequently done, rub over the advancing
tooth with a thimble or a piece of crust sugar.

Aconite may be given when there is much fever, restlessness, and pain, manifested by the child's crying and starting.

Belladonna. — Especially when there is great derangement of the nervous system; hot head, flushed face, inflammation and swelling of the gums; or, when there are convulsions, the child starts from sleep as if frightened, and stares, the pupils of the eyes are dilated, the whole body becomes stiff; convulsions followed by sleep. See " CONVULSIONS."

Chamomilla. — Perhaps this is the most generally called-for remedy for the difficulties of teething children. It is especially called for while the child is restless and uneasy at night, twitches and jerks while asleep; starts at the slightest noise; with general heat, redness of *one* cheek, moaning, groaning, and general uneasiness; diarrhœa, with watery, slimy, and greenish evacuations; worse at night. May be given in alternation with *Belladonna.*

Cina. — When, during teething, there is a dry, spasmodic cough, resembling hooping-cough; also when there are worm-symptoms present, such as distension of the abdomen, rubbing the nose, grating the teeth, wetting the bed, &c.

Coffea. —When the child shows restlessness and cannot sleep, with some fever.

Ignatia. — Should there be, in connection with symptoms of convulsions, frequent flushes of heat, sudden starting from sleep with piercing cries. Consult " CONVULSIONS."

Lycopodium. — When the child rolls its head from side to side; sleeps with its eyelids half open, and moans while asleep.

Magnesia-c. — Diarrhœa with stools like scum of a frog-pond, green and frothy.

Mercurius. — Diarrhœa with greenish evacuations, and great straining; profuse flow of saliva from the mouth, redness and soreness of the gums.

Ipecacuanha. — Should there be nausea, vomiting, and diarrhœa, fermented stools, or mixed stool of different colors.

Nux-vomica. — For obstinate constipation. May follow, or be given in alternation with Bryonia.

Calcarea-carb. — For fat children, with light complexion, and in whom the process of teething is slow; also when it is accompanied by diarrhœa with yellow stools, or stools like clay.

Sulphur. — For diarrhœa with sour, white, or hot stools, which excoriate the parts.

Should constipation prove obstinate, give injections, as directed in article on " CONSTIPATION."

When teething is complicated with convulsions or other diseases, consult such complaints, under their respective heads.

ADMINISTRATION OF REMEDIES. — Of the selected remedy a dose may be given, every one, two, three, or four hours, according to the urgency of the case.   When the globules are given, three will be a dose; when given in water dissolve six pills in as many spoonfuls of water and give one spoonful of the solution for a dose.

DIET AND REGIMEN. — A judicious restriction in diet will often be all that is necessary in most of these cases, especially if the child is kept in a cool room, and allowed plenty of cool water to drink.

THE TEETH.

GENERAL REMARKS. — The proper culture and preservation of the teeth of children is a subject demanding the attention of every thoughtful parent.   When taking into consideration the importance of sound and regular teeth, alike in regard to health, comfort, and appearance, the little care and attention requisite to keep them in a proper state seems almost insignificant.   What adds more to the beauty of an individual than a handsome set of teeth? and what detracts more from the appearance of a child, than a mouthful of blackened, irregular, and half-decayed teeth?   They not only present an unsightly appearance, but are very injurious to the health.

The teeth are a part of the digestive apparatus, and in a great measure their soundness depends upon a healthy state of the stomach and bowels, so you will readily observe that whatever tends to derange these organs will exert a deleterious effect upon the teeth.   Children are often refused candies because they are said to rot the teeth.   Now that sugar itself ever directly injures the teeth, is a matter of doubt, but certainly the confectioners' preparations, together with the thousand other little, fancy fixings, which children have given to them, or by some means procure, excite a direct and injurious effect upon the stomach, deranging the bowels, producing dyspepsia, flatulence, and gassy eructations which blacken and corrode the enamel of the teeth, thus laying the foundation for their decay and speedy destruction.

As a general thing healthy persons have sound teeth, while

sickly, feeble ones have decayed teeth. It therefore becomes us well to study and practice the few simple rules that promote health.

To preserve the teeth they must be kept clean; and to do this it is not necessary to use any of the thousand and one dentifrices, such as tooth-pastes, powders, tinctures, washes, etc., sold by chemists and perfumers. They are all more or less injurious and should therefore be avoided.

Many persons are unable, or at least they think so, to keep their teeth clean without some kind of a dentifrice, and for such Dr. Hering recommends a charcoal made by burning stale bread quite black, and reduced to a fine powder, by pulverization, after which it should be washed, to free it from salts, and then dried.

The most pleasant and efficacious way of cleaning the teeth is to wash or rub them with sour milk, after which the mouth should be rinsed with warm or tepid water.

The mouth should be cleaned, washed with water, *every night* and *morning*, and the teeth brushed with a soft brush, both on their anterior and posterior surface. The teeth should also be cleaned after every meal, either with the brush, or a piece of soft flannel; this will prevent the collection of tartar. Care should also be taken, that all particles of food, that may have lodged between the teeth, and are inaccessible to the brush, are removed; for this purpose toothpicks are found necessary; these should be made of wood, ivory, or the common goose-quill; a fine thread is at times convenient, it may be drawn backward and forward between the teeth. Metallic toothpicks are highly objectionable, and ought never to be used; they injure the enamel.

The pernicious habit of children, in picking their teeth with pins, needles, or penknives, should be peremptorily forbidden.

Sudden changes of temperature, produced by the introduction of very hot or very cold substances into the mouth, crack the enamel, and eventually produce decay. All articles of diet should, therefore, be of a medium temperature when partaken of.

The temporary teeth should be removed as they become loose, but not till then, unless they are crowded and irregular, or when a permanent tooth makes its appearance before the temporary ones are shed; and, in such cases, the milk-tooth, though sound, should be removed without delay. This is necessary, that the first or temporary teeth may not interfere with the permanent set, for these latter are to last the child its lifetime; and it is, therefore, desirable that they should present a uniform and beautiful appearance.

If, when the permanent teeth make their appearance, they are irregular and crowded, in consequence of the jaw being narrow and short, or from other causes, it may be necessary to remove one or more of them, in order to give the remaining ones a chance for free development, so that they will not present a pinched or crowded, and therefore unsightly, appearance.

When it is necessary that a tooth should be extracted, do not have it done by a botch. It requires just as much skill and knowledge to extract a tooth *well*, that is, properly, as it does to amputate a limb; therefore be particular to select a well-educated, competent dentist, a skilful operator; and, for fear that you are not aware of the *fact*, — for fact it is, — perhaps I had better inform you, that nine dentists out of every ten are entirely ignorant of the first principles of dentistry. It is with dentists, as it is with most other men, — the less they know, the more pretentions they make; many a one who does not know a bicuspid from a molar, hangs out his sign of " Surgeon Dentist," with a presumption only equalled by that of quack doctors.

The teeth of children, and no less those of adults, too, should be frequently examined, and wherever the enamel has become broken, and the body of the tooth commenced decaying, it should be immediately filled or plugged with *gold foil;* this, in many instances, will arrest the further decay. Never allow your children's teeth to be filled with an amalgam of any description; most of them contain mercury, and all are injurious, not only to the teeth, but to the general health. If you do not think it advisable to fill the milk or temporary teeth with gold, rather let them go unfilled than have recourse to any of the pastes or cheap patent fillings. Keep the cavities clean, and filled with white wax; it is far preferable to any amalgam, and has the advantage of being innoxious.

The practice of cracking nuts with the teeth, of lifting heavy bodies, of biting threads, etc., is injurious, because it cracks the enamel; and where the enamel is cracked, and the body of the tooth exposed, decay is sure to commence.

TOOTHACHE, OR ODONTALGIA.

DEFINITION. — CAUSES. — This troublesome affection, over which children shed so many tears, and adults sigh for want of sympathy, may arise from many causes; some are hereditarily predisposed to it, while others suffer from every exposure; again, it may arise from disturbances going on elsewhere in the system, or it may be purely nervous. It is often rheumatic; often arises from carious teeth; also from abuse of coffee or of calomel. Many are its causes, and as numerous are its forms; it may be confined to one tooth, or it may extend to many; one side of the face, both, or even the whole head may be affected. The pain may be of any, and of all forms imaginable, from a dull, heavy ache to a sharp, shooting pain.

TREATMENT. — Do not allow yourself to be too easily persuaded into the belief that, because a tooth aches, it necessarily ought to be extracted, for toothache, in its severest forms, is often cured with homœopathic remedies. It is not advisable to extract teeth when you can save them, and this can generally be done, unless they are ulcerated at the roots; in which case, extraction affords the only reliable and prompt relief. Here, again, let me caution you in the selection of a dentist. Do not run to the first " tooth-puller," for no other reason than because he is near at hand. Choose your dentist as you should your minister and physician, when in health, and then, when trouble comes, you will know where to seek relief.

Many of the domestic remedies for toothache are objectionable; creosote, laudanum, clove-tincture, and the like, afford but temporary relief at best; the pain soon returns with redoubled violence; besides, the majority of them are injurious to the general health, as well as to the teeth themselves. It is better, far better to obtain permanent relief from some remedial agent that will remove the *diseased condition*, of which the toothache is the *result*.

The principal remedies for toothache are, — *Aconitum, Arnica, Antimonium-crud., Arsenicum, Belladonna, Bryonia, Chamomilla, Kreosote, Mercurius, Nux-vomica, Pulsatilla,* and *Sulphur.*

Aconitum. — When there is feverishness, with great anxiety and restlessness; violent throbbing or beating pain; rheumatic pain in the face and teeth; congestion of the head; heat, redness, and swelling of the face; toothache, occasioned by cold. When the relief afforded is but transient, follow it with *Belladonna* or *Chamomilla.*

Arnica. — When the pain is the result of mechanical injuries, as from extraction or plugging. Children often fall and injure the teeth, at the same time bruise and cut the lips or cheeks; in such

30

cases Arnica may be used as a lotion, as well as taken internally; when used as a lotion, one part of Arnica tincture should be mixed with five or six parts of water; a linen cloth dipped in this mixture may be laid upon the injured part, and renewed every three or four hours, according to the extent and severity of the injury. When the injury is on the interior of the lips, or the teeth alone are affected, the mouth should be rinsed or gargled out with a similar mixture.

Antimonium-crud. — For pain in hollow and decayed teeth.

Arsenicum is useful when everything cold aggravates the pain.

Belladonna. — When there is a sensation of ulceration at the roots of the teeth; drawing pain in the face and teeth, extending to the ears, aggravated in the evening on getting warm in bed, or on applying anything hot; heat and throbbing in the gums.

Bryonia. — Drawing, jerking toothache, with a sensation as though the teeth were loose and elongated, especially during and after eating; pain in decayed teeth; toothache caused by wet weather, or accompanying rheumatic affections; pains relieved momentarily by cold water held in the mouth. *Bryonia* is serviceable for pains through the face generally; for pains which shoot from one tooth into another.

Chamomilla. — Violent, boring, and throbbing pain, extending through the jaws to the ears, also into the temples and eyes; the child is cross and feverish; complains of pain in all the teeth; cannot tell which aches the most; worse at night, when the patient is warm in bed; also after eating anything warm; swelling and redness of the cheeks. *Chamomilla* is serviceable for toothache before menstruation.

Kreosotum. — For pain in decayed teeth, with swelling and congestion of the gums.

Mercurius. — For pains in hollow teeth; tearing pain through the roots of the teeth; shooting pain, passing over through the sides of the face, extending to the ears, especially at night, aggravated by cold food or drink; swelling and inflammation of the gums.

Nux-moschata. — Especially for pregnant women; also sometimes for children, when the pain arises from taking cold.

Nux-vomica. — Toothache arising from cold, with throbbing, boring, or gnawing pain throughout the teeth and gums, aggravated by eating, or exposure to the open air; tearing pain on one side; rheumatic pains deep down in the nerve of the tooth, with pain as though the tooth were being wrenched out.   May be given in alternation with *Mercurius*.

Pulsatilla. — Is most suitable for young girls, or children of a mild or timid disposition; shooting pain, that extends to the ear of the affected side; jerking pain, as though the nerve were tightened, and then suddenly relaxed, particularly of the left side; the pain increased by warmth and rest, better when walking about, especially in the open air; toothache accompanied by earache and headache.

Sulphur. — Tearing and pulsative pain, particularly in carious or decayed teeth, extending to the upper jaw and into the ear; pain worse at night, when warm in bed; swelling of the gums, attended with shooting pain.   Suits well after *Mercurius*.

GENERAL INDICATIONS.

Toothache in Children. — Aconite, Bella., Cham., Coffea, Pulsat., Merc.

" " *Females.* — Acon., Bell., Cham., Chin., Coff., Hyos., Puls., Nux-m.

" *during Nursing.* — Acon., Bell., Chin., Nux-vom.

" " *Menstruation.* — Calc., Cham., Puls., Bry., Lach.

" " *Pregnancy.* — Bell., Bry., Nux-v., Puls., Staph., Rhus.

" *from Calomel.* — Carb.-v., Hepar., Puls., Sulph., Lach.

" " *Taking Cold.* — Acon., Bell., Bryo., Dulc., Hyos., Merc., Nux.-vom., Rhus., Phos., Puls.

" *with Swelled Face.* — Cham., Merc., Nux-v., Puls., Bryon.

" " *Swelled Gums.* — Acon., Bell., Merc., Nux-v., Sulph.

" " *Swelled Glands.* — Merc., Bell., Nux-v.

" " *Faceache.* — Merc., Acon., Bell., Bryo., Cham.

" " *Earache.* — Cham., Merc., Puls., Calc., Sulph.

" " *Headache.* — Bell., Glon., Nux.-v., Lach., Puls.

" *of a Nervous Nature.* — Acon., Bell., Coff., Ignat., Hyos., Cham., Nux-v., Spig.

" " *Rheumatic Nature.* — Cham., Merc., Bryo., Bell., Sulph., Puls., Rhus.

" " *Congestive Nature.* — Acon., Bell., Cham., Puls., Chin.

" " *Hysterical Nature.* — Ignat., Cham., Hyos., Sep., Bell.

" *on the Left Side.* — Acon., Cham., Nux-m., Phos., Sulph.

" " *Right Side.* — Bell., Bry., Staph.

" *in the Upper Jaw.* — Bell., Calc., Bry.

" " *Lower Jaw.* — Caust., Nux., Staph., Sulph.

ADMINISTRATION. — After having made a careful selection, dissolve, of the chosen remedy, twelve globules in twelve teaspoonfuls of water; of this solution, give one teaspoonful for a dose. Repeat the doses from fifteen minutes to an hour or two hours apart, according to the severity of the pain. When a remedy given has afforded some relief, do not change it for another until it has had time to show its full effect.

SORE THROAT, OR QUINSY.

DEFINITION. — This common disorder has several appellations: quinsy, angina faucium, cynanchia tonsillaris, amygdalitis, tonsilitis, and laryngitis. These are but a few, and most of them are quite expressive to the professional reader, denoting the precise nature and locality of the difficulty. However, the treatment, ignoring names, and being governed entirely by the symptoms present, all may be summed up and described under the more common name of sore throat. The disease consists in an inflammation of the back part of the throat, including the palate and tonsils. It appears in different degrees of intensity, from the slightest irritation, causing but moderate inconvenience, and lasting but a short time, to the highest degree of inflammation, ending in suppuration, or the formation of abscesses in the tonsils or adjacent parts.

CAUSES. — The exciting cause is not always easily ascertained; but, in the vast majority of cases, I believe that exposure to cold produces the attacks: we continually meet with them in the cold months of the year, and during cold, damp weather.

This disorder is not strictly limited to any particular age; the prattling babe, the boy at school, the young lady in her teens, and manhood in its prime are alike subject to its invasions.

SYMPTOMS. — Ordinary quinsy, of moderate severity, generally begins with restlessness, irritability, fever, sometimes a slight cough, and more or less soreness in the throat, especially when swallowing; the older children complain of this pain and refuse all diet except drinks and soft food, while the infant betrays it by refusing to nurse, and wincing its face whenever swallowing is attempted.

At first, there is but a slight sense of constriction and soreness; or, at times, a pricking sensation in the throat, which becomes decidedly manifest when an attempt is made to swallow. This soreness increases as the disease progresses. The constitutional symptoms, in mild cases, are not often decidedly marked; the face is generally flushed; fever moderate, and respiration somewhat accelerated; the voice is thick, and at times speaking is difficult or painful. Young children are often drowsy, but seldom sleep quietly on account of the fever and irritability, which produces a restless, uneasy disposition. Pain is not invariably present, especially in young children; and when there is a sudden rise of fever, with rapid respiration, slight dry cough, and more or less pain, this affection may be mistake for inflammation of the lungs; but upon placing your ear to the chest, you will readily mark the difference in the two diseases by the entire absence of all physical signs of pneumonia. Should any doubts remain as to the true nature of the disease, a thorough examination of the throat will soon decide it.

To examine the parts well, the head should be thrown back, the mouth widely opened, and the root of the tongue depressed with the handle of a spoon; by this means the whole interior of the throat will be brought into view.

Severer forms of this disease than that above described are not of unfrequent occurrence; not, however, in children under ten or twelve years of age, unless it be malignant or putrid sore throat in connection with scarlet fever; but of that we shall speak hereafter.

Ordinary sore throat, in its severer forms, is quite a serious affair, and is at times ushered in with vomiting, fever, and great nervousness; there is considerable thirst; the pulse is high, strong, and frequent; the cheeks are swollen; the glands about the neck are enlarged and painful; swallowing is difficult, sometimes almost impossible; the inflammation is extensive, frequently ending in suppuration, or in the formation of abscesses in the tonsils or adjacent parts. The tonsils are enlarged, sometimes enormously swollen, presenting serious obstruction to respiration. When the tonsils gather, relief may not be looked for until the abscess bursts.

This disease is not regarded as dangerous, and seldom amounts to but a trifling inconvenience if taken in season and properly treated.

TREATMENT. — The following are some of the principal remedies for sore throat. *Aconitum, Belladonna, Bryonia, Chamomilla, Hepar-sulph., Ignatia, Lachesis, Mercurius, Nux-vomica, Pulsatilla, Rhus, Sulphur.*

Aconitum and *Belladonna* are generally the most appropriate remedies to commence the treatment with, and, in the majority of cases, will effect a cure without other aid; particularly when the following symptoms are present: violent fever; pulse full and bounding; great thirst and restlessness; deep redness of the parts affected; constant desire to swallow; swallowing produces spasms of the throat, which forces the liquids partaken of out through the nose; burning and pricking sensation, with dryness of the throat; pains shooting into the tonsils, and up into the ears; swelling of the outside of the throat; profuse salivation; red and swollen face; skin hot and dry; pain in the forehead.

During the prevalence of scarlet fever, *Belladonna* should commence the treatment of almost every variety of sore throat.

Bryonia. — Especially after taking cold, or after getting overheated; hoarseness; oppressed respiration; pricking and painful sensibility of the throat; pain on turning the head; dryness of the throat, with difficulty of speech; swallowing painful; some fever, either with or without thirst; chilliness; pain in the limbs, back, and head. In alternation with *Rhus*, or *Rhus* may follow *Bryonia* when that remedy fails to afford complete relief.

Chamomilla. — Especially where sore throat has been induced by taking cold from exposure to a draught of air, while in a state of perspiration; swelling of the tonsils; tingling in the throat; hacking cough; hoarseness; fever in the evening, with flushes of heat; flushed cheek, or one cheek flushed and the other pale; the child is cross and restless, wishes to be carried in the arms, and wants things which, upon obtaining, it throws away.

Hepar-sulph. — In cases where the abscess in the tonsils is determined to break, this remedy will hasten the process; it promotes suppuration. It may also be given when there are several small ulcers, which appear slowly and are not painful. In the beginning of the disease, when there are lancinating pains in the throat, may be given in alternation with *Mercurius*.

Lachesis. — This will be found a useful remedy where *Belladonna* or *Mercurius* has been used without effect; also, when there is a constant disposition to swallow; dryness of the throat; extensive swelling of the tonsils with threatened suffocation; a sensation as of a tumor or a lump in the throat; sensitiveness to the slightest noise or touch, even to a handkerchief or the bedclothes about the neck. All the symptoms are worse during the evening.

Mercurius. — This is a valuable remedy and may often be given at the commencement of an attack, especially when sore throat arises from taking cold, accompanied with rheumatic pains in the head and nape of the neck; violent throbbing in the throat and tonsils, extending to the ears and glands of the neck, especially when swallowing; disagreeable taste in the mouth; profuse discharge of saliva; chills in the evening, or heat followed by perspiration; swelling and inflammation of the parts affected; ulcers, and tendency to suppuration in the .throat. *Mercurius* may, at the commencement of an attack, be given in alternation with *Belladonna;* if, however, there are strong symptoms of suppuration, it should be alternated with *Hepar-sulph.*, and should be continued sometime after the abscess has broken.

Nux-vomica. — In cases similar to those mentioned under Chamomilla; also, when there is soreness, with a feeling of excoriation or as if the throat had been scraped, and when there is a sensation as if there was a plug in the throat.

Pulsatilla. — For females or persons of a mild character; throat feels swollen inside; tonsils and palate have a dark-red appearance; shooting pain in the throat towards the ear when swallowing; scraping sensation in the throat; chilliness in the evening, followed by heat.

Rhus. — For symptoms similar to those of *Bryonia.*

Sulphur. — For frequent or continued sore throat, especially in vitiated constitutions. Sulphur is a valuable remedy to hurry forward the suppuration process, when an abscess seems certain to burst; also, after the discharge of an abscess, when the cavity is slow in healing, or when many abscesses form in succession. It may be given in alternation with *Silicea.*

ADMINISTRATION OF REMEDIES. — For ordinary cases dissolve, of the selected remedy, twelve globules in twelve teaspoonfuls of water, and of this solution one teaspoonful may be given every two or three hours, until relief is obtained. In severe cases, where swallowing is difficult and very painful, three globules may be given, dry, upon the tongue, every two hours, or even oftener, every hour, until a change takes place. In all cases, lengthening the interval between the doses, as the severity of the symptoms subside.

DIET AND REGIMEN. — The diet will have to be regulated according to the degree of inflammation. If the inflammation is extensive, the throat much swollen, and swallowing difficult, of course solid food cannot be taken. Custards, panadas, gruels, light soups, and the like, are about all that can be swallowed with any degree of comfort, and even these at times produce great pain; in fact, the mere *act* of swallowing is almost impossible.

In no disease perhaps is the beneficial effect of cold water more marked than in sore throat. When going to bed at night put a wet bandage around the throat, and cover it with a dry cloth. If the patient is confined to the house, repeat the same through the day. The application of water should be made at the commencement of the attack; if, however, the disease continues in spite of the treatment, and suppuration is about to take place, which may be suspected when there is a pulsating or throbbing sensation attended with stitches in the parts affected, the suffering may be relieved, and the bursting of the abscess hastened by the repeated external application of warm linseed poultices, and gargling the throat with warm water. When much pain is present the inhalation of vapor from boiling water will often afford great relief.

All medicinal gargles are injurious, and all external applications of blisters, leeches, mustard drafts, etc., are worse than useless.

A predisposition to sore throat exists in some persons. *Sulphur*, *Graphites*, and *Silicea* have been found useful in overcoming this constitutional difficulty. When taken for this purpose, a dose every second or third night, until six doses are taken; then discontinue the medicine for one week; after which, take it again, as above.

MALIGNANT OR PUTRID SORE THROAT.

DEFINITION. — CAUSES. — Malignant sore throat constitutes a part of that terrible scourge, malignant scarlet fever. It also forms an independent disease, generally occurring in damp autumnal seasons, attacking children of vitiated, impoverished, or delicate constitutions, weakened by some previous disease. It is also more apt to attack children living in low, damp, cold, mouldy, or ill-ventilated houses, and in want of warm clothing and healthy food. Epidemic sore throat, under these circumstances, readily assumes a malignant type.

This species of sore throat, however, is by no means exclusively confined to the class above described; but, as I have already stated, they are most liable to it. Still, those living in the very lap of luxury, where want never enters, are not exempt from its invasions. It is an exceedingly dangerous disease, wherever and whenever it appears; therefore the treatment should be prompt and energetic, and should never be attempted by domestic practice, unless it be impossible to obtain the services of a homœopathic physician.

SYMPTOMS. — This disorder commences with a chill, not always, however, distinctly marked; sometimes, indeed, amounting to but a slight shivering, followed by fever and languor; oppression at the chest, with or without vomiting; cheeks of a crimson hue; more or less inflammation of the throat and tonsils, with an acrid discharge from the mouth and nose, excoriating the parts with which it comes in contact. Pulse, weak and very quick, almost imperceptible; throat, and glands about the throat, much swollen; face bloated; patient very restless.

Upon examination of the throat, you will perceive numerous small ulcers, covered with an ashy-gray crust, while the surrounding tissue is of a livid or dark red color. These ulcerated spots vary in different cases. Sometimes they are few in number, and confined to the throat and tonsils; in others, they are numerous, the whole mucous membrane of the mouth being thickly studded over with them; they even extend through the opening into the windpipe.

These small spots, or patches, of ulceration are of a yellowish-gray color, looking more like spots of lard than anything else. They may remain isolated and circumscribed; but, in severe cases, they run together, and present a gangrenous appearance, become soft, and slough away. At this stage of the disease, there is excessive prostration; the teeth and tongue are covered with a blackish incrustation, similar to that seen in typhus fever; there is more or less delirium; the breath is fetid, the countenance sunken; vomiting and fetid diarrhœa supervene; the pulse grows feebler; the skin, which was previously harsh and dry, now becomes covered with a cold, clammy sweat; stupor sets in, and the patient dies.

The milder cases, and those which respond readily to appropriate treatment, generally yield on the third or fourth day, terminating in profuse perspiration. The breathing becomes easier; the pulse less frequent, but stronger; the ulcers in the mouth become cleaner, lose their ashy-gray cast, and are surrounded by a bright redness; the breath loses its bad smell; swallowing is less difficult; and the general expression of the face becomes more lively.

TREATMENT. — The remedies are *Belladonna, Arsenicum, Carbo-veg., Lachesis, Mercurius, Nitric-acid, Secale,* and *Sulphur.*

As a general thing, the treatment may begin with *Belladonna*

31

and *Mercurius* in alternation, especially should there be much dryness of the mouth, with restlessness, or even delirium.

It is often advisable, at the outset of an attack, when the skin is hot and dry, and the fever appears of an inflammatory nature, to give a few doses of *Aconite;* but as soon as dryness of the throat, difficulty of swallowing, and a sense of constriction or choking in the throat, manifest themselves, immediate recourse should be had to *Belladonna.*

Mercurius. — This remedy is specially indicated when there is a profuse secretion of saliva from the mouth, with ulceration of the throat, particularly in the early part of the disease. If, however, *Mercurius* fails to produce any decided relief, or, what is still worse, if the disease progresses, while the ulcers increase in size, and become painful, *Nitric-acid* may be given, either alone or in alternation with *Mercurius.*

Arsenicum. — This is one of the principal remedies for gangrenous sore throat, and should be given when there is extreme prostration of strength, rapid sinking of the patient; also when the ulcers present a dark-red appearance, or where they are covered with dark scabs, and surrounded with a livid margin; teeth and lips covered with blackish incrustations; tongue dark and cracked; constant muttering and delirium; breathing difficult; acrid discharge from the mouth and nose, excoriating the parts with which it comes in contact. This remedy can be given alone, or in alternation with *Lachesis,* which is also another valuable remedy, and is particularly indicated should the neck be much swollen or discolored and tender, or rather painful to external pressure.

Carbo-veg. — When the discharge from the ulcer is thin, copious, and fetid, accompanied with great prostration.

Secale. — When the patient is disposed to sleep a good deal, or when he lies in a drowsy, half-stupefied state, if Secale does not have the desired effect, give *Opium.*

Nitric-acid. — When the patient is out of danger, but the ulcers are slow in healing, one dose may be given night and morning.

ADMINISTRATION OF REMEDIES. — The remedies may be given dry, upon the tongue, or they may be dissolved in water. If dissolved, put twelve globules in twelve teaspoonfuls of water; of this solution, give one teaspoonful for a dose. At the commencement of an attack, or where the symptoms are severe, the remedy indicated may be given as often as every hour; always lengthening the interval between the doses as the severity of the symptoms subsides.

As a general thing, the dose may be repeated at intervals of from one to four hours. In this, as in many other cases, you will have to depend a great deal upon your own judgment in the administration of remedies.

DIET AND REGIMEN. — The first thing to be done, is to place your patient in a dry, airy room : plenty of pure, fresh air is the best adjuvant in the treatment of this or any other disease. Stagnation of air is an abomination greatly to be feared.

The food, as a matter of course, will have to consist of rice, arrow-root, corn-starch, thin flour gruel, broths, and the like. When the mouth is very hot and dry, it is advisable to moisten it with a little warm milk and water. The mouth should be frequently washed out; and this must be done very gently, so as to produce no irritation. As a wash, warm water appears to be the most desirable. During convalescence, care should be taken that the patient does not overload the stomach, as this would tend to produce a relapse, or at least excite some gastric derangement, which would retard an otherwise speedy recovery.

TONSILITIS, OR INFLAMMATION OF THE TONSILS.

DEFINITION. — CAUSES. — Acute tonsilitis has already been considered, in the article on sore throat. I will, therefore, here only make a few remarks in regard to chronic enlargement of the tonsils. You will frequently hear the ignorant speak of children having tonsils in their throat, as though *all* children, and adults too, did not have them there.

The tonsils are two oblong, somewhat rounded bodies, placed between the arches of the palate. In some they can scarcely be said to exist; while in others they fill up the throat to such an extent as to impede swallowing, or even respiration. The use of these glands is to secrete a fluid which makes smooth and slippery the passage to the stomach, for the easy transmission of the food we swallow.

Enlargement of these glands from chronic inflammation, or enlargement either congenital, or arising from excessive nutrition, constitutes the disease under consideration.

SYMPTOMS. — The first symptom that attracts attention, is the habitually loud breathing of the patient during sleep, or in other words, his snoring. This is caused by the enlarged tonsils pressing upon the palate, which partially closes the passage through the

nose; the air being forcibly drawn through this narrowed opening produces that horrid noise. The voice becomes thick and in-
articulate. These symptoms and that of snoring become aggravated upon the slightest attack of cold or catarrh.

Deafness is also another symptom and originates from a similar cause, — the enlarged tonsil pressing upon the Eustachian
tube, which is a small canal leading from the throat to the internal ear. But the most serious consequence of long-continued
enlarged tonsils is the effect it produces upon the chest; enlargement of the tonsils, and the " pigeon breast," usually go together.
The obstruction preventing the free entrance of air into the lungs,
these organs are but imperfectly developed; *how* this imperfect
expansion of the lungs produces the prominence of the breastbone, would take too much room for me to explain, but you can
rest assured that such is the fact.

Enlargement of the tonsils, you will therefore see, though
seemingly of slight importance, may lead to serious results.

A weakly child, with slight enlargement of the tonsils, will
often get rid of the ailment as he gains strength, and at the age
of fourteen or fifteen have entirely outgrown it.

TREATMENT. — The application of Nitrate of Silver, — caustic, —
Iodine, Alum, or to cut the tonsils out, as allopathic physicians too
frequently do, is worse than useless, for — and they too must have
noticed the fact — in the majority of cases, after such barbarous
operations, the lungs become affected, and sooner or later, as the
result, the patient dies of consumption. If there be, in a patient, a
hereditary taint of, or a predisposition toward consumption, I know
of no other way to light up that latent spark so certain and so rapid
as by the application of Nitrate of Silver to the throat.

The only rational way of curing these enlarged tonsils, is to
put the patient under a strict course of homœopathic treatment;
it will take some time, it is true, for their reduction; but, even
though it takes a long time, it is much better than to run a miniature guillotine down a child's throat and clip off his tonsils, though
it is done expeditiously; for one is safe, and the other is not.

The appropriate remedies are *Belladonna, Causticum, Calcaria,
Graphites, Hepar-sulphur, Lachesis, Mercurius, Nux-vomica,* and
Sulphur.

When an aggravation of the symptoms occurs, caused by slight
cold, the application of cold water to the throat, in the shape of
wet bandages, over night, frequently gargling the throat with cold

water, together with a few doses of Mercurius will be sufficient to remove the difficulty.

When a course of treatment has been commenced, the remedy given should be continued for at least six weeks, a dose every other night.

FALLING OF THE PALATE.

DEFINITION. — CAUSES. — TREATMENT. — This is entirely an imaginary difficulty; there is no such thing as falling of the palate. Some persons, however, after a slight cold or an attack of indigestion, suffer from a trivial inflammation of the palate, which produces from its thickened and elongated state, a sensation as if it had fallen. I have often seen old ladies, in such cases, put a little pepper, salt, or mustard upon the palate, — to make it jump up, I suppose.

The best remedy that I know of for this condition of things, is *Nux-vomica*. If it be a recent attack, a dose of three globules may be given every two hours, until the unpleasant feeling subsides.

Should this fail to effect a cure, *Mercurius*, *Belladonna*, or *Sulphur* may be tried in the same manner.

Cold water is here very beneficial, applied both externally and internally.

DIET. — Avoid all stimulating articles of diet, such as fancy, high-seasoned dishes, and the like.

DIPHTHERIA, OR DIPHTHERITE.

DEFINITION. — Diphtheria is a term used to designate a specific and peculiar form of inflammation of the throat. Unlike ordinary inflammations of these parts, this is attended with an exudation of false membrane upon the mucous surface, which is developed after a variable amount of constitutional disturbance, attended usually with a low grade of fever, and is mainly confined to the throat, tonsils, and nasal cavities.

Diphtheria is a constitutional disease, presenting, as its characteristics, local manifestations in the throat. This is not, as many suppose, a new disease. A consultation of the best authorities discloses many severe epidemics to have visited both this country and Europe. We have a record of its ravages in Rome as early as A.D. 330. We read of it in Holland in 1337; in Paris in 1576;

in Naples in 1618; again in France in 1818 and 1835. A severe epidemic passed over England during the years 1858, '59, and '60. — *Brit. Med. Journal.*

It seems to have first visited the United States in 1771—*Trans. Am. Philos. Soc.* vol. 1. — when it was but imperfectly recognized, being confounded with membranous croup, putrid, and other forms of sore throat. It reappeared here in the latter part of 1856 or the fore part of 1857, when it was at once distinctly recognized and successfully combated.

In the year 1857 there were two cases of diphtheria reported in the city of New York; in 1858, there were five; in 1859, fifty-three; in 1860, four hundred and twenty-two; in 1861, four hundred and fifty-two; in 1862, five hundred and ninety-four.

In this city, there were thirteen cases reported in 1859; one hundred and thirty-five in 1860; one hundred and sixty-five in 1861; two hundred and nineteen in 1862; two hundred and fifty-four in 1863; three hundred and sixty in 1864; one hundred and seventy-four in 1865.

I would here remark, that many diseases of the throat are called diphtheria, which have no analogy to the disease whatever.

CAUSES. — It is abundantly proven by long and repeated observation, that diphtheria is propagated by two causes, — epidemic influence and contagion.

Scrofulous children, — those subject to glandular enlargements, to catarrhal and croupous affections, — are usually the ones that are first affected when the disease rages as an epidemic.

I am well convinced that the disease is contagious, and for the following reasons: It habitually spreads in those families which it invades. I cannot call to mind a single family of children in my practice, where the disease entered, but that more than one suffered, while, in numerous instances, it spread through the household, affecting all, both adults and children, and those who were most closely in communication with the patient, were the first to be taken sick, while those who were early removed, sent away from the sick person, usually escaped. If this be not the result of direct contagion, then we must say that it becomes epidemic in a household to an extent sufficient to cause its spread from one member to another. *Reductio ad absurdum.*

Surgeons have been seized, after a portion of saliva or false membrane had fallen upon the lips or mucous membrane of the nose while engaged in examining the throat, and have died from the effects of it. — *Monograph on Diph.*, G. F. SNELLING, M. D., p. 10.

I have known a child from the country, merely carried through one street in Albany, a short time since, who sickened with the disease in six hours, and died within a few days. — *Loc. cit.*

Dr. Giddings, of S. C. — *Am. Jour. Med. Science,* vol. xxiv. — remarks: "Under particular circumstances, as when many persons are crowded together; when ventillation is imperfect, and cleanliness is neglected, there can be no question of the generation of a contagious influence, capable of transmitting the disease from one person to another." Dr. Bard, of N. Y., in the same journal asserts the disease to be infectious, and that the infection depends not so much upon any prevailing disposition of the air, as upon effluvia received from the breath of the infected person.

After reviewing the facts of the question, as presented by observation, and the written opinion of eminent authors, it may be concluded that diphtheria arises from a specific poison, taken into the system, which, acting through the blood, produces a true constitutional disease, exhibiting its local manifestations in the formation of false membrane upon mucous and abraded cutaneous surfaces, and becomes capable of transmission from one to another, without any recurrence to the original source of the poison.

If these views be accepted as correct, the importance of removing all the children in a family, as soon as any one is taken with the disease, will be obvious to all. This precaution, if early observed, will, I am fully convinced, be the means of saving many lives.

In the majority of instances children are the subjects of diphtheria.

There has been quite a difference of opinion among physicians with respect to diphtheria and scarlet fever, some intelligent observers contending for the identity of the two, maintaining that diphtheria was but an altered form of scarlatina, or, those cases where the eruption has not come well out, or coming out, has receded. Indeed, when we come to compare the symptoms, we find a strong resemblance existing between the two diseases. They are each preceded by a fever of about the same duration. There is headache, flashes of heat, chilliness, sore throat, enlargement and tenderness of the glands of the neck. Many cases of pure diphtheria are attended with an eruption, which, at times, resembles measles; at others, it is of a bright red, like scarlet fever. "Indeed, the eruption in diphtheria would seem to be a sort of a cross between that peculiar to measles and scarlet fever." — Dr. Ludlum, *on*

Diphtheria. Besides, the sequelæ, or secondary affection which follows the diseases, are almost identical.

Scarlet fever and diphtheria appear as contagious epidemics in the same neighborhood, at the same time; indeed, we have had the two diseases existing at the same time, in the same family.

"In the same house, the father and mother had well-marked scarlet fever severely, while the three children had all the marked symptoms of diphtheria, without much feverishness and *no rash*, though attended by the same premonitory symptoms, the cases occurring at the same time." — *Brit. Med. Jour., June 8th,* 1858, p. 449.

For my own part, though acknowledging the similarity of the two diseases, I do not by any means consider them identical. That scarlet fever *invites* diphtheria is quite manifest; it has been the painful experience of every physician to have scarlet fever patients swept off by diphtheria.

It is a matter of general observation and remark, that all diseases occurring during an epidemic of diphtheria, and especially those affecting the throat, have a tendency to become complicated with that disease. Sore throats, that in ordinary seasons would amount to but a slight inflammation during such seasons, almost certainly become diphtheritic, just as diarrhœa during an epidemic of cholera, has a tendency to pass into that disease.

SYMPTOMS. — Some authors, both for reasons real and imaginary, have divided diphtheria into numerous species, forms, and varieties. We can see no good reason for this division. On the contrary, there are many objections to the arrangement. Diphtheria is, in fact, with all its degrees of severity and apparent differences, a single and distinct disease, produced by one cause, inducing similar results, however much they may vary in gravity, and no more requiring than does dysentery to be divided into the many forms too often ascribed to it.

According to my observation, diphtheria, in the majority of instances, commences as does an ordinary cold or influenza, with slight chills and flashes of heat; some little irritation of the throat, but no great amount of pain or difficulty in swallowing; stoppage of the nose or fluent discharge; aching in the bones; general prostration and weariness, occasionally with high fever and severe pain in the head; disordered stomach and loss of appetite; followed in the course of twenty-four or forty-eight hours by a more or less decided aggravation of the throat trouble, the glands about the neck becoming sensitive and swollen, with an increased flow of saliva or water into the mouth. In many instances, the onset of disease is so insidious that its true nature would hardly be suspected, were the patches of false membrane not *seen* in the throat; and not unfrequently it is with no little difficulty that the parents can be persuaded that anything which may prove serious is the matter with the child.

In cases somewhat more severe, the patient early complains of soreness of the throat and stiffness of the neck; externally, the tonsils are found enlarged and tender; internally, the inflammation is plainly visible, sometimes appearing bright and glassy; at others, almost purplish, and dotted over with spots of false membrane. These spots may vary in size from a split pea to a half inch in diameter. When the membrane becomes detached, it leaves the surface beneath in appearance not unlike a piece of raw meat. There is more or less fever, with headache, in many cases, almost unbearable; the breath is extremely offensive. A characteristic symptom, and one not accounted for by the amount of local mischief going on in the throat, is the extreme prostration with which all these cases are attended. However, under judicious treatment, convalescence can usually be established in from eight to ten days, although it may be weeks, or, indeed, months, before the debility and nervous depression can be removed, and the vigor and elasticity of the system entirely restored.

The above may be taken as a fair example of diphtheria, as it ordinarily appears. But it must not be imagined that the disease always presents itself in this benignant manner, because it does occasionally assume a malignity truly terrific to behold.

It may be the patient is suddenly seized with rigors, and vomiting of a thin, white, yellowish matter of a very offensive nature; then purging of a fluid of a similar appearance and smell. These symptoms, subsiding after an hour or two, are followed by prostration and stupor. After a period varying from six to sixteen hours, the stupor passes off, and delirium, often of a violent character takes its place. — "*Lancet*," 1859, p. 183. When the onset of the disease is thus sudden, it runs its course with great rapidity; a few hours, often, being sufficient to place the patient beyond all hope of remedial assistance. In all cases of marked severity, the most striking symptom is the extreme prostration, at once indicating that the system has been overwhelmed by some powerful morbific influence. These sudden attacks explode a train of symptoms, which startles the most stoical observers. Whatever may be the particular feature of the assaults, the characteristic exudation quickly appears in the throat, and rapidly spreads. The glands about the neck become enlarged and tender; there is high fever, great excitement of the pulse, and severe headache; or the surface may be cool and clammy; swallowing becomes difficult, or even impossible. Though threatened strangulation is sure to be produced by every effort to swallow, the attempt must frequently be made to get rid of the saliva and other fluids, which collect in the throat. Whenever food or medicine is taken, it is violently ejected from the nose and mouth. The breath, which for some time has been extremely offensive, now becomes horribly so; indeed, the case becomes so repulsive that even the patient's best friends, — those who are most anxious to assist him, — cannot even come near him without feelings of aversion.

As the disease progresses breathing becomes difficult, from an extension of the membrane into the air passages, and all hopes of the case become futile. The countenance assumes a leaden hue; the skin is cold and shrivelled; the patient throws himself from side to side, fighting for breath, — a heart-rending spectacle to all whose duty compels their presence, — until death finally closes the scene.

A no less fatal, but more insidious, form of the disease is occasionally met with. It steals upon the patient without sounding a single note of alarm, until it has, so to speak, gotten the whole system entirely within its grasp. The general symptoms attending ordinary cases may be entirely wanting. The local soreness of the throat, which, for a day or two, attracts but little attention, by degrees begins to cause some inconvenience; when, all at once, from an extension of the diphtheritic deposit, the child is taken with croupy breathing, and is either in a few hours beyond all hope of recovery, or he may live after this alarming symptom sets in, for two or three days, at times brightening up, exciting hopes in the minds of his parents that he may possibly escape; but it is all illusive, for these radiant moments are but the last flickerings of the expiring lamp.

A careful internal examination of a diphtheritic sore throat displays a condition varying in appearance, according to the severity of the attack, and the stage at which it is observed. If observed early, the tonsils, palate, and back part of the throat present a red, shiny appearance, as though the parts had been brightly painted, and then varnished. To the casual observer, or one unaccustomed to diphtheria, this condition of the throat would occasion no anxiety, and might very readily be considered an ordinary case of tonsilitis. Not so, however, with the experienced man; he expects it; and is well aware that unless the disease is arrested, the condition of his patient, in a few hours, will be very materially changed.

If the disease has made some progress, before attention is called to the throat, a careful examination will disclose spots or patches of false membrane dotting over the palate and tonsils; these spots may be, at first, few in number, small and indistinct, but they are quite sufficient to warn one who is acquainted with the disease to arm himself for a conflict which the inexperienced would hardly anticipate. This false membrane, first observed in small, apparently insignificant spots, scattered over those parts hitherto so brilliantly red, may in a wonderfully short period of time conglomerate into one thick, plastic deposit, covering the palate, tonsils, and fauces, and sometimes extending into the nasal cavities. Indeed, it has been observed to cover the whole mucous membrane of the throat and mouth, within twelve hours from the first complaint, so that, on looking into the mouth, it appeared as though lined with gray velvet. — *Lon. Lancet, Aug.* 20*th*, 1859, p. 183. It has also been observed that the skin, when abraded, has become covered with false membrane, and blistered surfaces are especially liable to become affected in this way.

In those cases which have come under my immediate observation, the membrane was first observed at one or more spots on the tonsils of about the size of a split pea; from these points it has spread to the surrounding parts. I have never had it become general and extend to the windpipe and bronchial tubes, although it frequently does so, and always, I believe, proves fatal.

The physical appearance of the membrane is similar to that thrown out in true inflammatory croup, except it is soft, and appears, as indeed it really is, saturated with fluids. In color it is described as a yellowish-white, gray, or light brown; some compare it to gray velvet, others to wet chamois. Portions taken from the windpipes are lighter colored than that found in the mouth. It adheres with moderate tenacity to the mucous membrane beneath; when its edges become loosened, as frequently happens, a ·bloody secretion of a fetid odor exudes from beneath.

TREATMENT. — In all cases of diphtheria, no matter how mild

it may appear, the patient should, without delay, be placed under the care of an intelligent homœopathic physician. Until such services can be secured, give *Kali-bichromicum,* and *Proto-iodide of Mercury* (Mercurius-iodatus), in alternation every hour, or every two or three hours, according to the urgency of the case. I usually dissolve about a grain of the first trituration of *Kali-bich.* in a glass half full of water — just enough to tinge the water yellow — and give one dessert-spoonful of the solution at a dose. Of the *Mercurius,* I give the third trituration, about as much of the powder as you could hold upon a five-cent piece. Many physicians prefer the high attenuation. These two remedies, repeated every hour or two, will suffice to cure nearly every case. — Marcy & Hunt's " *Theory and Practice,*" vol. I. p. 764.

Aconite. — In many cases this remedy will be called at the commencement of an attack, especially, if there should be considerable fever; heat of skin; rapid and full pulse; dry tongue; offensive breath; the inflammation of the throat being of a dusky redness. If, in addition to these symptoms, there is a patch or two of wash-leather exudation upon the tonsils and far back in the throat, give *Aconite* and *Bryonia,* in alternation, every hour.

Belladonna. — Inflammation of a bright, scarlet redness, extending uniformly over the mucous membrane; enlargement of the tonsils; fever, and severe headache.

Rhus-tox. — When the inflammation is of a dark red, with dark crimson patches scattered over the surface.

Arsenicum. — " The breath fetid; the lining of the nostrils discharging a viscid, foul secretion; great and increasing prostration of strength. After the separation of the false membrane, it may remove the extreme tenderness which remains, as well as keep up the vital energies."

Kali-chlor. — " Is especially indicated if there be extreme depression, imperfect vitalization of the blood, a septic condition generally, and a tendency to stupor." — Dr. Snelling.

Numerous other remedies are recommended and used for diphtheria; among them, we have *Capsicum, Colchicum, Tartar-emetic, Nitric-acid, Amonium causticum, Cantharis, Lachesis, Borax, Spongia, Hepar-sulph., Bromine, Lycopodium, Muriatic acid, Iodine, Croton.*

Administration of Remedies. — Where directions have not already been given, dissolve twelve globules in six spoonfuls of water, and give one spoonful of the solution at a dose. Repeat it every six hours.

DIET AND REGIMEN. — The main danger in uncomplicated cases of diphtheria evidently arises from debility; and hence it has been insisted on, by nearly all practitioners, that a good, nourishing diet is essential, and stimulants often desirable. In every genuine case, at the outset, there is more or less febrile action going on: you will find heat of skin and acceleration of the pulse. During this stage, stimulants would be highly injurious; but this period is usually of short duration. The patient should be closely watched; and the moment the pulse begins to flag, the skin to get cool, and the characteristic weary prostration to show itself, we should at once adopt a sustaining regimen. An early resort to a good nourishing diet, judiciously combined with stimulants, I am satisfied, will very much diminish the period of convalescence, which, in too many instances, drags its weary length along through weeks and months. The amount of both stimulants and diet will, of course, depend upon the peculiar circumstances of each respective case; but, in order to insure a sufficiency, they should be various, administered in small quantities, at regular and frequent intervals. A great deal will depend upon the digestibility of the articles used, and their adaptation to the wants of the stomach. It must ever be borne in mind, that the digestive apparatus shares in the extreme depression of the system, and that, as it can no longer convert crude materials of aliment into the life-giving current, it will not do to choose food or stimulus hap-hazard, and to gorge the patient with it, blindly supposing that this will afford him support. Not by any means. That which is taken into the system, and not digested, becomes a source of irritation, from which untold mischief may result. In the majority of instances, beef-tea will be better adapted to the wants of the patient, perhaps, than any other article that can be selected. Beef-tea should be made after the following receipt: "Take a pound of perfectly lean, juicy beef; cut it in little squares, and put it into a wide-necked bottle, with a suitable quantity of salt; tie a piece of muslin over the mouth, and place it in a kettle of hot water, and let it simmer over the fire for six hours; then remove the juice. When done, the meat should be quite white and tasteless." The tea should be seasoned palatably to the sick persons, and they may be allowed to take as much of it as they choose. In some cases, it will be necessary to give it to the patients, especially when they loathe all nourishment, by the spoonful, every half-hour or hour, telling them that it is medicine. In cases of extreme prostration, where the beef-tea is rejected by the stomach, it must be given by enema.

In cases where the stomach refused all nourishment, rejecting whatever was administered, I have often found clam-broth to act like a charm, settling the stomach and giving the patient a relish for other kinds of food.

The soft parts of oysters, either raw or stewed, will often relish, and make a good substitute when the patient gets tired of beef-tea.

As stimulants, port wine, claret, champagne, milk-punch, and brandy and water, are all, when judiciously administered, of vast benefit. Eggs, beaten up with brandy, hot water, and sugar, make a good, nutritious stimulant, next, perhaps, to milk-punch.

For children, the best stimulant is wine-whey, or beef-tea mixed with port wine, or port wine and arrow-root.

As a beverage, when the patient is thirsty, barley-water or toast-water, acidulated with a little lemon-juice, is not objectionable. Perhaps, however, cold water, to which has been added a little raspberry or strawberry syrup, is to be preferred.

PYROSIS. HEARTBURN. WATER-BRASH. SOUR STOMACH.

DEFINITION. — SYMPTOMS. — Under the above names are usually arranged the following symptoms : a gnawing or a burning sensation at the pit of the stomach, accompanied with or followed by sour, acid eructations, or belchings, attended with nausea, coldness of the extremities, and often with faintings.

As you will readily observe, these are but symptoms of a disturbed digestion, the forerunners or accompaniments of dyspepsia. These disagreeable symptoms are always aggravated when anything is taken into the stomach which does not exactly agree with it.

When these symptoms are isolated, that is, apparently unconnected with any other derangement, the following remedies will be found of service.

A more detailed account has been given under the general head of " DYSPEPSIA," which see.

TREATMENT. — For water-brash — *Nux-vomica, Pulsatilla, Chamomilla, Arsenicum, Silicea, Carbo-veg., China, Belladonna, Sulphur.*

For heart-burn. — *Nux-vomica, Pulsatilla, Arsenicum, China, Sepia, Sulphur.*

For flatulency, or frequent rising of wind. — *Nux-vomica, Carbo-veg., China, Graph. Phosphorus, Pulsatilla, Sulphur.*

For sour stomach, — *Chamomilla, Nux-vomica, Pulsatilla, Phos-phorus, Sulphur.*

For flatulence, when it occurs after eating, and is accompanied with hiccough and sour risings, take *China.* When such symptoms arise from eating fat food, take *Pulsatilla.* If attended with colic, *Nux-vomica.*

When sour stomach occurs in nursing infants, *Chamomilla, Ipe-cacuanha,* or *Nux-vomica* will be of service. Sometimes a little sugar-water will afford relief when nothing else will.

Sour stomach of pregnant females may sometimes be remedied with a few teaspoonfuls of lemonade.

ADMINISTRATION OF REMEDIES. — Dissolve, of the selected remedy, twelve globules in twelve teaspoonfuls of water, and give the child one spoonful of this solution every hour, or oftener, if necessary. An adult may take six or eight globules every hour. — See " COLIC."

NAUSEA, VOMITING, AND REGURGITATION OF MILK.

DEFINITION. — Owing to the imperfect development of the infant's stomach, vomiting and regurgitation takes place with great readiness. Nausea and vomiting, as a general thing, in children, is simply an act of nature, kindly intended to rid the stomach of any excess of food it may have received. Nursing infants, when in the very best of health, are very apt to vomit or regurgitate, after having nursed abundantly. This, of course, arises from having overloaded the stomach, and is salutary. The milk is ejected just as it was drawn from the mother, or perhaps slightly curdled.

Older children also have their spells of nausea and vomiting, which not unfrequently follow an expedition to the grape-vine or apple-orchard ; or especially among city children, from having partaken of too great a variety of rich things, or from eating some indigestible substance. This kind of vomiting always affords relief, and it is fair to conclude that the act has been a beneficial one. We should thank nature for her kind assistance, instead of misinterpreting her, as people too frequently do, and concluding that the child needs a little " doctoring."

Sometimes, however, vomiting arises from other causes, and, instead of only a *portion,* the *whole* of the food is thrown up ; and not only that, but mucus and bile may be ejected, either with or after the contents of the stomach. This, of course, is *not* salutary, and, therefore, *does* need attention.

TREATMNET. — *Ipecacuanha.* — This is the first remedy, and will generally be all that is required.

Pulsatilla. — Should there be much flatulence, and distention of the abdomen, after *Pulsatilla*, give *Antimonium-crud.*, if only partial relief is afforded.

Chamomilla. — When the disease is attended with diarrhœa or with convulsions; diarrhœa, with greenish stools; pain in the stomach; great restlessness.

Nux-vomica or *Bryonia* — May be given where vomiting is attended with constipation. When vomiting arises from a natural weakness of the stomach, *Nux-vomica*, followed by *Bryonia*, will often be of service. Chronic cases, or cases of long standing, call for *Calcarea* or *Sulphur*.

Cina. — For vomiting caused by worms. Should no relief follow its use, give *Mercurius* or *Ferrum*.

ADMINISTRATION OF REMEDIES. — The medicine may be given dry, three globules at a dose, for an infant ; or you may dissolve twelve globules in twelve teaspoonfuls of water, and give, of the solution, one spoonful every four hours. In severe cases of vomiting, the remedy may be repeated every fifteen minutes, or every half-hour. — See "DYSPEPSIA."

BILIOUSNESS.

DEFINITION. — The common term "biliousness" is, to my mind, rather indefinite. Some people call everything "biliousness." Nine out of every ten patients that I prescribe for, among their other complaints, usually inform me that they are a "little bilious." If a child loses his appetite, has a cough, or any other slight ailment, it is because he is "so bilious." Should there be sickness at . the stomach, tongue coated, accompanied with a giddy, dizzy headache, it is because "there is too much bile on the stomach," and the patient must have an emetic. The contents of the stomach are thrown off, and, at last, from the very bottom, up comes the "villanous bile." Against this irrational mode of practice, permit me to enter my protest.

The common belief, that bile collects in the stomach, is erroneous. The bile-duct, or canal, leading from the gall-bladder, enters the *intestines* more than six inches *below* the stomach. Now, do you suppose that this fluid is going to travel up-hill, for the sake of tormenting a poor patient? Not a bit of it. But perhaps you will say, how is it that bile is thrown off the stomach, if there is none there. I will tell you. Suppose a patient is complaining. Aunt Susan, or some other good old lady, is sent for. She at once perceives, or, what amounts to the same thing, thinks she does, that *bile* has taken possession of the stomach. So *Mr. Ipecac.* is sent down with a writ of ejectment, and, at his command, everything is turned topsy-turvy, and heaved out; but no bile is found. "Oh! but, Mr. Ipecac," says the old lady, "I know there is bile on his stomach. You must try him again." So away he goes, with the hearty determination not to leave the premises again till the invader is found and ejected. Immediately after his second visit, the poor stomach, almost worn out with exhaustion, gets an inkling of what is wanted, and forthwith sends to her next neighbor below — the duodenum [1] — for a little bile, which she feebly drops at the feet of her conqueror. Now behold Aunt Susan, as, with a twinkle of triumph, she points to her conquered foe, really believing, and therefore positively asserting, that, now the invader is taken, the danger is past. That is the way in which bile is obtained from the stomach.

To show you that I am not alone in this belief, I shall quote a few lines from a work on physiology, by Dr. Calvin Cutter, p. 126: "If bile is ejected in vomiting, it merely shows, not only that the action of the stomach is inverted, but also that of the *duodenum.*[1] A powerful emetic will, in this way, generally bring this fluid from the most healthy stomach. A knowledge of this fact might save many a stomach from the *evil* of emetics, *administered on false impressions of their necessity, and continued, from the corroboration of these by the appearance of bile, till derangement and perhaps permanent disease are the consequence.*"

Now, do you not see the folly of attributing every little ailment to "biliousness," and the worse than folly, the barbarousness of administering emetics? In nine cases out of every ten, you do more harm than good. As Dr. Cutter intimates, in removing some trifling *temporary* ailment, you produce a serious *permanent* one. But, perhaps, after all, you do not belong to the "bilious" school, or, at least, when you speak of biliousness, you do not mean that there is "an overflow of bile," or that there is an actual collection of fetid bile within the stomach, but rather that the patient is suf-

[1] Duodenum, from the Latin word *duodenus,* meaning *twelve,* as if twelve fingers in breadth. The first part of the small intestines.

fering from a number of symptoms, originating perhaps from some
gastric derangement, and which, for want of a better name, you
call "biliousness." If so, then we exactly agree.

Now, in regard to emetics, under certain circumstances I do not
doubt their utility. I do not see how we could get along without
them; but still their use is very limited. I do not hesitate at all
to say that we ought *never* to make use of them, except when we
wish to remove some foreign substance from the stomach, which
has been *swallowed* either through indiscretion or by accident.

SYMPTOMS. — The patient at first appears dull and languid,
complains of headache, or, rather, a giddy sensation in the head;
great oppression, and a fulness at the pit of the stomach, with
nausea, sometimes vomiting; eructations of offensive gas, smelling
like stale meat or rotten eggs. The tongue is covered with a
thick, slimy, yellowish coating; there is a disagreeable, bitter,
putrid, or slimy taste in the mouth, especially in the morning.
Bowels are either constipated or quite loose; the passages are
dark, very offensive, and accompanied with a great deal of fetid
wind. The eyes are dull and heavy; at times they have a yellowish cast; also, the skin, particularly around the mouth and nose,
looks yellow.

TREATMENT. — The remedies are *Bryonia, Ipecacuanha, Mercurius,* and *Pulsatilla.*

Bryonia. — This remedy is called for, when, in addition to the
above symptoms, there is chilliness, followed by fever, rapid pulse,
and headache.

Pulsatilla. — When the disorder is occasioned by eating fat
meat, or greasy substances, with offensive eructations.

Ipecacuanha and *Mercurius* are valuable remedies, and will be
found sufficient in the majority of cases. They may be given in
alternation.

ADMINISTRATION OF REMEDIES. — Give, of the selected remedy,
one dose of six globules every hour or every two hours. Should
no improvement take place in the course of ten or twelve hours,
choose a second remedy and use in the same way.

DIET AND REGIMEN. — All meats and soups are strictly forbidden; nothing should be taken but gruel, — and oat-meal makes
the best, — dry toast, or milk toast, crackers, plain bread, with
but little butter, Graham bread, oranges, and cold water. Some
persons are in the habit of taking a few drops of lemon-juice;

this is not objectionable, unless there should be diarrhœa, and then the coarse bread should also be abandoned. — See "DYSPEPSIA."

OFFENSIVE BREATH.

CAUSES. — TREATMENT. — This unpleasant affection arises from one of several causes; for instance it may arise from decayed teeth; from inflammation or other disorder of the gums; from ulcers in the mouth, or from want of careful attention to cleanliness, allowing particles of food to collect and remain between and around the roots of the teeth. Where bad breath arises from decayed teeth, or from the accumulation of tartar about the roots of the teeth, consult a good dentist immediately; or, if this is not convenient, clean out the hollow teeth yourself, either with cotton or rolls of paper, and fill the cavities with white wax, or, what I think is better, gutta percha; this you can easily procure, and, after making it soft by immersion in hot water, plug the cavities. Before filling, either with wax or gutta percha, the cavity should be thoroughly dried with bits of cotton, or rolls of paper.

The mouth and throat should be rinsed with cold water, and the teeth thoroughly brushed with a soft brush, after *every* meal.

When offensive breath arises from a deranged stomach, or from other diseases, the proper treatment will be found under the head of such disorders.

In other cases, where it is the chief symptom, and its origin can be traced to no apparent or perceptible cause, the following remedies may be employed.

If it appears only in the morning, *Nux-vomica*, *Belladonna*, or *Sulphur*.

In the morning and at night, *Pulsatilla*.

If after a meal, *Sulphur* or *Chamomilla*.

If in young girls, at the age of puberty, *Aurum*, *Pulsatilla*, *Belladonna*, *Sepia*, *Sulphur*.

If caused by worms, *Cina* or *Sulphur*.

If caused by previous salivation with calomel, *Carbo-v.*, *Hepar-sulph.*, *Nitric-acid*.

ADMINISTRATION OF REMEDIES. — Of whichever remedy chosen, one dose may be given every night and morning, either dry, upon the tongue, or dissolved in water. For a dose, when given dry, six or eight pills. When dissolved, put about the same number in twelve spoonfuls of water, and give one spoonful of the solution at a dose.

Bad breath, caused by eating onions or garlic, may be removed by taking a little wine, or drinking a glass of milk, or eating a pear, or a piece of boiled beet. — See, also, "Dyspepsia."

COLIC.

DEFINITION. — All severe pains in the abdomen not dependent upon inflammation are called colic, but from the different causes, and attendant circumstances of this disorder, it is variously denominated, and divided into several varieties. When its principal symptoms are sharp and griping pains, it is called *spasmodic colic* When the pain is accompanied with nausea and vomiting, it is called *bilious colic*. When caused by wind, or when the abdomen is much distended, and relief is afforded by the passage of wind, then it is called *wind colic*. We will consider these three varieties, commencing with

SPASMODIC COLIC.

SYMPTOMS. — Sharp, spasmodic, cutting, griping, crampy pains. Sometimes the pain commences gradually, and continues to increase until it becomes so violent that it seems one cannot bear it; then it will gradually die away, to return again in five or ten minutes, with renewed violence. The patient writhes and twists himself around; is covered with perspiration; presses his abdomen with his hands; curls himself up, or lies across the edge of the bedstead, or presses against any hard substance; lies upon his face with a tightly folded pillow under him. The pain is principally confined to the region of the navel.

For colic, presenting these symptoms, give *Colocynth;* one dose of five or six pills every few minutes, until relief is obtained. If this does not answer, try *Chamomilla, Belladonna,* or *Nux-vomica.*

BILIOUS COLIC.

SYMPTOMS. — In addition to severe cutting, writhing pain of spasmodic colic, we have nausea and vomiting, with thirst and great anxiety. These symptoms usually come on after the patient has been indisposed for a few days, with a "bilious attack;"

tongue coated ; bad taste in the mouth, and other symptoms of a disordered stomach. If diarrhœa supervenes, the evacuations consist of bilious matter.

The remedies for this form of colic are *Nux-vomica, Colocynth, Mercurius, Pulsatilla, Chamomilla,* and *Plumbum.*

The treatment may commence with *Nux-vomica;* a dose of which may be given every five or ten minutes. If no relief is afforded after five or six doses, another remedy should be chosen, and given in the same manner.

WIND COLIC, OR COLIC OF INFANTS.

DEFINITION. — CAUSES. — This a very frequent and very troublesome disorder of young children. It arises from various causes, but most frequently from cold ; or it may, and frequently does, arise from some sudden or violent emotion of the mother ; such as a fit of anger, grief, or chagrin, improper food, or a confined state of the bowels. Some parents, and not unfrequently, experienced nurses, appear to think that a new-born infant should take a dose of something. Each one has her little innocent, domestic remedy, which is soon concocted, and the child forced to swallow it down. Now this is frequently the cause of stomach and intestinal derangement.

SYMPTOMS. — Flatulent complaints of children do not always terminate in colic. The only noticeable symptom may be a disturbed sleep; the child rolls its eyes, distorts its features, kicks out its feet, draws up its knees, moans; its sleep is broken and uneasy. It may amount to nothing more than this, or it may increase in severity; the child commences to cry out; writhes its body, draws up its knees, kicks out its feet; the abdomen becomes tense and swollen, with rumbling in the bowels. These attacks of pain sometimes become so severe that the poor child seems to be in the greatest anguish. It writhes and screams; nothing can appease it; it will not take the breast; or, if at times it does, it soon lets go, with another fit of screaming; the face turns pale ; the little sufferer trembles all over; a cold sweat breaks out, and the child seems entirely exhausted. Sometimes the belching up, or passing of a little wind affords momentary relief. The child appears easier when carried about in a sitting posture. Severe attacks of this kind, unless speedily relieved, may end in spasms or convulsions.

TREATMENT. — *Chamomilla* may be given in most cases, especially if there be distention of the abdomen; excessive crying; writhing and twisting of the body; drawing up of the knees; and coldness of the extremities. If, in addition to these symptoms, there is nausea, vomiting, and diarrhœa, give *Pulsatilla*, and particularly if there is rumbling of wind, shivering, paleness, and tenderness of the abdomen. If the passages are fermented, and have a putrid odor, give *Ipecacuanha*. In case the bowels are constipated, *Nux-vomica* should have precedence.

Grief and sadness on the part of the mother frequently injure the child's digestion. *Ignatia* will be found the most effectual remedy when colic arises from such causes. Should it, however, fail to afford relief give *Chamomilla*. In cases where the pain apparently has all subsided, but still the child cannot go to sleep, is restless and uneasy, give *Coffea*.

For colic, caused by worms, give *Cina*, *Sulphur*, or *Mercurius*. — See " WORMS."

For colic in pregnant women, *Chamomilla*, *Nux-vomica*, *Pulsatilla*.

Menstrual colic, — *Pulsatilla*, *Coffea*, *Belladonna*, *Cocculus*.

ADMINISTRATION OF REMEDIES. — For young infants, put two or three of the globules dry upon the tongue, or dissolve the same in a few teaspoonfuls of water, and give it to them often, part of a teaspoonful at a dose. For adults and the older children, you may give of the selected remedy, five or six pills every fifteen minutes, half hour, hour, or two hours, according to the severity of the case.

Most cases of colic are attended with constipation : a free evacuation of the bowels often affords instant relief. In severe cases it is desirable, therefore, that we should get a movement from the bowels as soon as possible, and the most simple and efficient way to accomplish this is by an injection of tepid water, or, perhaps, it may be advisable to add a little salt to the water. If the first injection should not produce a movement, a second one should be tried. Whenever you do give an injection, give one large enough to have some effect. Hot applications to the abdomen are also serviceable, and should always be made use of.

CHOLERA MORBUS.

DEFINITION. — This disease is characterized by great anxiety; painful and violent gripings; with copious and frequent vomiting and purging; coldness and cramps of the extremities. The griping pain evidently proceeds from violent spasmodic contractions of the alimentary canal, causing the repeated and frequent ejection of their contents by vomiting and purging.

CAUSES. — Intense heats of summer, especially when the days are hot and the evenings cool, with heavy dews; sudden atmospheric changes; cold drinks, when the body is overheated; and the incautious use of ice. Sudden suppression of habitual discharges; diarrhœa; cutaneous eruptions; vexation; fits of anger; errors in diet; partaking of unhealthy food, or in an improper quantity or quality; unripe or indigestible fruits, particularly melons, cucumbers, pine-apples, green apples; poisonous or irritating food of any kind. Large doses of cathartic drugs, not unfrequently produce it by their irritating qualities.

SYMPTOMS. — Cholera morbus generally attacks suddenly, without any premonitory signs, with vomiting and purging, accompanied by severe griping pains in the stomach and bowels; great anxiety; the patient tossing from one side of the bed to the other in quest of rest. The discharges from the bowels consist, first, of fæces; afterwards of watery, bilious matter; each evacuation is preceded and accompanied with violent burning and cutting colicky pains, especially in the region of the navel. In severe cases, the spasms extend to the arms and hands, with pinched features; paleness of the surface; sunken eyes; cold, clammy skin; great anxiety, and general depression.

The substance vomited consists at first of the contents of the stomach mixed with bilious matter; afterward of a watery liquid; and, finally, nothing is thrown up, but still the gagging and retching continue.

The severe symptoms often continue for some hours without the patient's strength being reduced, so that it is no uncommon thing for a patient to suffer a severe attack of this disease at night, and by morning have entirely recovered.

TREATMENT. —The principal remedies are, *Ipecacuanha, Chamomilla, Colocynth, Arsenicum, Veratrum, Cuprum, Chinchona.*

Chamomilla. — When the attack has been induced by a fit of anger; also, when there are severe pains, or colicky pains in the region of the navel; greenish evacuations; tongue coated yellow; sour vomiting, and watery diarrhœa; cramps in the calves of the legs.

Ipecacuanha. — Especially when vomiting predominates. It

may be given at the commencement of an attack, in alternation with *Veratrum*. When there is severe pain in the abdomen; frequent and small evacuations, with severe pressing-down pain, *Nux-vomica* may be given in alternation with *Ipecacuanha*.

Colocynth. — For extreme pain in the abdomen, as though the bowels were jammed between two stones; green vomiting, with violent colic; crampy pains and constrictions in the bowels, or cutting pains, as from a knife.

Arsenicum. — For cases attended with rapid prostration; insatiable thirst; great restlessness; violent vomiting and diarrhœa of greenish or blackish matter; tongue and lips dry, cracked and bluish; severe cramps in the fingers and toes; clammy perspiration; burning sensation at the pit of the stomach.

Veratrum. — Violent vomiting, with severe diarrhœa; excessive weakness; cramps in the calves of the legs; coldness of the extremities; countenance pale; eyes hollow or sunken; shrivelled appearance of the skin; violent pain in the region of the navel. This remedy will cure almost every case of cholera morbus without other assistance.

Cuprum. — When there are severe spasms of the limbs; violent cramps in the fingers and toes.

Chinchona. — Especially for the debility remaining after severe cases; also, when there is vomiting of undigested food, and when the evacuations contain undigested matter. — See, also, " COLIC."

ADMINISTRATION OF REMEDIES. — Immediately upon an attack, a remedy should be selected, and promptly given a dose of six pills, and repeat it every few minutes, until relief is obtained. If, after a reasonable length of time, no relief is afforded, select and administer another remedy. As the patient grows easier, the interval between the doses should be lengthened.

DIET AND REGIMEN. — For some little time after an attack of this disorder, the patient should be very careful of his diet, avoiding all vegetables, and all other articles of diet that he has the least reason to suspect will disagree with his stomach. — See " COLIC."

CHOLERA INFANTUM.

DEFINITION. — The common name of this affection is summer complaint. A correct definition of it can scarcely be given, except by a complete enumeration of its symptoms and characteristic features. The chief seat of lesion appears to be in the secretory,

— that is, the glandular apparatus of the stomach and intestinal canal,— but is chiefly confined to the large and lower part of the small intestines. It seldom, except in severe cases, extends to the stomach. By many physicians this disease is looked upon as an inflammation of the glandular structure of the intestinal tract. But I can hardly consider it as such, except in its last stages, or in very severe cases. Ordinary cases of cholera infantum, presenting little or no febrile movement, seem to me to consist of simple catarrh, or simple irritation, arising either from the evolution of the teeth, or,— what is more probable — the effect of an improper diet. However, in severe acute cases, and those terminating fatally after lingering a long time, we generally find traces of inflammation and its results, ulceration, softening, and thickening.

CAUSES. — This disease is seldom met with, except during the three summer months,— June, July, and August.

All writers agree in considering the heat of summer to be a powerful predisposing cause. It is true we but seldom meet with well-marked cases of cholera infantum in the cool seasons of the year. Nevertheless, heat alone cannot be charged with all the mischief; for, at the south, where the seasons are much hotter, and of longer duration, the disease is less frequent than with us. Besides, we seldom hear of it in the country, even within a few miles of the city, where it prevails extensively. It would, therefore, seem that in order to have heat produce the disease to any great extent, it must be combined with close, unwholesome air. Here we have just such an atmosphere, loaded with impurities, exhaled from the gutters, lanes, and dirty streets of our cities.

Teething may also be set down as a prolific cause of this, as well as of all other diseases of the digestive apparatus. During the whole period of teething, the child's system is in a peculiarly excitable condition; so that apparently slight causes, which at other times might produce no perceptible effect,— as slight errors in diet, exposure to night-air, fatigue, vexation, etc., may now usher in alarming symptoms of bowel disorders. But the chief causes are, in my estimation, errors in diet. As I have already asserted at more than one place in these pages, the deprivation of the breast, and the early resort to an artificial diet, is the chief predisposing cause of *all* gastric and intestinal diseases. Indeed, we can often date the commencement of an attack to the period of weaning, and the resort to artificial food. The fact of a child's

34

being weaned at a very early age, or fed upon artificial diet from birth, is a most unfortunate circumstance in every case of cholera infantum. On the contrary, where the child has not been weaned, and can depend upon the breast for a sufficient supply of nutriment, the disease seldom reaches an alarming state of severity; and when convalescence does begin, the case progresses much more rapidly toward a happy termination.

Wonderful it is how perfectly reckless some parents are in regard to the diet of their children, letting them eat anything and everything. There is not a particle of doubt but this is a frequent source of the disease in question. One of the worst cases I think I ever saw was a teething child, eighteen months old; and was occasioned by eating *new potatoes* for dinner, and a quantity of apples and of watermelon through the afternoon. How in the world any but a crazy woman could ever allow a child to eat such articles is entirely beyond my comprehension. I have frequently known the disease to arise from eating the smallest quantity of unripe fruit. Over-feeding may also be enumerated as another cause of the disease; also the preparation of the food in too thick and too rich a manner; for, by this means, the stomach is overtasked, and thereby disarranged. The child's stomach is incapable, especially during the early months, of digesting any but the simplest kinds of food, and that should be prepared very thin, and not too rich.

In fact, the diet of small children, till after the completion of first dentition, should resemble as much as it is possible for us to make it, the proper aliment supplied by nature. Its chief constituent should be milk, to which can be added fine rice, or barley-flour.

It is astonishing, what slight causes will excite the disease in some children, or rather in some families of children, compelling one almost to believe in the existence of a hereditary predisposition to it. Certainly the disease is more prone to attack children whose constitutions are feeble and delicate, or of a nervous, irritable tendency; children of scrofulous or consumptive parents. — See "GENERAL REMARKS," at the head of this chapter.

SYMPTOMS. — This disease is extremely variable in its mode of invasion. It may be sudden or gradual. A child, who to all appearance is enjoying good health, may be suddenly attacked with severe diarrhœa, accompanied with, or soon followed by vomiting, great exhaustion, anxious and contracted countenance, coldness and paleness of the skin, — symptoms indicative of real cholera in adults. As a general thing, however, its mode of attack is gradual, commencing with a diarrhœa of no great importance, but which proves obstinate, and after running a few days, more or less, becomes associated with nausea and vomiting. Attacks like these are usually preceded a few days by feverish restlessness; the child is unusually fretful and indifferent about eating or nursing. The mother lays it to the teeth, or calls the child cross, when in reality it is sick.

The most important symptom is the diarrhœa. The dejections becoming more frequent and abundant than is natural, they begin to be spotted, and streaked with green. As the disease increases, the green color predominates, and the passages look like chopped up greens, or spinach, and are mixed with particles of undigested food. At first they are thick, mushy-like, sometimes mixed with water of a yellow or greenish cast; they are often fluid, running directly through the diaper; occasionally they contain some blood and mucus. Their odor is bad, decidedly so, very fetid; sometimes it is so offensive that you will be compelled to raise the windows, and open the doors, and even then you can scarcely get rid of it. The passages are not as frequent as in dysenteria; sometimes they will not number over six or eight in twenty-four hours; in severe cases they may run up as high as twelve or even eighteen.

During each evacuation, there is more or less pain, the child will fret, toss about, and be restless and uneasy; some time before an evacuation, and when the passages have a dysenteric appearance, that is, — contain blood and mucus, — there is generally severe straining and pressing-down pain when at stool.

The vomiting, which is generally present at the commencement of an attack, is sometimes very frequent and distressing; so much so that everything, no matter how slight it may be, is thrown off with great violence as soon as taken into the stomach. The substances vomited consist of the contents of the stomach, — undigested food, mixed with phlegm, mucus, and bilious matters. At other times, there are frequent retchings and efforts to vomit, but nothing is thrown off. The frequency and severity of the vomiting, like all the other symptoms, depend upon the violence of the attack; sometimes, as just stated, it is very severe; again it may occur but two or three times throughout the day, or only when food is taken, and it may be absent entirely. It seldom continues throughout the sickness, but, as a general thing, subsides after the first few days.

At first, the tongue is coated with a dirty-white or yellowish-brown fur, with the exception of the tip and edges, which are generally red. In cases of long standing, the tongue acquires a dry, smooth appearance; the mouth is hot and dry; thirst is intense; the appetite is variable; sometimes the child will eat voraciously everything you will give it; again it is diminished, and in some cases entirely wanting.

There is always more or less fever; in slight cases it is hardly perceptible, except at night, when there may be quite a flush. In cases of great severity, the fever is high, usually of a remittent type. In all cases, the fever is higher during the afternoon and evening.

The abdomen is seldom tender; but you will always find it distended and tense.

The temperature of the surface is not generally even throughout; the head and abdomen being hot, while the extremities are cold.

Emaciation is rapid, so that, in a short time, if the disease is not checked, the countenance and whole appearance are so changed, you would scarcely recognize it as the same child, who, but a few days before, with its bright face and merry laugh, made the whole house happy. In cases that have been neglected or improperly treated, this emaciation is one of the most marked symptoms. The skin becomes dry and harsh; it has a withered appearance, and hangs in folds about the face, neck, arms, and thighs. The child has an old look: the eyes are hollow, the nose pointed, and the chin prominent. In fact, it looks as though some one had made a mistake, and the child had got into the skin of old age.

The sleep is disturbed and unrefreshing; the eyes are never entirely closed, but partially open, leaving the white of the eye exposed. The child always wakes up crying.

DURATION. — The duration of cholera infantum depends, in a great measure, upon the treatment pursued. Old-school physicians have but little control over it; and, when treated by them, it runs an indefinite course of from six weeks to six months. The severity of their treatment soon undermines the delicate constitution, making it, in the last stages, a difficult matter to distinguish from which the child is suffering most, —the original disease, or the medicine given for its removal. As a last resort, the physician advises a prompt removal to the country, wisely concluding that a change of air will do more towards the restoration of the little sufferer than a continuance of his treatment.

The homœopathic treatment of cholera infantum, when promptly applied at the commencement of an attack, is productive of the most satisfactory results, cutting short the disease before it has had time to reduce the child to any great extent.

TREATMENT. — Dentition being, as we have already observed, a powerful predisposing cause of the disease, we should never neglect, at the commencement of our treatment, closely to examine the state of the gums ; and if they are found swollen, hard, hot, and shiny, they should be freely lanced. This can be done just as well with a penknife as with a surgeon's lancet. Place the point of the blade directly over the tooth, and make a free incision down to it. This operation should never be performed unless you can see or feel the tooth through the gum.

This simple operation is often of great service : it sometimes affords immediate relief, allays all irritability, and renders the disease more tractable to the remedies administered, by removing one of its prominent and important causes.

The principal remedies are, first, for recent or acute attacks, *Ipecacuanha, Veratrum, Arsenicum, Mercurius, Chamomilla, Bryonia,* and *Chinchona ;* second, for chronic or cases of long standing, *Calcarea, Sulphur, Arsenicum, Mercurius,* and *Carbo-veg.*

Ipecacuanha — will be the first remedy called into use in nine cases out of every ten, — that is, recent cases, — and generally will arrest the disease at once. It is specially indicated when the following symptoms are present: nausea and vomiting of food or drink, or of mucous or bilious looking matter, attended with watery diarrhœa, green or fermented stool, with white flocks ; coated tongue ; great thirst, and loss of appetite. The next important remedy is *Veratrum*; it may be given for cases that have lasted some time, or when the attack has been violent, with great exhaustion from vomiting, especially when the vomiting comes on in paroxysms, while drinking, or when the slightest movement produces retching ; loose, brownish, or watery evacuations, coldness of the extremities ; pale face, with sunken eyes ; great thirst for cold water.

Chamomilla. — Mucous or sour vomiting; diarrhœa ; evacuations, looking like stirred eggs ; or green and slimy, with colicky pains in the bowels. This remedy is specially adapted for teething children, when they are very cross, fretful, and uneasy. Fever, with nightly exacerbations.

Magnesia-c. — Diarrhœa, with stools like scum of a frog-pond, green and frothy. Chronic sour diarrhœa.

Podophyllum. — Diarrhœa, with cramp-like pains in the abdomen, light-colored stools, exceedingly offensive, frothy mucous, and slimy stools. The child moans while asleep, sleeps with its eyes half closed, and rolls its head from side to side.

Mercurius. — For cholera infantum, attended with colic, and straining when at stool; evacuations scanty, greenish, and sour; frequently the evacuations are mixed with blood and slime. The child smells sour. Diarrhœa is worse at night.

Chinchona. — For diarrhœa occurring immediately after eating, with fetid stool, containing undigested portions of food; loss of appetite.

If the stools are very thin, have a putrid smell, and are attended with burning pain, give *Bryonia* or *Carbo-v.*

Arsenicum. — For extreme cases, where there is great prostration, nausea, and vomiting, after partaking of the least food or drink. The child nurses with avidity, probably from intense thirst. Evacuations green, brown, or yellowish, very offensive, putrid, and undigested. If the lips and tongue become dry, cracked, and black ; skin dry, like parchment, or cold and clammy ; abdomen hard and distended; sleep disturbed by moaning and grating of the teeth.

Calcarea. — is especially indicated at a late period of the disease, or in cases of long standing, where there are swelling and hardness of the abdomen ; great emaciation and debility ; diarrhœa of mushy, clay-colored stools ; the skin is dry and withered ; hair looks dry and dead ; nervous system becomes very sensitive ; child is cross, easily vexed. *Calcarea* generally acts better *after Sulphur.*

Sulphur. — is a valuable remedy for protracted cases, especially when the evacuations are greenish, watery, and frequent; distension of the abdomen ; countenance pinched ; skin shrivelled ; great emaciation.

Should head symptoms manifest themselves, give *Aconitum, Bryonia,* or *Helleborus.*

ADMINISTRATION OF REMEDIES. — In sudden attacks of small children, you may dissolve twelve globules in twelve teaspoonfuls of water, and give of this solution one teaspoonful at a dose.

The dose may be repeated as often as every fifteen minutes, or every half hour, until the severer symptoms have subsided, when the intervals ought to be lengthened. In chronic cases, the dose should not be repeated oftener than once in two hours.

DIET AND REGIMEN. — Upon this point, it is necessary to be explicit; for a successful treatment of cholera infantum depends, in a great measure, upon a proper regulation of the diet.

If the disease appears in an infant at the breast, and the nurse has enough for it, no change need be made; and nothing else should be given, except it be cold water, of which the child can have as much as it wants to slake thirst.

If possible, a child should not be weaned until after the second summer. If, however, you are forced to adopt an artificial diet, either partially or wholly, you will have to be governed, in the selection of food, in a great measure at least, by circumstances or previous habits. No specific articles can be named which will agree with the peculiarities of every child's stomach.

In all cases, it is equally important to regulate the quantity as it is the quality. Overloading the stomach with good food would prove just as injurious as small quantities of bad. I have often found it necessary to restrict the quantity of food given to the smallest amount, especially in those cases where everything taken into the stomach excites nausea and vomiting. There is no other way to succeed in keeping anything upon the stomach but to give it often; and even then it will not always stay.

I am fully persuaded that fresh cow's milk should form the principal ingredient of a child's diet. It should be diluted with about one-third water, boiled for ten or fifteen minutes, and moderately sweetened with loaf-sugar. For a change, to this may be added rice-flour, arrow-root, sago, tapioca, or wheat-flour.

Rice-flour gruel makes a very good diet for children with bowel-complaints. It should be prepared as follows: take one table-spoonful of flour and one table-spoonful of milk; stir them together; then add a little salt and nearly a pint of warm water; stir well, and boil for fifteen minutes; when cold, this is about the thickness of starch. Add a little finely powdered white sugar when feeding it.

Though a child in good health may be able to digest milk, either pure or diluted with a third part water, it may fail to do so when its stomach is weakened by disease. In such cases, a further reduction should be made, or, what is sometimes better, you may take one part *cream* to five of water, and to this add a little arrow-root, rice-flour, or almost any other farinaceous article. A still better article is that recommended by Dr. Meigs, which I have quoted on page 142.

In some cases of cholera infantum, when there is excessive vomiting, you should give nothing but a little gum-water, or rice or arrow-root water, one teaspoonful at a time, until the vomiting ceases.

Fresh air is just as important as good diet. This is so well understood now, that those who can, without waiting for the disease to attack their children, remove to the country before the hot months, and thus avoid it altogether. But then all cannot do this. It is very important that the child should spend a large part of its time in the open air; and this can be done even by those whose circumstances will not allow of their going into the country. The child can be carried about the yard, or in the street; or, what is still better, you can make short trips with it into the adjoining country, or out upon the water. New Yorkers need never want for places to make such excursions to. Our beautiful bay and rivers are threaded with steamers, that, in half an hour, will land them upon shady spots, as free from city dust and air as though they were a thousand miles away.

If the child is too sick to be taken out, it must be carried through the house upon a pillow, when the doors and windows are open and a free ventilation established.

Cool, fresh air and bathing are important, and cannot be too highly recommended. The bath — not cold, but tepid — should be used frequently. Sponging may be preferred to bathing in severe cases, where there is great exhaustion, or where the bath annoys or worries the child.

The dress should be adapted to suit the weather, and changed to suit the changes of temperature, care being taken not to clothe the child too warmly.

DYSPEPSIA, OR INDIGESTION.

DEFINITION. — To understand aright the pathology of dyspepsia, it is first necessary to know the physiology of digestion; or, in other words, rightly to understand what constitutes digestive disease, it is first necessary to know what constitutes digestive health.

The theory of digestion is simple and easily understood. The first preparation of food for its introduction into the system, consists in its proper mastication, or its reduction into fine particles by the act of chewing; and while undergoing this process, the food is moistened, or rather mixed with a considerable quantity of saliva from the salivary glands, which are situated within the mouth. This facilitates its easy passage into the stomach, where it immediately comes in contact with the gastric juice.

The food, on reaching the stomach, is subjected to a double process; first, the solvent power of the gastric fluid; second, the churning process of the stomach. The churning is produced by the presence of food, which excites a contractile action of the muscular coat of the stomach, and by this means the position of the contents of the stomach is changed from one part of this cavity to another. Each particle of food, thus brought in direct contact with the mucous coat of the stomach, becomes saturated with gastric juice, the action of which, together with the constant agitation that it is subjected to, reduces the whole to a homogeneous mass, of a creamy consistence, called *chyme*. As fast as the food becomes converted into chyme, it is passed into the duodenum, which is the upper portion of the small intestines; its presence here creates an action not only in the duodenum, but also in the liver and pancreas. The liver secretes the bile, which first comes in contact with the food we swallow here within the duodenum, and not within the stomach, as many suppose. The pancreas, which is a small gland, situated just behind the stomach, secretes the pancreatic fluid, which enters the duodenum with the bile; the mucous surface of the duodenum also throws out a secretion. And thus the chyme from the stomach being mixed with, and acted upon, by these three fluids, is converted or transformed into a milk-like liquid, called *chyle*. This is the substance from which blood is made. The chyle thus formed, now passes along the tract of the intestines where it comes in contact with the lacteal vessels. These are little delicate tubes, spread over the mucous surface of the small intestines, and whose office it is, to imbibe or take up the chyle and transfer it through the mesenteric glands into the thoracic duct, through which it is conveyed into a large vein at the lower part of the neck. In this vein the chyle is mixed with the venous blood, and soon reaches the heart, to be distributed throughout the system. The veins of the stomach and intestines also act as absorbents.

35

The residuum, or excrementitious matter left in the intestines after the lacteals have absorbed the chyle, is conveyed into the large intestines, through which it is passed along and excreted from the system as effete matter.

In the process of digestion you will observe the food is subjected to five different changes. 1st. The chewing and admixture of the saliva with the food. This process is called *mastication*.

2d. The change through which the food passes in the stomach by its muscular contraction, and the secretion from the gastric glands; this is called *chymfiication*.

3d. The conversion of the homogeneous chyme by the agency of the bile and the pancreatic secretion into a fluid of milk-like appearance; this is *chylification*.

4th. The absorption of the chyle by the lacteals, and its transfer, through them and the thoracic duct, into the subclavian vein at the lower part of the neck.

5th. The separation and excretion of the residuum.

Perfection of the second process of digestion requires *thorough* and *slow mastication*. The formation of proper chyle demands appropriate mastication and chymification; while a healthy action of the lacteals requires that all the anterior stages of the digestive process be as perfect as possible.

Having fully comprehended the physiology of digestion, we have now arrived at a stand-point from which we can take an intelligent view of dyspepsia.

The term dyspepsia, in its literal sense, means difficult digestion; or, if you like it better, indigestion. Any condition or state of the stomach, in which its function of digestion is disturbed or suspended, giving rise to a train of multifarious symptoms, such as want of appetite, sudden and transient distension of the stomach, eructations of various kinds, heart-burn, water-brash, pain in the region of the stomach, uneasiness after eating, rumbling noise in the bowels, sometimes vomiting, and frequently constipation or diarrhœa, with an endless string of nervous symptoms, usually passes for dyspepsia.

Chronic inflammation of the stomach is quite a common disorder; it deranges the function, and perverts the feelings of the stomach to such an extent that digestion is suspended, or but imperfectly performed, and the patient rendered miserable from the dyspeptic symptoms which are sure to follow. Perhaps this state of things might not properly be called dyspepsia; but, nevertheless, the larger part of all the causes of indigestion originate from some similar condition.

Dyspepsia may consist in a derangement of all or any one of the five several different processes whose combined *healthful* action goes to make up the phenomenon of digestion. For instance, imperfect mastication, or the introduction of coarse and indigestible substances into the stomach ; the absence of the natural quantity or quality of the gastric juice ; an imperfect churning or agitation of the contents of the stomach from · muscular debility, caused by too frequent or over-taxation.

Indigestion may be, and no doubt frequently is, simply debility, or a defect of muscular power in the stomach ; a want of vital power and strength.

Some persons are fond of attributing dyspepsia to some derange-' ment of the liver, thinking perhaps that a perverted secretion from this organ is thrown into the stomach. This is incorrect ; bile is the natural stimulus of the intestines ; and when this secretion is obstructed or perverted, the bowels may become sluggish, and the general health disturbed, but still stomach digestion remain good.

CAUSES. — " Indigestion, although not confined to any period of life, occurs most commonly between the ages of twenty and forty-five, and in *its simple form* more frequently in the female than in the male sex. The upper classes of society and the middle ranks of life are most subject to this variety of the complaint. It is more prevalent in cold and temperate than in warm climates, and in the winter than in the summer ; but whatever may be the temperature of the climate or of the season, damp weather and a moist atmosphere may be regarded as among its most active *predisposing* causes. The predisposition to this disorder is sometimes hereditary, particularly in persons of a weak, relaxed fibre, with high nervous susceptibility, and general debility of constitution. Those in whom the functions of the stomach are naturally weak, and feebly performed, the circulation languid, the temperature of the extremities, below the natural standard, and the secretion generally disordered, or more abundant than usual, are also constitutionally predisposed to dyspepsia. Sedentary occupations, especially when carried on in close rooms and factories ; indolent habits either of body or mind ; long and intense study ; insufficient exercise in the open air ; addiction to debilitating excesses and injurious indulgences, luxurious modes of living, indulgence in sleep

or in bed, breathing impure air, and confinement to close or ill-ventilated apartments, remarkably predispose to this complaint. In
persons thus predisposed, the slightest excess or irregularity, or the
most trivial exciting cause, is often sufficient to bring on an attack
of indigestion ; while a repetition of such causes, or long exposure
to their action, in those of a stronger habit and more vigorous constitution, cannot fail to have a similar effect." — COPELAND's *Medical Dictionary*.

Whatever tends to impair the condition of the digestive function is an active cause in producing dyspepsia.

"Indigestion," says an eminent writer, "is the prevailing malady of civilized life.

"The principal exciting cause of indigestion is imperfect mastication.   The fact is, we as a nation have not time to eat; business or pleasure is too pressing."

Tom Moore, the poet, truly says, —

> "No digest of laws like the law of digestion; "

but of this we seem to be oblivious.   Seven eighths of the persons
one meets in society, display an ostrich-like willingness to devour
the most indigestible substances, and to imbibe fluids which tend
first to excite and afterwards to paralyze the stomach.   That dyspepsia has become a "National Disease" with us, is not to be
wondered at, when we come to consider our "bill of fare," and
how we partake of it.   We are the swiftest eaters in the civilized
world, and the most dyspeptic of all people.   From childhood to
old age we are in the habit of "bolting" our food, as if our teeth
were in our stomach and we could masticate it at our leisure, like
a ruminating animal.   The stomach, especially of children, is
unable to digest solid lumps, or tough masses of food, and whatever passes it undissolved receives but little digestive aid from the
duodenum.   Of course the lacteals — whose duty you will remember it is to absorb the milk-like liquid, chyle, as it comes from the
duodenum — refuses to pick up crude chunks of bread, meat, or
potatoes.   And thus the food partaken of passes through the
whole digestive apparatus without even having undergone the first
natural change.   And instead of nourishing the system, it but
taxes the vital energies, and finally either becomes impacted in
the large intestines, producing constipation, or acts as a constant
source of irritation, producing, and, if continued, maintaining an
exhausting diarrhœa.   All this, you will observe, arises from im

perfect mastication. Again, a weak, dyspeptic stomach acts slowly, or not at all, upon solid lumps and tough masses of food. The delayed morsels undergo spontaneous changes, promoted by the warmth and moisture of the stomach; gases are extricated; acids are formed; perhaps the half-digested mass is at length expelled by vomiting, or it passes undissolved into the duodenum, and becomes a source of irritation and disturbance during the whole of its journey through the intestines. — WATSON.

Festina lente was inscribed on the walls of the old Roman banquet-halls. It signifies " Hasten slowly," and should be printed in letters of gold upon the border of every dinner-plate in the United States. In order that digestion may be perfect, it is requisite that the food be in a state of minute division, so that when it enters the stomach it may be readily incorporated with the gastric juice and reduced to chyme.

We should eat slowly, not only that we may reduce the food to a state of comparative fineness, but that the salivary glands in the mouth may have an opportunity to pour out their secretion. Now, no doubt but one great source of indigestion arises from the fact, that we eat so rapidly that the food passes through our mouth and is swallowed before the salivary glands are even excited to action, and so the food passes into the stomach scarcely moistened. This tends to induce disease, not only in the salivary organs, by leaving them in a state of comparative inactivity, but in the stomach, by the deficiency of salivary stimulus.

" Persons who eat rapidly generally drink large quantities of water, tea, or coffee; this retards digestion by partially suspending salivary secretion, also by diluting, and thus lessening the energy of the gastric juice. Besides, all liquids taken into the stomach while eating must be removed by absorption *before* digestion proper can commence." — CUTTER.

As we have before remarked, chronic inflammation of the stomach is not unfrequently the cause of indigestion. This chronic inflammation is, in the majority of cases, caused by indigestible substances or irritant condiments which we have swallowed. There are certain things upon which the gastric juice has no power: the green coloring matter of certain vegetables, the cores of apples, the skins of many fruits. No doubt you have observed that dry currants, the husks of apple-seeds, and raisin-skins, swallowed entire, reappear unchanged among the egesta. These things while in the stomach, subjected to the agitation which everything in transit there receives, must necessarily, from their harsh nature, irritate the coats of the stomach, and excite at least a low degree of inflammation. Food prepared for the dainty palates of the people of this generation is so saturated with condiments of various kinds, including black pepper, red pepper, allspice, cloves, mustard, horseradish, ginger, nutmeg, cinnamon, and other irritating substances, that the wonder is, not that we suffer from dyspepsia, but that our stomachs are not entirely destroyed. If any doubts that to put such irritating substances into the delicate organization of the stomach is wrong, just let him put a little pepper or mustard into his eyes or nose, or retain them a little while in contact with the mucous membrane of the mouth. Why, apply mustard, pepper, horse-radish, or almost any of these articles to the external gross organization of the skin, and you will soon excite an inflammation and even ulceration. Is it reasonable to suppose, that such substances can be put into the stomach with impunity, especially into the stomachs of children?

Another fruitful cause of indigestion is the habit we have of going immediately from severe mental or bodily exercise to our meals. All organs, while in action, require and receive more blood and nervous fluid than when at rest. The increased amount of blood and nervous power supplied to any organ, during extra functional action, is abstracted from other parts of the system. Now, of course, those parts of the system which supply this demand do so at their own expense, and thereby, to a certain extent, become enfeebled and prostrated. Suppose a child comes in after severe bodily exercise. Perhaps she has been jumping the rope, rolling the hoop, or otherwise actively engaged, so that the muscular system has demanded and received an extra amount of blood and nervous force from the balance of the system. The stomach has, of course, supplied its share, and is now in a state of comparative debility, and *consequently* unfit to digest food. She eats a hearty meal. What is the consequence? It is not digested, — that is quite certain, and will not be, either, until the increased action of the arteries and nerves abate, and a due supply of blood and nervous fluid is sent to the stomach, or until an equilibrium of action in the system is re-established. The result generally is, the child suffers from a fit of indigestion. The mass of food may pass the stomach, and become a source of disturbance during the whole of its journey through the intestines; or, as we before remarked when speaking of other causes, the delayed morsels may undergo spontaneous changes, promoted by the warmth and moisture of the stomach : gases are extricated, acids are formed, and the whole digestive apparatus thrown into a state of derangement. It is equally as injurious to digestion that we should enter upon severe mental or physical toil immediately after as immediately before eating. If you wanted your boy to help you grind your knife, you would not expect to find him turning the stone, would you, when you had just set him to cutting wood ? But, still, you will set your brain to work to supply the nervous power to carry on some mental or manual labor, and, at the same time, expect digestion to progress. You might as well expect to grind your knife without the stone revolving. An English gentleman once fed two dogs upon similar articles of food. He permitted one to remain quiet in a dark room ; the other he sent in pursuit of game. At the expiration of one hour, he had them both killed. The stomach of the dog that had remained quiet was nearly empty ; the food had been properly changed, and carried forward into the alimentary canal. In the stomach of the dog that had used his muscles in chasing game, the aliment remained nearly unaltered.

Now, from what has previously been said, you will readily see why one dog digested his food, and the other did not. This applies as well to children and men as to dogs. If our brain or muscles are intensely engaged soon after eating, the stomach will not be sufficiently stimulated by blood and nervous fluid to change the food in a suitable period of time ; and the result will be, a fit of indigestion, just the same as in the case where the patient eats a hearty meal immediately after severe exertion, either mental or physical.

Another cause of indigestion is eating late at night or just before retiring. If the boy goes to sleep and stops turning, of course you cannot grind your knife. So, if the brain goes to sleep and withholds its power, of course digestion stops. That is plain enough, I am sure. It is no unusual thing for those persons who have eaten heartily just before going to bed, to have unpleasant dreams, to be disturbed in their sleep by all sorts of visions. The brain has gone to sleep, or at least has become partially dormant, and does not impart to the digestive organs the requisite amount of nervous influence. The nervous force or stimulus being deficient, or entirely withheld, the food remains in the stomach unchanged, causing irritation of this organ.

Here we have a common cause of dyspepsia and an easy preventive.

Among the numerous causes of indigestion, Dr. Cutter, in his valuable school-book upon " Anatomy, Physiology, and Higiene," remarks: " The condition of the skin exercises an important influence on the digestive apparatus. Let free perspiration be checked, either from uncleanliness or from chills, *and it will diminish the functional action of the stomach and its associated organs.* Restricting the movements of the ribs and diaphragm impairs digestion. . . . . It is noted of individuals, who restrain the free movements of the abdominal muscles by tight dresses, that the tone and vigor of the digestive organs are diminished. The restricted waist will not admit of a full and deep inspiration, and so essential is this to health, that abuse in this respect soon enfeebles and destroys the functions of the system."

There is not the slightest doubt but that many a young lady has died of consumption from tight-lacing. Not that tight dresses directly caused the disease, but they ruined the digestive function by compressing the digestive apparatus into an unnatural and constrained position. The result, indigestion, enfeebled and wasted away the system, the latent seeds of consumption shot forth, and the patient was soon carried away ; whereas, had the functions of digestion and assimilation remained perfect, consumption might never have shown its hideous form. Its seeds would have remained dormant through life. Hereditary diseases are like weeds : when the shell that contains the germ or seed becomes broken, or commences to decay, they sprout forth and grow luxuriantly, overtopping and destroying all other plants whose fruits we wish to cherish.

Any sudden intelligence, a violent fit of passion, or of great joy, sometimes instantly brings on an attack of indigestion. Grief, anxiety, envy, jealousy, indulgence in tender feelings, repeated disappointment, reverses in fortune, night-watching, etc., are active causes in exciting the disease.

Another very frequent cause of dyspepsia, in this country, is the excessive use of cathartic medicine in the shape of pills. No wonder that patent medicine-venders, who advertise specifics, drive a lucrative business in the sale of these pretended remedies. " Were we to give the amount of the latter, — cathartic pills, — annually swallowed in the United States, the statement would not be believed, and yet we have it from good authority, namely, *that of the manufacturer himself*, that one establishment in New York city turns out by the aid of steam no less than ten barrels per

day — 3130 bbls. per year! — and this is by no means so extensive as some others of a similar kind. These pills, which are
highly drastic, are used by immense numbers of people, not only
in cases of actual illness, but in time of health, as prophylactic,
preventive remedies. The consequences are easily predicted. In
addition to this, great quantities of bitters are used, in which
brandy, wine, or some alcoholic liquor forms the principal ingredient; and, on the occurrence of the least feeling of discomfort,
recourse is had to the panacea, till at length the powers of the
stomach are exhausted, and derangements, either functional or
structural, take place. We could wish that the epitaph of the
Italian count could be placed so as to be seen by every man,
woman, and child: ' *I was well, wished to be better, took physic, and
here I am.*'

" Much of this evil is doubtless owing to physicians, who have
been too much in the habit of pouring down drugs empirically in
every case of illness, slight or severe, in order to humor a popular
notion, that the materia medica must furnish a remedy for every
disease, and a popular prejudice, that want of success is a sure
indication of poverty of resource on the part of the practitioner."

The above quotation is from an eminent allopathic physician,
Dr. Copland. And I have no doubt it will be endorsed by every
physician in the land. The open assertion, however, which the
doctor makes in regard to physicians themselves consenting to
dose their patients, for some trivial disease, until their digestion is
ruined, *simply to humor a popular prejudice*, must be rather humiliating to his colleagues, and certainly will not go very far toward
establishing that feeling of mutual confidence which should exist
between physician and patient.

I do not know of a physician, neither do I believe there are
such, who would intentionally ruin a patient's constitution, simply
that he ever after, or at intervals, might *be a patient.* I have not
the least doubt, however, but that many a stomach has been ruined, while acting as a distributing reservoir for large, complicated, and nauseous prescriptions. How, for instance, can you
expect to take with impunity into your stomach a prescription
which the physician cautions you against permitting to come in
contact with your teeth, as it will injure them. Is your stomach
less susceptible to injury than your teeth? I think not. We frequently hear patients complaining of dyspepsia, which they have

30

had ever since an attack of fever, years before, or ever since Dr. So-and-so salivated them.

A transient attack of dyspepsia, to which all children are liable, occasioned, as it generally is, by a surfeit of what people call "good things," and which a judicious restriction in diet would remove in a short time, without any medicine at all, is frequently rendered permanent by this eternal habit of drugging.

What has been said applies almost exclusively to adults and the older children. Almost exclusively, I say, because indigestion in nursing infants is frequently caused by these very excesses in the mother. We frequently see children suffering from indigestion, attended with vomiting, acid eructations, and diarrhœa, in consequence of the mother's having indulged in a very rich diet, and particularly in vegetables and fruits. Who has not seen a nursing infant suffer from indigestion, the next day after its mother had attended a party or a ball, where she had partaken of a *variety* of *choice* cakes, fruits, ices, etc., danced considerably, returned home late, and nursed her child?

Indigestion in infants is frequently caused by an unhealthy state of the milk of the nurse.

I would here remark that indigestion of young infants is a very common disorder. Dyspepsia, in fact, is a very frequent affection during the whole period of childhood. As the infant's stomach is very delicate, it takes but the least thing to derange it.

What has been said in the article upon " DIET DURING NURSING," might not inappropriately be quoted here ; but instead, the reader in referred to it on page 124. Also to the article on the " DIET OF INFANTS, " and to " GENERAL REMARKS," at the commencement of this chapter.

SYMPTOMS. — I shall describe the symptoms of indigestion, first, as it occurs in adults and in children as young as ten or twelve years; secondly, as it occurs in younger children, and in the nursing infant.

Dyspepsia may be either transient or habitual. By the former, I mean what is usually called " a slight attack of indigestion," such as arises from over-eating, or from partaking of some indigestible or unwholesome article of food, occasioning a temporary derangement of digestion. By habitual indigestion, I mean what would perhaps be called chronic, — those cases, from whatsoever cause originating, which are continued in consequence of a persistence of the cause. The stomach loses its digestive power, and perhaps months, even years, will elapse before the patient will be able to digest any but the simplest kind of food. This state of things not unfrequently arises from the quantities of drugs taken to remove some little gastric derangement caused by eating too much rich or unwholesome food. Frequently, some persons, in anticipation of a "bilious attack," will take a few doses of anti-bilious pills as a preventive; but this just produces what they wanted to avoid. You will find that those individuals who are al-ways taking bitters, pills, and patent medicines, are just the ones who look sallow and unhealthy; and are always complaining of being "bilious." . They think that, unless they keep drugging themselves, they will not be able to keep about; whereas the very medicine they are constantly taking is the very thing that produces all their ill feelings.

An occasional attack of indigestion is characterized, chiefly, by a sense of distention of the stomach; by acrid or acid eructations and flatulence soon after a meal; by loss of appetite, or loathing of food; and occasionally by nausea or vomiting. These symp-toms, however, vary with the nature and quantity of the food. Some persons, though possessing weak and debilitated stomachs, will manage to dispose of a light, easily-digested meal about as well as ever; but a meal of rich meats, or a hearty meal of any description, will invariably produce heart-burn, putrescent eructa-tions, and a feeling of weight or oppression in the region of the stomach.

The tongue is generally pale, flabby, or slimy; or it becomes dry, clammy, or loaded with a thick coating, especially on rising in the morning.

There is generally present, headache, languor, and a general indisposition to look upon the bright side of anything. Rancid, oily, indigested substances are eructated, or brought off the stom-ach with nausea or retching. If nausea and vomiting take place, the contents of the stomach are thrown up either partially or alto-gether undigested. The matters thus thrown up are most usually sour. Vomiting does not always take place. Most-generally, I think, there is repugnance to food; sense of weight, and fulness at the pit of the stomach, and some pain, aggravated by pressure; fre-quent gaseous eructations, sometimes very offensive; also eructa-tions of sour or acrid fluids. When vomiting does take place, instant relief is afforded.

Attacks of this description, as we have before observed, arise from some errors in diet, and, for their prompt removal, only need a proper restriction of diet.

Habitual indigestion, or chronic dyspepsia, may come on gradually, almost imperceptibly, as a consequence of the foregoing. The *acute* attacks are liable to pass into the *confirmed*, or chronic state of the complaint, especially when they occur frequently, or are improperly treated or neglected.

As dyspepsia becomes confirmed, various additional symptoms and sympathetic affections appear. The patient at first complains of local and general debility. All the physical and mental functions betray more or less inactivity. The sleep is disturbed or unrefreshing, sometimes heavy and prolonged. The appetite, in the morning, is impaired and capricious, savory articles being chiefly relished ; and a sense of soreness or relaxation in the throat is complained of. A full meal is followed by heaviness, yawnings, stretchings, and by an almost irresistible disposition to sleep ; by a sense of fulness, weight, flatulence, or by rancid or acrid eructations, etc. As the disorder continues, the appetite is more impaired and more capricious. The bowels become costive and irregular ; the discharges being scanty, offensive, discolored, or more copious, or frequent, and sometimes containing imperfectly digested portions of food. — COPLAND.

Flatulence is troublesome, particularly when the stomach is empty ; the mouth is clammy, and the tongue coated or furred, especially in the morning. The countenance becomes pale or unhealthy, and the body occasionally enlarges about the trunk or abdomen. Vertigo, loss of memory, lowness of spirits, apathy, indifference, and numerous associated and sympathetic disorders ensue, according as the weakness of the stomach extends to the duodenum and intestinal canal, or the secreting organs of the viscera.

The symptoms referred directly to the stomach are often very severe. Indigestion is, in many instances, attended with scarcely any pain ; while, in others, the pain is very tormenting. In many, it does not amount to a pain, or, rather, they do not describe it as such, but complain of great discomfort, and a sense of a load ; of an uneasiness, or a sensation of gnawing in the stomach. Others complain of a burning sensation, which is greatly aggravated by a full meal or by pressure. One form of pain connected with indigestion is popularly called *heart-burn*. This appears to be more of a permanent uneasiness than an actual pain. A second form of pain in the stomach is when it occurs *immediately after* taking food, and continues during the whole process of digestion, or until vomiting ensues, which gives instant ease. In such cases, we have reason to suspect the existence of chronic inflammation of the mucous membrane of the stomach. A third form of pain in the stomach comes on at uncertain intervals in most violent paroxysms. This is properly called *cramps* or *spasms* of the stomach. It is often accompanied by a sensation of distention, much anxiety, and restlessness. In females, it is very frequently combined with hysterical symptoms. I have lately had under my charge a case of this description, which had been treated by a neighboring physician, over six months, as " womb complaint." A fourth variety of dyspeptic pain makes its appearance in from two to four hours after a meal, and continues for several hours. This is the most common form of the complaint.

Water-brash is another modification of pain or uneasiness and disorder of the stomach, of which the distinguishing feature is the vomiting, or rather the eructation of a thin, watery liquid, sometimes sour, but usually insipid and tasteless, and often described by the patients themselves as being cold. In some cases of dyspepsia, the only observable symptom is a loss of the natural appetite. The patient refrains instinctively from certain kinds of food, or feels, perhaps, absolute repugnance and disgust at the very thought of eating. The appetite may even be morbidly craving and ravenous, or capricious and uncertain.

Nausea and vomiting are, in some instances, the most distressing results and symptoms of dyspepsia. Sometimes nausea comes on soon after the food is swallowed. Sometimes there is no nausea; but, after the lapse of a certain period, an hour or two, generally, the food is rejected by vomiting; the matters thus thrown up are usually sour, and not unfrequently mixed with bile, especially if the retching has been violent or long continued, and then the patient is ready to ascribe the whole of his complaint to an " overflow of bile," although, in fact, the secretion of the liver has nothing whatever to do with it. If bile is ejected in vomiting, it merely shows that the action, not of the stomach only, is inverted, but also that of the duodenum. A powerful emetic will, in this way, generally bring bile from the most healthy stomach.

Belching and flatulence are not only distressing, but exceedingly annoying symptoms; they are produced by gas, which is evolved from the undigested food which is detained in the stomach and undergoing fermentation. Gases are sometimes generated apparently by the stomach itself; for the flatulence and eructations frequently come when the stomach is entirely empty of food.

Dyspepsia is almost always accompanied by a sluggish state of the bowels. The evacuations are most commonly dry, scanty, and deficient in healthy color and odor.

" There are innumerable sympathies of distant parts with a dyspeptic stomach, in respect to which I can do little more than barely enumerate a few. Thus, indigestion is often accompanied by pain in the head, with some confusion of thought, or, at all events, with a loss of mental energy and alertness; together with a violent headache, there are frequently nausea and vomiting; and the complaint popularly known by the name of ' *sick headache*,' or, in the fashionable jargon of the day, as a ' *bilious headache*.' " — WATSON.

While an immense number of diseases originate in neglected or protracted indigestion, various disorders are entirely sympathetic with it. Diseases of the urinary organs, of the liver and bowels; palpitations of the heart; irregularities of the pulse; fits of asthma; menstrual irregularities; womb affections; nervousness; hysterical symptoms; eruptions on the skin, and many others thus arise.

The brain and the organs of sense are often much affected by indigestion. Some writers argue that functional disorders, thus sympathetically induced in the brain, may pass into organic disease. Headache is one of the most common and severe affections sympathetically excited by this complaint. The manifestations of the mind are often more or less disturbed; memory is almost sure to be impaired; attention is unsteady, and cannot be long continued; the disposition is more fickle, and the temper more irritable than natural; there is often confusion of thought or of ideas; lowness of spirits; despondency and vertigo, particularly in old, chronic cases. Sight becomes weakened; specks appear before the eyes; hearing is frequently impaired. These symptoms depend upon a weakness of the nerves. — COPLAND.

Dr. Cullen, in speaking of Hypochondriasis, — a disease characterized by languor or debility, depression of spirits, or melancholy, with dyspepsia, a species of insanity, — says: " In certain persons there is a state of mind distinguished by the occurrence of the following circumstances; a languor, listlessness, or want of resolution and activity with respect to all undertakings; a disposition to seriousness, sadness, and timidity; as to all future events, an apprehension of the worst or most unhappy state of them, and, therefore, often upon slight grounds, an apprehension of great evil. Such persons are particularly attentive to the state of their own health, — to the very smallest change of feeling in their bodies; and from any unusual feeling, perhaps of the slightest kind, they apprehend danger, and even death itself. In respect to all these feelings and apprehensions there is commonly the most obstinate belief and persuasion."

Cases like this are of no uncommon occurrence. We meet with them every day. Your " one idea " people belong to this class of afflicted mortality. Pick out the advocates of " Woman's Rights," " State Rights," " Spiritualism," and other like crazy notions, and you will find them to be long-faced, bilious-looking individuals, whose mental faculties are as warped and withered as their external features. All their functions and faculties are blunted and despoiled. Everything they look upon is as " through a glass darkly," and if they do not see " trees as men walking," they behold other things equally strange.

The influence of dyspeptic complaints, in producing affections of the lungs, was referred to in the first part of this article. The debility, caused by protracted disorders of the digestive organs, calls latent tubercles into activity, or rapidly develops them.

In females, dyspepsia not unfrequently occasions difficult, too frequent, or delayed, or irregular menstruation, leucorrhœa, chlorosis, hysteria, and painful affections of the spinal nerves, with tenderness and soreness of the back.

We come next to consider the symptoms indicative of indigestion in infants and young children. Indigestion in infants, like that of adults, may be either transient or habitual.

All children are liable to occasional attacks of indigestion. The symptoms come on soon after the child has nursed freely, or after a hearty meal of artificial food. The child becomes restless and peevish; it moans and cries; it turns pale; contracts its face; shows unmistakable signs of nausea; occasionally it retches and perhaps vomits; there is distension of the abdomen, with wind, eructations, and in many cases, diarrhœa. When vomiting takes place, the milk thrown up is curdled, and its rejection is followed by immediate relief.

In some cases, there is a complete loss of appetite; the infant cares neither for the breast nor for any other food that may be offered it. It nurses but little, is soon satisfied, and even the

small quantity taken is soon regurgitated, or thrown up. Sometimes there is an unnatural craving for food; the child wishes
to nurse all the time. But, though it sucks much, the milk
evidently does not set well upon the stomach, for, soon after
nursing, it begins to worry and cry, and appears to be in much
pain until it has vomited. *Habitual* indigestion produces a train
of symptoms similar to those just described, except that they
are more severe. Nausea and vomiting are present after every
meal; diarrhœa is a common accompaniment; there is constant
restlessness and discomfort; the child frets and worries all the
time, especially at night, when it ought to be asleep; it is never
contented, except when dragging at its nurse. The milk it gets
does not agree with it, or, at least, it produces pain and suffering,
and is soon ejected by vomiting. The milk thrown up is curdled
and sour. In some cases there is no vomiting and no pain after
eating, but the child is distressed by frequent acid or offensive
eructations; its breath has a sour or nauseous smell, and its
evacuations have a most fetid odor. —WEST.

The child has "constant fits of most violent screaming from
colic, sometimes lasting for hours;" there is also a "dull and
languid expression of the countenance, or else an uneasy, contracted look, like that produced by continued suffering; more
or less emaciation; failure of the natural growth in stature and
size, so that the child is small and puny for its age." The child
"suffers unusually from cold, as shown by the coldness of the
hands and feet." — MEIGS.

Indigestion in children that are "brought up by hand," and
those who are fed considerably upon an artificial diet, as well as
those who have been entirely weaned, is a very common disorder.

Children, who have completed their first dentition, and are
allowed to partake of a varied and rich diet, not unfrequently
suffer from severe attacks of transient indigestion. The attacks
usually begin within a few hours, or a day, after the child has
partaken of some indigestible substance, with languor, perhaps
chilliness, headache, pain in the stomach, nausea, and very often
a disposition to drowsiness and sleep. When these attacks end
in vomiting, which they not unfrequently do, the child appears
perfectly well from the time the offending substance is ejected.
If, however, vomiting does not take place, fever usually sets in,
the skin becomes flushed, hot, and dry; tongue, coated; thirst,
considerable; the child is restless and uneasy; keeps tossing from

side to side, or lies in an uneasy, drowsy state; there may be also frequent starting or jerking of the limbs or crying out. Symptoms of this description are frequently followed by convulsions. Attacks of indigestion, in children of from three to five years of age, are not unfrequently followed by convulsions, and, when symptoms indicative of a tendency in this direction are present, the patient should be carefully watched and the accident guarded against.

Gastric derangements, such as we have just described, generally continue until nature relieves the stomach by vomiting or diarrhœa, or until the proper remedies have been administered. A judicious restriction of diet and an occasional dose of the appropriate remedy will soon restore the healthy action of the stomach, whereas, when the case is left entirely to nature, and the child permitted to indulge its appetite, a perfect restoration will take place slowly, or not at all. *Transient* attacks of indigestion, when neglected, or what is still worse, improperly treated, are very apt to leave the digestive organs in an irritable or debilitated condition, from which it will require time and great care on the part of the physician and nurse to make a permanent cure.

Habitual indigestion in children, who have completed their first dentition, "is a condition analogous to, if not identical with, the dyspepsia of the adult. The symptoms of this form are the following: the general appearance of the child is delicate, as shown by a pallid or sallow tint of the skin, instead of the ruddy complexion of health; as, also, by thinness and flaccidity of the muscular tissue. There is an habitual air of languor and listlessness, with absence of the usual gayety and disposition to play natural to its age, and the child often complains of being tired. The appetite is feeble or uncertain, being sometimes absent, and at other times too great; or it is peculiar, there being a willingness to eat of dainties, but a refusal of food of a simple character. The tongue presents nothing peculiar; it is, however, more frequently somewhat furred than clean and natural. The temper is usually irritable and uncertain. The child rarely sleeps well; on the contrary, the nights are restless and much disturbed, the sleep being broken and interrupted by turning and rolling, by moaning or crying out, and by grinding of the teeth. These latter symptoms, together with picking at the nose, which is a frequent accompaniment, are almost always referred by the parents and nurses to worms, and it is often impossible to convince them

37

to the contrary, even though frequent and violent doses of vermifuges have failed to show the existence of entozoa (worms). . . . . This form of indigestion, like dyspepsia in the adult, is generally a very chronic affection, seldom lasting less than several weeks or months, and sometimes for years."—MEIGS' *Diseases of Children.*

TREATMENT.—To facilitate as much as possible the treatment of this complicated disease, we make the following divisions with the remedies attached to each variety. Before selecting a remedy, consult its details below.

For dyspepsia of *Adults.*—Acon., Ant.-c., Arn., Bell., Bry., Cal.-c., Carb.-v., Cepa, Cham., Chin., Hepar-s., Ipecac., Lach., Merc., Nux-v., Phos., Puls., Sepia, Sulph., Verat.

Of *Children.*—Acon., Bry., Calc., Cham., Ipecac., Puls., Sulph.

Transient, or acute dyspepsia.—Acon., Arn., Ant.-c., Bell., Bry., Ipecac., Merc., Nux-v., Puls.

Habitual, or chronic dyspepsia.—Ars., Bell., Cal.-c., Chin., Hepar-s., Lach., Merc., Nux-v., Phos., Puls., Sepia, Sulph.

Arnica.—When the disorder is caused by a fall, a blow upon the stomach, or by lifting heavy weights, with pain, and a sensation as if the small of the back was broken. Also, when there is great sensitiveness and nervous excitement; frequent eructations, with a putrid or bitter taste; tongue covered with a thick, yellowish coat; some nausea, with inclination to vomit; head full and giddy; also, a heaviness of the limbs. Should *Arnica* not suffice, try *Nux-vomica.*

Aconitum.—When, at the commencement of the attack, there is considerable fever, with thirst and nausea. Or, at any time during the disease, when the fever runs high, or partakes of an inflammatory nature. Also, where there is much heat, redness and soreness of the mouth or throat.

Antimonium crudum.—This remedy is particularly useful when the disorder arises from an overloading of the stomach, and the following symptoms are present: frequent eructations, which taste of the food last partaken of; or a gulping up of particles of undigested food soon after eating, either with or without sickness at the stomach. The tongue is coated with a white, or yellowish mucus; the stomach feels distended, and is tender to the touch; there is, besides, flatulency, accompanied with griping pains, or diarrhœa.

Belladonna.—When there is painful distention of the abdomen, with griping pains, as if the bowels were grasped with the fingers; flatulent colic; hiccough; bitter eructations; nausea, with loathing of food; vomiting of water, or bile. Also, when, in addition to the derangements of the stomach, there is dulness of the head, or congestion of blood to the head.

Arsenicum. — In serious chronic cases, where there is great prostration of the vital powers; countenance sunken; extremities cold; face blanched; dark circle around the eyes; nose pointed;· tongue white, or of a brownish color, dry and trembling; pulse irregular, small, frequent, and weak. Also, when there are severe cramps in the stomach, with a sensation of coldness or much heat; when vomiting becomes excessive, everything taken is returned from the stomach; the skin hot and dry; the patient becomes emaciated, and the countenance cadaverous. If *Arsenicum* does not soon produce a favorable change, give *Lachesis*.

Bryonia. — This is an important remedy for dyspepsia, especially when it occurs in summer, or when the weather is warm and damp; also, when it is accompanied with chilliness, headache, pain in the limbs, small of the back, etc., — symptoms such as follow a cold.

Tongue dry and red, or coated over with a whitish-yellow fur; loss of appetite; great aversion to food, — even the smell of food causing great disgust; sometimes there is a great craving for food, an unnatural appetite, the child literally gorging itself if allowed to have its own way; or there may be loss of appetite alternating with unnatural hunger; craving for acid drinks; much thirst; insipid, clammy, sweetish or bitter taste in the mouth; eructations; gulping up of particles of food after every meal; vomiting of food, particularly at night; inclination to vomit after every meal; morning nausea; water-brash; distention of the stomach, especially after a meal, even though ever so little has been eaten; burning in the stomach; constipation of the bowels, especially in infants; temper restless, irritable, and obstinate; also, when anger excites or aggravates the derangement.

Bryonia may follow *Aconite* when the patient complains of dryness of the mouth and considerable thirst. Should *Bryonia* produce little or no improvement, it may be followed by *Rhus*.

Cepa. — When there is no hunger, but considerable thirst, fulness of the head, pain in the bowels from wind, tongue coated, especially near the root.

Carbo-vegetabilis. — Loss of appetite; aversion to meat; distention of the stomach after eating. Empty eructations, or belching up of air tasting of the fat and food which had been eaten. Flatulence, fetid and offensive; wind-colic; rumbling in the abdomen; water-brash; hiccough; contractive or burning pain in the stomach; nausea in the morning; offensive diarrhœa.

Calcarea-carbonica. — Particularly for children of a scrofulous tendency, or those who are backward about learning to walk, or very fat children, whose circulation is torpid and sluggish. Frequent eructations; acid stomach; water-brash; sour eructations and sour vomitings; fulness and swelling in the region of the stomach, with tenderness to touch; pressure in the pit of the stomach; gnawing or griping pains; enlargement of the abdomen, particularly in scrofulous children. Sour-smelling diarrhœa of children. Chronic dyspepsia of adults, with heart-burn after any kind of food; vomiting of food; eructations; morning nausea; aversion to meat and warm food; loss of appetite, at other times canine hunger; distention of abdomen; pain in the stomach after eating, followed by nausea and vomiting; desire for wine, salt things, or dainties.

Chamomilla. — Especially for gastric derangements brought on by a fit of passion, which sometimes happens even in females, or by standing in a draft when perspiring. Bitter or sour eructations; regurgitation of food; nausea; vomiting of food, green phlegm, or bile; cramps in the stomach; distention of the stomach; sometimes constipation, but generally diarrhœa, especially in children, while the evacuations are green and watery. Headache, fulness, giddiness, and staggering in the morning when getting up. Sleep disturbed, tossing about, frequent awaking. Face red and hot; obscuration of the eyes; mind very sensitive.

China. — This remedy is especially applicable for that form of dyspepsia which is caused by an impure atmosphere, — an atmosphere that is overloaded with exhalations of decayed vegetable matter; also for indigestion which precedes or accompanies chills and fever, and is caused by the same miasmatic influence which produces that disease.

Pressure in the stomach, as if from a load; a constant feeling as if one had eaten too much; constant eructations; gulping up of particles of undigested food, especially after supper; aversion to food and drink, with feeling of fulness; flat or bitter taste in the mouth; desire for a variety of things, — this or that dainty, — without knowing which; after eating, drowsiness, oppressive fulness in the stomach and abdomen; heart-burn, with flow of water in the mouth ; ineffectual retching ; morbid craving for something strong, sharp, or sour. Weakness and tired feeling ; a disposition to lie down, without being able to remain quiet. The patient yawns, bends, and stretches his limbs, from a sense of weariness ; is melancholy and morose.

Hepar-sulphur. — This is an important remedy for correcting a chronic derangement of the stomach, caused by taking blue-pills or other preparations of mercury ; or in those cases where the stomach appears to be very sensitive and easily deranged, though the patient may be healthy, and very correct in his general habits. Nausea in the morning, with eructations, or vomiting of sour, bilious, or mucous substances. Appetite only for something sour and piquant; aversion to fat ; desire for wines. Distention in the pit of the stomach, as from wind ; one cannot bear tight clothes. Bowels constipated ; stools hard and dry, or, in children, a sour, whitish diarrhœa.

Ipecacuanha. — This is the principal remedy for indigestion of children, arising from imperfect mastication or from partaking of improper food. It should be given at the commencement of an attack, especially when the following symptoms are present : tongue coated with a white or yellowish coating ; nausea, with empty eructations ; vomiting of undigested food, or of bile, or of bitter, acrid-smelling water and jelly-like mucus ; vomiting with diarrhœa ; easy vomiting, generally attended with coldness of the face and extremities ; diarrhœa, with fermented stools ; diarrhœa, with nausea, colic, and vomiting.

Also for adults, when the tongue is not coated, although the patient is sick at the stomach and vomits ; aversion to food, particularly to fat, rich food, such as pork, pastry ; or for dyspepsia, caused by eating such things. Headache, attended with nausea and vomiting ; stitching headache, with heaviness of the head ; pressure in the forehead ; nausea and vomiting ; vomiting, with diarrhœa ; bilious headache — that is, sick headache.

Lachesis. — In cases where *Hepar* has been insufficient, or for indigestion, which is worse immediately after eating, and for severe cases, such as described under *Arsenicum.*

Mercurius. — Acrid, bitter eructations ; putrid, sweetish, or bitter taste in the morning ; bilious vomiting ; repugnance to solid food and meat ; pressure at the pit of the stomach after eating ; weak digestion, with constant hunger. Suits well before or after *Lachesis.*

Nux-vomica. — This remedy is specially adapted to those persons who lead a sedentary life, seamstresses, school-girls, and the like; also those who possess a lively, restless, irritable temperament, in whom anger, chagrin, or anything which crosses them is apt to induce a fit of indisposition. And for those who are fond of dissipation, keep late hours, sip wine, etc.

The following symptoms call for *Nux-v.*: head confused; reeling, giddiness, or dulness in the head; heaviness in the back part of the head; headache, unfitting one for, and increased by, mental exertion; tearing pain in the head and checks; ringing in the ears; drawing in the teeth, sometimes above and sometimes below. The headaches are deeply seated in the brain, often confined to one side, or the back part of the head, coming on chiefly in the morning, after a meal, or in the open air. Tongue coated white; mucus collects in the mouth; metallic, bitter, sour, or putrid taste, chiefly in the morning or after eating. At times, there is no taste at all, or all kinds of food taste insipid. Heart-burn; bitter eructations; nausea and inclination to vomit, especially early in the morning and after eating; periodical attacks of vomiting; vomiting of undigested food. Distention and pressure in the stomach and pit of the stomach after eating. Distention and pressure, as from a stone in the abdomen; a feeling of tightness of the clothes around the waist; wind-colic; the abdomen is sensitive to pressure or contact. Constipation; ineffectual urging to stool; and hard, difficult stool, streaked with blood; blind piles.

Pulsatilla. — This is another important remedy for dyspepsia, and is peculiarly adapted to persons of a lymphatic temperament, — those ardent, enthusiastic, good-natured individuals, who easily laugh or weep, and have pale faces, blue eyes, and blonde hair. *Pulsatilla* should be given in all transient, or recent, cases of indigestion, especially when it has been caused by eating any kind of food which produces flatulency; by over-eating, or by the use of pork, mutton, or pastry, or any greasy substance; and when there are eructations tasting of the food which has just been eaten; or sour, bitter eructation; inclination to vomit, especially after eating or drinking; taste flat or putrid, resembling bad meat or tallow. Aversion to food, especially to meat, bread, butter, milk, and anything warm. Pressure in the pit of the stomach, especially after eating; wind-colic after supper or at night; rolling and rumbling in the abdomen; slow, small evacuations or diarrhœa; frequent urging to stool; water-brash. The patient feels chilly, is weak, cross, sad, melancholy, annoyed at every trifle, not inclined to talk.

Phosphorus. — Empty eructations, especially after eating; sour regurgitations; vomiting after eating; burning in the stomach; pressure and fulness in the stomach; acidity and sour taste in the mouth; drowsy and lazy after eating; chronic looseness of the bowels.

Sepia. — For chronic dyspepsia, with or without sick headache; eructations, sour, bitter, or tasting of the food; loathing of food; putrid or sour taste; nausea before breakfast, also after eating; nausea of pregnant women; distention of the stomach, with pressure as from a stone. This remedy suits well for nervous, hyster ical persons.

Sulphur. — This remedy acts well after *Nux-vomica* and *Mercurius*, in cases of long standing, or when there is loss of appetite; aversion to meat; difficulty of breathing; nausea after eating; belching or vomiting of food; colic immediately after eating; water-brash; sour stomach; flatulency and constipation. Mental depression; morose irascibility; dissatisfied with everything and everybody.

Veratrum. — When *Ipecac.* has proved insufficient, or where, after the use of *Ipecac.*, there is still left diarrhœa, with griping pains in the bowels. Also, for cases attended with vomiting and diarrhœa; thirst for cold drinks; empty, sour, or bitter eructations; nausea, with great prostration; coldness of the hands, and shuddering all over; vomiting of the food; burning in the whole abdomen.

Rhus, Sulphuric-acid, Natrum, Cocculus, Thuja, and a few other remedies, in addition to what we have enumerated, are sometimes, but not often, called for.

ADMINISTRATION OF REMEDIES. — In all transient or recent cases of indigestion, especially if there should be much pain and sickness at the stomach, a dose of the selected remedy may be given every half-hour, until relief is obtained. As soon as the severity of the symptom begins to abate, the interval between the doses may be lengthened. For recent attacks of indigestion of ordinary severity, the doses may be taken from one to three hours apart. In chronic cases the remedy may be repeated about three times a day. Dose, for an adult, ten globules; for an infant, two.

DIET AND REGIMEN. — Perhaps there is no class of diseases in which it is more important to adhere strictly to dietetic regulations than in those which consist in some derangement of the digestive apparatus.

Our opponents are very fond of attributing our cures to the strict attention which we pay to all the laws of Hygeia. This I consider one of the highest compliments ever paid to our branch of the profession, and at the same time one of the severest censures ever applied to the other schools, even by their most bitter enemies. If we can, and do, as they assert, by a simple regulation, and restriction of diet, cure our patients of serious disorders, then how, in the name of suffering humanity, can they justify themselves in giving such enormous quantities of drugs for slight ailments?

Of course, we maintain that our remedies are active in this or in any other disease; but, at the same time, are perfectly willing to admit that we do not, by any means, place our whole reliance for the cure of disease upon the administration of "infinitesimal doses" of medicines, because we know full well that unless the diet of the dyspeptic patient be duly regulated, medical means will be employed in vain. We think that a large number of all the cases of indigestion, if put under proper dietetic regulations, would get well without any medicine at all; not as promptly, however, as they would if a judicious hand administered proper doses of the proper remedy at the proper time. Therefore, considering the subject of diet in connection with digestive diseases to be an important one, our remarks upon this point will be somewhat extended.

In speaking of diet with reference to indigestion generally, we shall remark upon, 1. The kinds and quality of food. 2. The quantity of food. 3. The time of eating, or the periods which should intervene between meals. 4. The kind and quantity of drinks. 5. The conditions deserving notice in connection with eating and drinking.

THE KINDS AND QUALITY OF FOOD. — Although a healthy person, leading an active life, may eat, with impunity, almost every variety of food, there are a great many articles of diet which ought to be preferred, and others which *must* be avoided, by the dyspeptic. *Vegetables* are slower of digestion than animal and farinaceous aliments, and more liable to undergo fermentation in weak stomachs, and to occasion acidity and flatulence. Fat and oily meats are also very indigestible, and give rise to acid or rancid eructations and heart-burn. Soups and liquid food are acted upon by the stomach with great difficulty, and if the diet consists chiefly of them, they furnish insufficient nourishment, and never fail of producing the more severe forms of dyspepsia, and the diseases of debility. Soups are hurtful when taken at the commencement of a full meal; but, when little or no animal food is eaten along with them, and rice or bread is taken with them, so as to promote their consistency, they are digested with greater ease. — COPLAND.

Animal food is easier of digestion than vegetable food, because there is less "conversion" required; it is already nearer, in its composition, to the textures into which it is to be incorporated by assimilation. It is well known that vegetable food, when the stomach is weak, produces more flatulence than animal; this is simply because digestion is slow and incomplete. The most digestible meal that can be taken by a dyspeptic, in my estimation, is a fair proportion of both animal and vegetable food, — a mixture of the two. This is better than a rigid adherence to either kind of aliment singly. Well-cooked animal food, either beef, mutton, venison, or game, eaten with a moderate quantity of bread, or with roasted, mashed, or dry, mealy potatoes, or with rice, will seldom disagree even with a very delicate stomach. The kind, however, of animal food, and the mode of cooking it, should depend much upon the peculiarities of each individual case and the stage of the disorder. One of the greatest evils of the present generation is modern cooking, — *the art of rendering food unhealthy.*

One of the best aids to digestion we know, is good cookery. By bringing out the flavor, it increases the nutritiousness of food, which bad cooking toughens and deprives in part of its nourishing properties.

Dyspeptics should avoid all cured meats, such as ham, tongue, salted, smoked, or pickled meats, sausages, etc. All raw vegetables, also, must be eschewed; salads, cucumbers, pickles, etc., *must* be banished.

Fish holds an intermediate rank between animal and vegetable food, as respects digestibility. It is less nutritious than mutton or beef; therefore a larger quantity is requisite to satisfy the appetite. Fish is most digestible when *boiled;* is less so when *broiled;* and the least so when *fried.* Shell-fish are slow of digestion; raw oysters are more digestible than crabs or lobsters; but oysters, when stewed or otherwise cooked, are heavier than either. Fruits in general are refreshing and wholesome, but not very nutritious;

when the stomach receives them, and suffers no inconvenience, there can be no reasonable objection to their use.

THE QUANTITY OF FOOD. — One great and indispensable principle in the treatment of indigestion is that of restricting the *quantity* of food at any time. Those who lead sedentary lives, and whose circumstances admit of free living, are peculiarly liable to dyspeptic complaints. This is owing chiefly to the quantity of food indulged in. The quantity of food should always be proportioned to the digestive powers of the stomach, and the wants of the system. When digestion is liable to be easily impaired, it is of great importance, not only to refrain from those substances which are known to be indigestible, but, also, to avoid mixing together in the stomach different substances which are of different degrees of indigestibility. You will here see the reason why it is salutary to dine off one dish. 1st. Because we avoid the injurious admixture just adverted to ; and, 2d. Because we escape that appetite, and desire to eat a *large quantity*, which is provoked by new and various flavors. One great cause of over-eating, is the hasty manner in which we partake of our food. We do not pay sufficient attention to the preliminary process of mastication. The consequence is, too much food is received in a short time, in a state of insufficient preparation, and the stomach is overloaded before the sensation of hunger is completely appeased. Of course it is impossible to lay down any rules respecting the quantity of food that should be taken. This, however, you can take for granted, that the *first* intimations of a satisfied appetite are *warnings* to stop eating. Dyspeptics, especially, should remember this, and heed it.

THE TIMES OF EATING, or the period which should intervene between meals. At least six hours should elapse between one meal and another. Even healthy stomachs require from three to four hours to digest an ordinary meal. Of course, dyspeptics require a much longer time.

" The stomach also requires an interval of rest, after the process is finished, in order to enable it to enter upon the vigorous digestion of the next meal. As a general thing, breakfast about half an hour or an hour after rising, will be found most beneficial. The dyspeptic, especially, ought never to travel, or to enter upon any exertion with an empty stomach, and never with an overloaded one."

As a general rule, not more than six hours should elapse from breakfast till dinner. For children and youth that are growing rapidly, as well as for convalescents and persons taking active exercise in the open air, the interval may be somewhat shortened ; but for sedentary persons it may be somewhat prolonged. The habit of eating between meals is especially injurious ; some children's stomachs are kept constantly at work digesting, or at least making an effort at all sorts of trash. This constant taxation must prove injurious. See what has been said, at the commencement of this chapter, under head of " GENERAL REMARKS ; " also, supplementary, " DIET OF CHILDREN."

THE KIND AND QUANTITY OF DRINKS. — Dyspeptic patients are very importunate to know *what* they may drink, as well as what they may eat. They may think a little bitters of some kind are necessary to aid digestion, and many are in the habit of taking malt-liquor, wine, or brandy, with or after every meal. Some allowance must be made, no doubt, for custom, but my impression is, that *teetotal* abstinence would be more conducive to health than the use of any of these articles. Dr. Beaumont says that the use of ardent spirits, wine, beer, or any intoxicating liquors, when continued for some days, *invariably* produces morbid changes in the stomach. Drinks which are followed by evident disturbance and discomfort, are manifestly injurious. It is very easy, indeed, for me to say that pure *spring water is the best possible drink*, either for a sick or a well person, but perhaps it would not be so easy for me to vindicate the assertion. Nevertheless, I do believe it to be true, that pure, cold water is the best drink, and the only one of which nature ever intended that we should make use. Persons suffering from transient attacks of indigestion should never drink anything but water, toast-water, or whey. Stimulating beverages will prove injurious to them, and greatly increase the liability of the disorder becoming chronic. Persons should always drink when thirsty, *and then only*. Frequent sipping, or drinking by mouthfuls, will quench thirst better than large draughts ; besides, large draughts suddenly distend the stomach and lessen the energy of the gastric juice by its dilution, and thus retard digestion. The dyspeptic ought never to drink largely, either during, or soon after, a meal. As we strongly advocate the constant use of cold water, perhaps we should say, that, by cold water we do not mean *ice water*. Ice-cold water is injurious, from the fact that it so suddenly reduces the temperature of the stomach. Dr. Beaumont says that a gill of water at the temperature of 55°,

when put in the stomach, reduces the heat of the organ from 99°
to 70°. The shock which the constitution thus receives, from
having the temperature of the most vital and central organ so
suddenly and remarkably depressed, paralyzes all the other vital
movements. Dr. Dunglison states that laborers in Virginia are
frequently killed by drinking copiously of cold water when overheated. Such examples are frequent in cities.

THE CONDITIONS DESERVING NOTICE IN CONNECTION WITH EATING
AND DRINKING. — Dr. Caldwell remarks that dyspepsia commences
perhaps as often in the brain as in the stomach. By this he means
that care, anxiety, envy, and excessive mental exertion impede
digestion; and no doubt this is true. In considering the causes
of dyspepsia, we had occasion to refer to this. It is important
that there should be rest of body and tranquillity of mind, both
before and just after eating. The Spanish practice of having a
" siesta," or sleep, after dinner is far better than the custom of the
Americans, who hurry from their meals to the store, shop, or study,
in order to save time, in holy horror of losing a single moment.
" The practice of the Spaniards may be improved by indulging for
an hour before resuming toil in moderate exercise of the muscular
system, conjoined with agreeable conversation and hearty laughter,
as this facilitates digestion, and tends to ' shake the cobwebs from
the brain.'" That is what Dr. Cutter says, — sensible man ! The
same author sums up his observations upon digestion, as follows: —
" Digestion is most perfect when the action of the cutaneous vessels is energetic; the brain and vocal organs moderately stimulated by animated conversation; the blood well purified ; the
muscular system duly exercised ; the food of an appropriate quality, taken in proper quantities, at regular periods, and also properly masticated."

The importance of the subject is my only apology for the length
of this article.

CONSTIPATION.

DEFINITION. — As a general thing, there should be one evacuation, at least, from the bowels every day. By constipation, or
costiveness, as it is also called, we understand a prolonged retention of the fæces, or slow, imperfect, or difficult evacuation of
them.

CAUSES. — Constipation of infants generally arises from an

improper mode of living, on the part of the nurse or child. It most frequently appears in those children who are wholly or partially fed upon an artificial diet, and in those whose nurses or parents are similarly disposed.

Constipation of the adults and the older children is frequently caused by improprieties in diet, by stimulating and astringent aliments and drinks, too long indulgence in sleep, inattention to the first intimation of a desire to evacuate the bowels, sedentary habits, impaired or torpid condition of the digestive function, and the habitual use of aperient medicine.

SYMPTOMS. — Constipation is always accompanied by a train of symptoms more or less definite. In fact, constipation itself is, in the majority of instances, but a symptom of some actual derangement, — usually indigestion, — and in such cases you will find the tongue coated at the root, while the sides and point are red; the urine high-colored; the pulse slower than natural, usually a little quicker an hour after a meal. There is also slight sallowness of the countenance and skin, and more or less distention and uneasiness about the lower portion of the abdomen; also, there is frequently much flatulence, and almost always more or less headache.

Constipation of small children is usually attended with distention of the abdomen. The child is restless, cries a good deal, and breathes heavily. In some cases constipation is followed by jaundice, and may be followed by convulsions.

TREATMENT. — If there is any point in the whole range of therapeutics, where allopathia and homœopathia stand in direct opposition to each other, it is just here. Allopathia, looking upon this abnormal condition as a simple obstruction of the intestinal canal, orders a cathartic, thinking that if the bowels are once opened, the case is cured. It is true, it is admitted that it may return again in a few days, but then, as drugs are plenty, take another dose. Homœopathia takes quite a different view of the case, and says, "Correct the system; remove the cause of the disorder; the bowels will take care of themselves."

Allopathia may be compared to a scavenger, who, with physic for his shovel, removes the collection, while Homœopathia goes to the fountain-head, to the original source, and administers a corrective, or turns the stream in another direction, well knowing that if the cause is removed, the difficulty itself, as a matter of course, must soon disappear. Which is the more rational mode of procedure, I will leave the reader to decide.

Now, that cathartics are not the proper medicines for constipation, is just as easily demonstrated as that two and two make four. Take a person in health, and give him a dose of physic; what is the effect? Diarrhœa at first, it is true; but does not constipation soon follow? The secondary, or lasting effect of all cathartics is the very reverse of their primary or first effect. They first produce a few free evacuations, but the final result is constipation, and constipation will be your eternal companion if you persist in taking aperient medicines.

Patients often come into our hands so thoroughly imbued with a sense of the necessity of an occasional downright scrubbing-out, that they look back and sigh for a good old-fashioned dose of physic. It is hard to deprive them of such a luxury, but then it is necessary for their health. Nevertheless, when we point out to them the absurdity of their former habit, they generally come to the conclusion that their late physician recommended it as but a temporary relief, well knowing that he would soon have an opportunity to prescribe for them again, and thus double the fee. In fact, that he would thus keep them vibrating like a pendulum — upon a one dollar tick — between cathartics and astringents.

Many a slight case of constipation, which, if left to nature, would have disappeared, leaving no ill consequences, has, by an ill use of cathartics, been converted into habitual constipation, fairly embittering existence, and predisposing the constitution to a variety of diseases in after life.

The principal remedies for constipation are *Nux-vomica, Bryonia, Platina, Lycopodium, Mercurius, Opium, Sulphur,* and *Plumbum.*

Nux-vomica. — When there is frequent but ineffectual urging to stool; disagreeable taste in the mouth; inclination to vomit; loss of appetite; distention of the abdomen; irritability; frequent complaining. Also when constipation is preceded by diarrhœa, or accompanied by a feeling of general depression.

Should *Nux* prove inefficient, give an occasional dose of *Bryonia,* and especially if the disorder occurs in warm weather, or is accompanied with disordered stomach.

Opium is sometimes useful in alternation with *Nux-vomica,* especially should there be a total absence of all inclination to stool. *Opium* is also useful when there is an inclination to evacuate, but a feeling as if the anus were closed. Determination of blood to the head; redness of the face; and headache.

Platinum. — When, after much straining, the fæces are evacuated in small, hard lumps. If not relieved, try *Magnesia-mur.*, especially if the stool consists of hard lumps, like small marbles. *Platinum*, when there is shuddering over the body after an evacuation. Should *Platinum* fail, give an occasional dose of *Lycopodium*.

Lycopodium is an excellent remedy when there is painful urging, severe bearing down, but inability to pass the hardened fæces.

Sulphur.—For habitual costiveness, particularly should the patient be troubled with piles ; or when there is a disposition to hæmorrhoids, either blind or bleeding ; or where there is a frequent inclination to go to stool, but without the desired result.

Plumbum.—For obstinate constipation ; tenacious, hard, difficult stools, sometimes in hard lumps or balls.

For constipation of pregnant women, *Nux-v., Opium, Sepia.*

For constipation of lying-in women, *Bryonia, Nux-v.*

For constipation of nursing infants, *Bryonia, Nux-vomica, Opium, Sulphur.*

ADMINISTRATION OF REMEDIES. — Give, of the selected remedy, a dose about once in four hours. The remedies may be given dry, or dissolved in water. When given dry, put six pills upon the tongue, and let them dissolve. For an infant, two or three pills will be a dose. When given in water, dissolve six globules in about as many teaspoonfuls of water, and give one spoonful for a dose.

INJECTIONS. — In addition to the remedies above mentioned, you will find that great assistance may be derived from a proper use of injections. Some people are disinclined to make use of them, for fear of establishing a habit from which it will be difficult to escape ; but no fear need be had upon this point, if you, at the same time, use the remedies above recommended. You are aware that an injection affords but temporary relief, at best ; but this is infinitely preferable to physic, because it does not affect the general system. It simply acts as an attendant, to prevent the further accumulation of fæces, while, in the meantime, the remedies taken are actively engaged in removing the cause of the disorder ; and, when this is accomplished, the injections may be discontinued. It is erroneous to suppose that an injection of tepid water — and nothing else should ever be used — can prove injurious. I have no doubt that injections, as some people make them, often do permanent injury, producing piles, weaknesses, etc. But this is never the case with simple water-injections.

I have often, by their timely use, succeeded in removing alarming symptoms, for which it never would have answered to wait the slower action of remedies. Injections are indispensable; and 1 would advise you, upon all occasions where prompt relief is desirable, to make use of them. They may be repeated as often as necessary, until a normal action of the bowels can be brought about by proper medication. But never give physic, nor incorporate any of it, with your injections.

DIET AND REGIMEN. — For a prompt and permanent cure of constipation, as much depends upon a proper mode of living as upon medical treatment; perhaps more. When constipation is caused by sedentary habits, of course the remedy is a larger amount of exercise. Children who spend most of the day in school, whose out-door exercise is restricted, and who are permitted to lie abed late in the morning, are generally troubled with this disorder; while those whose out-door sports are not so limited, — those who are provided with hoops, ropes, and balls, and are *permitted to use them*, — who spend the larger part of their time out of doors, are not troubled with constipation. Their systems are not sluggish; their appetites are keen; bowels regular; they sleep well at night; are up with the lark in the morning, bright and happy. But then fashionable people call them rude, and imagine they look coarse, because their cheeks are plump and red, and their skin sunburnt, instead of being white, or rather, pale and delicate.

Those persons who are habitually constipated, should be careful to avoid all articles of diet of a binding nature, such for instance, as animal food; especially salted meats, cheese, wheaten flour in any shape, stimulating drinks, high-seasoned dishes, etc. And, on the other hand, take a liberal allowance of all kinds of fruits and vegetables, soups, coarse bread, and such other things as experience has taught us to be of a laxative nature. Above all other things, see that children masticate their food well; do not allow them to do, as most grown people do, bolt it.

Make free use of cold water as a drink. A good drink of cold water, on going to bed, has been found beneficial.

Another very important point demanding your attention, as parents, is to see that the children attend *regularly* to the calls of nature. The best time, probably, is in the morning. Once accustom them to attend to this at a certain hour every day, whether there is a desire or not, and you will be surprised to see what uniformity will be soon established.

DIARRHŒA.

DEFINITION. — By the term diarrhœa is understood a too frequent evacuation from the bowels, without any perceptible symptoms of an inflammatory action; without fever; such as physicians would call "a mere functional derangement." That is, a diarrhœa existing without any anatomical or structural lesion; and this constitutes the great difference between diarrhœa and dysenteria. The first is simply a failure on the part of the digestive apparatus to perform its accustomed duties, and an effort to rid itself of some offending substance; while the second consists of an actual injury to the parts affected; as inflammation; the mucous membrane becoming swollen, thickened, red, softened, and ulcerated.

CAUSES. — Diarrhœa is a frequent ailment of children, and may arise from an unhealthy diet, or from over-feeding, from taking cold, from sudden atmospheric changes, fright, vexation, too close confinement in unwholesome or ill-ventilated apartments, and dentition. — See causes of " CHOLERA INFANTUM," and " GENERAL REMARKS," at the head of this chapter.

TREATMENT. — Although diarrhœa affords sufficient evidence of a disordered action going on within the system, it is not always best to interfere with it; for we should ever remember that this is one of nature's ways of relieving herself, by expelling some offensive or irritating substance; and, when this is accomplished, the diarrhœa will pass away without requiring any assistance on our part. We should not be in too great haste to give medicine. Most people are very fond of giving *something,* and usually astringents of some kind are chosen. They little think or little know how much risk they are running by throwing back upon the system that matter which *may* prove most injurious.

Nature is wise in all her dealings; and these changes are often salutary and conservative. Would it not, therefore, show more wisdom on our part to assist by judicious treatment, rather than to interfere with one of nature's laws.

In the proper treatment of this disease, as much depends upon a judicious regulation of diet as upon the administration of remedies. — See " GENERAL REMARKS," at the beginning of this chapter; also " CHOLERA INFANTUM."

In commencing the treatment of a case of simple diarrhœa, first forbid the use of all food that is not perfectly easy of digestion;

39

also acids, coffee, everything highly seasoned, vegetables, fruits fresh or dried, as well as fresh meats, and meat-soups of every description.

The patient may be allowed toast, rice, boiled milk, oatmeal, hominy, arrow-root, sago, etc. When the patient begins to improve, and the appetite begins to return, a little mutton-broth, thickened with rice or wheaten flour, may be allowed.

The remedies are *Chamomilla, Ipecacuanha, Dulcamara, Chinchona, Mercurius, Pulsatilla, Nux-vomica, Rheum, Podophyllum.*

Dulcamara. — For diarrhœa which is caused by taking cold, evacuations watery, worse at night, attended with no great pain. If this fails to afford relief, give Bryonia.

Ipecacuanha. — For thin mucus; frothy, fermented evacuations, or small, yellow stools, with pain in the rectum; very offensive evacuations, with great weakness; dysenteric stools, with white flocks, and subsequent pains, as though more would pass; great prostration, inclination to lie down, paleness of the face. Where Ipecacuanha gives but partial relief, follow it with Rheum. Ipecacuanha may be given to nursing infants, when diarrhœa arises from overloading the stomach, accompanied with nausea and vomiting, frequent crying, stool yellowish, or green and streaked with blood, and very offensive.

Chamomilla. — Especially for infants; evacuations slimy, green or yellowish, or of undigested matter, looking like chopped straw, and smelling like rotten eggs; distention of the belly; tongue coated; thirst; want of appetite; rumbling in the bowels. The child draws up its legs, frets and worries, wants to be held or carried all the time.

Chinchona. — For profuse, watery, and brownish diarrhœa, intermingled with portions of undigested food. For painless diarrhœa, with a great deal of wind; undigested milk in the stools.

Podophyllum. — Diarrhœa with cramp-like pain in the abdomen; stools light-colored and exceedingly offensive; frothy mucus, and slimy stools. The child moans, and rolls its head from side to side while asleep. Sleeps with its eyes half open.

Magnesia-c. — Diarrhœa with stools like scum of a frog-pond, — green and frothy.

Rheum. — For sour-smelling evacuations; thin, slimy, fermented diarrhœa, common to small children; sour smell proceeding from the child, which washing will not remove; diarrhœa from acidity of the stomach; distention of the abdomen; colic; crying, both before and after an evacuation : ineffectual urging before and after stool. If *Rheum* does not relieve, give *Chamomilla*.

Mercurius. — This remedy suits for almost any diarrhœa, especially when accompanied with griping in the bowels before, and burning in the anus after, a passage; diarrhœa with ineffectual urging; cold perspiration and trembling; evacuations bilious, slimy, or frothy, or mixed with blood; violent colic; fetid breath; loss of appetite. When *Mercurius* does not relieve, give *Nux-vomica*.

Pulsatilla. — Diarrhœa from indigestion, with pap-like, or watery and offensive evacuations; green, bilious, slimy stool, or when each stool is of a different color from the preceding one; nausea, disagreeable eructations, or vomiting. When the slimy evacuations are mixed with blood, and attended with great straining, give an occasional dose of *Mercurius*.

Nux-vomica. — When there are frequent but scanty evacuations, accompanied with much straining and pressing-down pain in the rectum.

When teething children are attacked with diarrhœa, do not be in too great haste to check it; rather wait a day or two, and if no other symptoms set in, it may not be necessary to give anything; or whenever loose evacuations afford relief from any disorder, from which the patient is suffering, wait a while before you give any medicines; and only where, by its long continuance, or its severity, it becomes necessary to check it, make a selection from the above list of remedies.

ADMINISTRATION OF REMEDIES. — Of the remedy best indicated, dissolve six globules in twelve teaspoonfuls of water; and of the solution give one teaspoonful for a dose, every half-hour, hour, or two, or three hours, according to the severity of the pain, and frequency of the evacuations. Should the liquid produce nausea, give the pills — three or four at a dose — dry, upon the tongue.

DYSENTERIA.

DEFINITION. — Dysenteria — sometimes called bloody flux — is characterized by frequent evacuations of scanty, bloody, and mucous stools, containing little or no fæcal matter. It is essentially an inflammation of the mucous lining of the large intestines, accompanied with an alteration or retention of the natural excretions, and general constitutional disturbance. The mucous membrane is found swollen, thickened, red, and softened, and in severe cases, ulcerated.

Dysenteria is not, as many seem to suppose, an aggravated form of diarrhœa. It is quite different from that disorder; in fact, it is just the reverse; it is constipation, with a constant desire for an evacuation. This constant desire and straining is not caused by the presence of fæces, as in health, but by the inflammation. The swollen and congested parts are tender and painful.

CAUSES. — Dysenteria most frequently makes its appearance in the autumn, when the days are hot and the evenings cool. It is frequently epidemic. It may be excited by cold, exposure to wet; unripe or sour fruit, stale vegetables or meat; drinking cold water when in a heated state, or when perspiring freely. It also frequently manifests itself without any obvious cause.

SYMPTOMS. —The symptoms of dysenteria are too well known to need any lengthy description. In mild cases, there is little or no fever; while in severe attacks, there is high fever during the first few days, marked by frequent pulse; hot, dry skin; excessive ing thirst, etc.

It often, but not always, begins with diarrhœa; the evacuations at first containing some fæcal matter, soon become thin, scanty, and streaked with blood. Blood sometimes passes in considerable quantities, either black, or of a dark-reddish color, resembling the washings of meat.

There is constant straining and desire to stool, with pain and burning in the lower bowels. There is also severe pain just before and after each evacuation; a painful constriction of the anus, termed *tenesmus*. Small children manifest this suffering by restlessness, drawing up of the lip, etc., about the time of an evacuation. There is always more or less fever, nausea, sometimes vomiting and headache.

The number of stools varies, according to the severity of the case, from three to thirty or even forty in twenty-four hours. When fæces reappear in the stools, or the stools increase in consistency, though they may be streaked with blood, it is an indication that the inflammation is subsiding, and the case decidedly advancing toward convalescence.

TREATMENT. — The treatment of dysenteria under homœopathic administration is very satisfactory indeed. It is often asserted by our opponents, that we never meet with the severer cases. This, I am inclined to think, is quite true, but attribute the fact to different causes from what they do. Dysenteria, occurring under severe epidemic visitations, makes sad havoc, when assisted by cathartics, alteratives, astringents, or other forms of *regular* treatment. In view of the happy termination to which this disease is so speedily brought, under what allopathic physicians choose to call the " do-nothing " treatment, one would think a shadow of doubt as to the utility of excessive medication would occasionally steal over them, and they would be tempted, if not induced, to put their patients under proper dietetic regimen and bread-pill treatment. Then, perhaps, they too would meet with a few of those mild cases which providentially fall to our lot to treat. And then, could they be induced to put these few under homœopathic remedies, they might have the happy satisfaction of seeing their patients promptly recover, without relapse or provoking sequelœ.

The principal remedies for dysenteria are, *Aconitum, Arsenicum, Belladonna, Chamomilla, Colocynth, Bryonia, Mercurius, Nux-vomica, Colchicum, Sulphur, Veratrum.*

Aconitum. — For the following symptoms: pain in the bowels; bilious, or thin, watery evacuations mixed with mucus, and sometimes streaked with blood, with rheumatic pains in the neck, shoulders, and limbs; the patient of a full habit; pulse strong and fast; face hot and red; great thirst; loss of appetite; urine hot and red. Inflammatory dysenteria, with high fever, and intense remittent colic.

Arsenicum. — In severe cases, where the stools pass involuntarily, and have a putrid smell; red or bluish spots appear on the skin; the patient is very weak; the pain in the bowels is burning; also, when the patient is restless and uneasy. If *Arsenicum* fails to have the desired effect, give *Carbo-v.*

Belladonna. — When *Aconite*, which seemed indicated, fails to afford relief; when there is dryness in the throat and mouth; tongue coated and red at the tip; tenderness of the abdomen; constant bearing-down pain; frequent small evacuations of blood.

Chamomilla — is sometimes of service after *Aconitum*, when there is still fever, with thirst, headache, nausea, foul tongue, and accompanied with great agitation and tossing about. *Chamomilla* is especially called for, when the disease has been brought on by a sudden check of perspiration.

Colocynth. — When there is severe pain in the bowels, the discharges mixed with green matter, or else slimy, sometimes mixed with blood. — See " Colic."

Podophyllum. — Diarrhœa, with cramp-like pains in the abdomen; light-colored stools, exceedingly offensive; frothy mucus and slimy stools; the child moans and rolls its head from side to side while asleep; sleeps with its eyelids half open.

Mercurius. — This is by far the most important remedy for dysenteria that we have, and is useful in all cases. Its special indications are, severe tenesmus or painful constriction of the anus; urgent desire to evacuate, as if the intestines would force themselves out; after much straining, there is a discharge of light blood, sometimes streaked with mucus or greenish matter; at other times the evacuations resemble scrambled eggs as much as anything; violent straining, both *before* and *after* an evacuation; frequent small mucous stools; violent colic; nausea; shivering and shuddering; great exhaustion and trembling; thirst for cold drinks; aggravation of pains at night.

When there are severe colicky, griping pains, which cause the patient to bend double, *Colocynth* may be given in alternation with *Mercurius.* *Mercurius* is useful in dysenteria of children, where there is much crying and screaming.

Nux-vomica. — Frequent small stool, consisting of bloody mucus; violent cutting pain about the navel; intense heat; great thirst. The pains calling for Mercurius are increased by an evacuation, while those of Nux are relieved by a movement. The symptoms of Mercurius are worse during the afternoon and evening, while those of Nux are worse after midnight and in the morning.

Colchicum. — Dysenteric stools, consisting of white transparent, jelly-like mucus.

Sulphur may be given in obstinate cases, and when the other remedies do not afford permanent relief. An occasional dose will be sufficient.

ADMINISTRATION OF REMEDIES. — The remedies may be either given dry or in solution. When given in solution, which is the better way, dissolve ten pills of the selected remedy in twelve teaspoonfuls of water, and give one teaspoonful every hour, or every half hour, if necessary, until eight or ten doses have been given, when the medicine may be withheld for a couple of hours to await its effect. If then necessary, repeat the same, or select another remedy. As soon as amelioration takes place, lengthen the interval between the doses.

DIET AND REGIMEN. — Dysenteria often arises from taking cold, and is most frequent when the days are hot and the evenings cool. Care should be taken that children are not exposed to the evening air too thinly clad. Children should never be allowed to sit upon the bare, cold stone steps of the stoops, as they are very liable to take cold there.

A patient, suffering from dysenteria, ought to lie in or upon the bed constantly; this is very important. Small children should not be put upon the floor, nor older ones allowed to run around, even during convalescence.

For food you can make use of water-toast, arrow-root, sago, gruels, and the like. When recovering, a little mutton-broth may be allowed. The patient should eat often and but little at a time. For a drink use cold water, or, if preferred, toast-water or barley-water. All kinds of animal food and wines should be avoided, even during convalescence.

Dr. Hering, of Philadelphia, when speaking of epidemic dysenteria, remarks: "If you have one patient in your house, there will soon be more from the use of the same privy. The surest, easiest, and cheapest way of disinfecting, is a solution of copperas, — sulphate of iron, — one part of copperas to twenty parts of water; some of this solution should be mixed with all the discharges of the patient, and a quart or two of it poured every few days into the privy."

PROLAPSUS ANI, OR FALLING OF THE BODY.

DEFINITION. — By prolapsus ani is understood a protrusion or falling down of the lower part or extremity of the bowels. This complaint, when first witnessed by a young mother, causes great and unnecessary alarm, — unnecessary alarm, because there is really nothing dangerous about it.

In health, every time there is an evacuation from the bowels, a small portion of the mucous or lining membrane protrudes and goes back as the parts contract. Now, properly speaking, prolapsus ani is a protrusion of this mucous lining beyond what is natural. It is very common in infancy, and, indeed, it is not uncommon at any period of life.

CAUSES. — No doubt the most frequent cause of this accident is a natural laxity of structure. It also arises from habitual costiveness, straining at stool, diarrhœa, hœmorrhoids, drastic purgatives, worms, and other causes.

TREATMENT. — The first thing to be done is to replace the protruded membrane. This should be accomplished as speedily as possible. Sometimes it will return itself, if the child is laid upon its back, with a pillow under the hips, so as to raise them up a little. If it does not return of its own accord, then, after protecting the protruded parts by laying over them a piece of soft, smooth cloth, wet with warm water, or sweet oil, embrace it with the ends of the fingers, and gently and steadily press it upward, not using a great deal of force until it slips in, which it will do in a minute or two if the operation is rightly performed. If it has become red, swollen, and inflamed, do not be in a hurry to reduce it, but place upon it rags wet in a weak solution of Arnica water, and give a dose — three pills — of *Nux-vomica* every half-hour. As soon as the inflammation subsides, the bowel may be returned. When once returned, great care should be taken to prevent a repetition of the trouble, and happily we have remedies which will diminish the tendency to this troublesome disorder.

Ignatia is one of our principal remedies. It may be given once every twenty-four hours for six or eight days; then discontinue for a few days, and, if not better, give a dose of *Sulphur* every other evening for one week.

Nux-vomica. — When there is a great deal of pain and straining, especially in young children, and those subject to constipation. Give same as *Ignatia*.

Sulphur. — This is an excellent remedy for both recent and chronic cases. A dose of three globules may be given every twelve hours. *Sulphur* may be given in alternation with *Nux-vomica*. Give *Sulphur* in the morning, and *Nux-vomica*, at night for one week, then discontinue for one week; if not better follow with

Calcarea. — A dose of three globules, every twelve hours. *Calcarea* is an excellent remedy for obstinate chronic cases where other remedies have failed.

Mercurius. — When the protruded intestine is much swollen, or is bluish, or bleeds and pains much when at stool. Follow *Mercurius* with *Ignatia*. Dose, same as of *Ignatia*.

This treatment will be greatly assisted if care is taken to keep the bowels in a regular condition. The child should be accustomed to use the chamber at regular intervals. The ordinary chamber is faulty in its construction, and unfit for children suffering from this disorder; the hole is too large and of an improper shape. Instead of being round, it should be oval and about half the ordinary size. The child should be watched to prevent its over-straining while sitting upon its chair, or remaining upon it too long, particularly if the bowels are any way constipated.

Cold hip baths, or sponging with cold water, and sometimes cold water injections, are of great service. The temperature of the water should be graduated according to the age and vigor of the child.

DIET. — The diet should be the same as that observed in derangements of the digestive organs in general. If possible the diet should be so governed as to prevent either constipation or diarrhœa. The child may be allowed as much cold water as it wants to drink.

RUPTURE, OR HERNIA.

DEFINITION. —VARIETIES. — FREQUENCY. — By the term rupture, or hernia as physicians call it, we are to understand a swelling formed by the protrusion or escape of a portion of intestine from the cavity of the abdomen. The places, at which these swellings most frequently make their appearance are the navel and the region of the groin.

The point of egress selected by the hernia gives it a particular name to express its position; as, umbilical, when it appears at the umbilicus or navel; inguinal, when it appears in the groin.

There are several varieties of hernia, but three only are especially met with in children, namely: *umbilical, inguinal,* and *oblique inguinal.* The latter variety is where the intestines have intruded into the scrotum.

Hernia is termed *reducible,* when it can at any time be returned into the abdomen, and *irreducible,* when, without inflammation or obstruction to the passages of fæces, it cannot be returned to the cavity of the abdomen, either owing to adhesions or entanglement of the intestines. *Strangulated,* when the protrusion is not only incapable of being reduced, from constriction of the aperture through which they passed, but the circulation is arrested; the passages of fæces towards the anus cut off; inflammation sets in; the tumor becomes hard, and tender to the touch; pain, nausea, and vomiting occurs, accompanied by other alarming symptoms.

The first two varieties of hernia are not of uncommon occurrence in children of all ages.

CAUSES. — Children whose muscular development is not compact, but, on the contrary, relaxed and flabby, leaving the natural outlets of the abdomen unusually large, or capable of easy enlargement, are more prone to accidents of this nature than those

40

who are robust and strong, having their muscular fibres closely and firmly knit together. The weakest parts are those at which the accident most frequently occurs. And in children where there is general or local muscular debility, either from imperfect development or recent indisposition, the most trivial circumstance, as crying, coughing, or straining, may produce hernia; but in other cases, where no such predisposition exists, the protrusion only takes place under great bodily exertion, or in consequence of external injury.

SYMPTOMS. — Umbilical hernia need not be mistaken for any other tumor. Those appearing at the groin, however, are sometimes so closely simulated by other diseases, that mistakes may readily be made by any but a physician. The general symptoms of hernia are an indolent tumor upon some part of the abdomen, usually, in children, at the navel or groin. The tumor appears suddenly, is developed above, and descends. It is subject to changes in size, being smaller when the patient lies upon his back, and larger when he stands upright. The tumor diminishes when pressed upon, and grows larger when the pressure is removed. It is larger when he is coughing, sneezing, or drawing a long breath. Vomiting, constipation, and colic are often present in consequence of the unnatural situation of the bowels.

TREATMENT. — In every case of hernia, no matter how slight or trivial it may appear to your inexperienced eyes, send immediately for your family physician, and ascertain from him its precise nature and probable termination. It is of the utmost importance that a cure should be effected during childhood; otherwise the individual will, in after years, suffer great inconvenience, be unfitted for any kind of manual labor, and may any day be in danger of losing his life.

Rupture at the navel is by far the most frequent form in which hernia appears in young children. It is generally first observed about two months after birth. You will readily recognize it by the unnatural protrusion of the navel. The navel, instead of closing, as it should have done, remains open, allowing a portion of intestine, covered by the skin and integuments, to escape from the cavity of the abdomen.

The hernia, or swelling, thus formed varies in size from a hazelnut to a walnut, always, however, increasing in size when the child strains, either by coughing or otherwise. It is not often painful, unless it becomes very large.

The first point to be obtained, in the treatment of these cases, is to keep the intestines permanently within the cavity of the abdomen, so as to give nature an opportunity of closing the opening. This can best be accomplished by covering a piece of cork or pasteboard with soft muslin, and then binding it over the opening with a broad bandage. You had better, for the first time at least, let your physician prepare your compress, and show you how to apply it. Physicians use different appliances, — some one thing and some another. The object to be attained by all, however, is as above stated; and whatever is best adapted to the particular case is always the best, no matter of what it consists.

You should take particular pains to inquire how, and precisely where, the compress is to be applied, and be careful to keep it always in place. Whenever it is removed, either for the purpose of bathing the child or to apply a fresh bandage, the nurse should invariably place her finger over the navel, to guard against the bowels coming through the opening. This will require constant care and watching; but perseverance will generally prove successful. The cure will be much facilitated by frequent bathing in cold water. The only remedies that it will be necessary to give, are an occasional dose of *Nux-vomica* or *Sulphuric acid*. I prefer the *Sulphuric acid*, and generally give a dose of three globules every evening for about a week.

Rupture at the groin is much more troublesome to treat than that at the umbilicus, as it is impossible to keep a truss, or any other mechanical means properly applied to the parts, especially with children under a year or eighteen months old. At least, such has been my experience. You may try a truss, if you feel disposed; but, be as vigilant as you may, the instrument will become displaced in spite of you; the straps constantly getting soiled, irritating the child, making it cross and fretful, until you will be compelled to abandon all hope of help from mechanical appliances, at least until a later period. "Moreover, the very fretfulness caused by wearing the instrument at this time and under these circumstances tends to increase the size of the rupture. You must be content to bathe the parts with cold water night and morning, and keep the child as tranquil as possible to avoid crying." *Nux-vomica* and *Sulphur* are highly recommended, not only for simple rupture, but also for the peculiar disposition or habit which leads to the formation of hernia.

The *Sulphur* may be given every morning for four successive days, to be followed every evening by *Nux-vomica* for the same length of time. Then wait eight days, and, if there is no manifest improvement, repeat the remedies.

If there should be diarrhœa, give a dose — three pills — of *Chamomilla*, every three hours.

When hernia results from a fall, or an injury of any kind, *Arnica* or *Rhus* will be the appropriate remedies, and may be given from two to four hours apart.

You will generally succeed, by this treatment alone, in effecting a cure long before the time arrives when a truss can be usefully resorted to.

Wherever and whenever a truss is used, care should be taken, that the bowel is properly returned to the cavity of the abdomen. If the patient perceives, after the truss has been adjusted, that something still protrudes, it is a sure sign that the truss is either good for nothing, or has been improperly applied.

It is of great importance to know how hernia should be reduced. In order to accomplish it properly, the patient should lie upon his back, with a pillow under his hips, so that the ruptured part will be higher than the rest of the abdomen; he should then incline a little to the ruptured side, so that the muscles of the abdomen may relax as much as possible. And then the reduction may be accomplished by gentle pressure upon the tumor with one hand, while, with the fingers of the other hand, the tumor is grasped, so as to direct it backward through the aperture from whence it protruded. The effort should be continued gently and patiently a sufficient length of time gradually to effect the reduction.

Alarming symptoms sometimes accompany hernial protrusions. Should there be violent burning in the abdomen, as from a hot coal, with tenderness of the tumor, the least touch giving pain; sickness at the stomach, with bitter bilious vomiting; nervousness and cold perspiration; give *Aconitum*, a dose every half-hour. In case Aconitum only alleviates the symptom for a short time, without any permanent good, try *Veratrum*, in the same manner, for two hours. If your efforts should fail to reduce the hernia, give *Sulphur*.

Now, I have given you the above synoptical directions, in regard to the reduction of hernia, not for the purpose of fitting you to take charge of such cases, but simply that you may understand the nature of the disease, and its appropriate treatment. Should you be so circumstanced, as to meet with a case where no physician can be had, you might in such extremity offer some assistance to the sufferer. Cases will occur, when necessarily three or four hours must elapse before a physician can be had; in such emergency you can, with perfect propriety, make use of the means which I have pointed out.

I do not deem it safe for any but a competent physician, or surgeon, to take charge of a case of hernia; therefore I would not have you assume the responsibility, unless it were in a case of extreme necessity.

JAUNDICE. — ICTERUS.

DEFINITION. — By physicians this disease is called *Icterus*, which is the Greek name for a bird with golden plumage, — the golden thrush, — the sight of which by a jaundiced person was death to the bird and recovery to the patient; at least so Pliny tells us. — WATSON.

Jaundice is not specially a disease of childhood; nevertheless, it appears sufficiently frequent in children to claim a passing notice in such a work as this.

It is characterized by yellowness of the eyes and skin; whitish stool; urine having the color of saffron, and communicating a yellow tinge to white linen; deranged digestion, and sometimes pain in the region of the liver.

CAUSES. — DURATION. — Jaundice depends upon various and very different internal causes. It may arise from inflammation of the liver, from obstruction of the gall-duct, from diseases of the bowels, and from fevers. And frequently we cannot ascertain, at all, even in the simplest cases, what the precise cause may be. It is not unfrequent among the studious and indolent — those leading sedentary lives, or lives of inactivity, — and is common to all ages and sexes.

The most frequently recognized, exciting causes are, severe mental emotions; as rage, fright, grief, anger, despondency, and irritability of temper. Also particular kinds of food, the excessive use of strong coffee, acids, unripe fruit, and indeed any error in diet, which has a tendency to disarrange the digestive apparatus. It is not unfrequently caused by the inordinate use of quinine, rhubarb, or calomel.

Its duration depends in a great measure upon the exciting cause, the constitution of the patient, and the treatment which he is under. It may last a few days, or it may last a week, and, indeed, if improperly treated, it may last for years. When it depends upon some mental or moral cause it is of much shorter duration than when it depends upon some organic disease.

SYMPTOMS. — This disease generally makes its appearance preceded by, or accompanied with, great languor, depression of spirits, slight chills, or rigors, and flushes of fever, loss of appetite, giddiness, constipation, flatulence, sour eructations, and sometimes there is nausea and vomiting; there is also a sense of weight and uneasiness about the chest and abdomen, with some pain in the region of the liver. There is frequently a disagreeable itching or tingling sensation in the skin before the discoloration appears. The yellow tinge begins in the eyes and extends to the temples, brow, and face, then to the neck, chest, and whole surface of the body. In some spots the color is deeper than in others, especially so in the folds and wrinkles of the skin. The color varies from a light yellow to a deep lemon or a greenish-brown.

Constipation is generally present; the evacuations are scanty, and of a pale-clay color, indicating an absence of bile. The urine is commonly high-colored, at first yellow, afterward of a deep-saffron color. Bilious sweat sometimes occurs, staining the patient's linen yellow.

The characteristic yellow hue of the skin is owing, no doubt, to the presence of bile, or its coloring matter in the blood, and the deep tint of the urine is derived from the same source.

There is, for the most part, but little fever attending the milder forms of this disease, but in bad cases there frequently is a high degree of fever accompanied with a stupid sleep from which it is difficult to arouse the patient. This latter symptom is caused, we presume, by the retained bile which acts upon the nervous system like a narcotic poison. When the disease assumes this aspect which fortunately is not common, it is regarded as dangerous, and · death may follow in a short time.

Jaundice, as we ordinary meet with it, is not a dangerous disease, and you will seldom find any difficulty in successfully combating it with the following

TREATMENT. — *Mercurius* will be found the best remedy in most cases, provided the patient has not already abused this drug. A dose of six pills may be given once in four hours. If not better in the course of three or four days, give *Hepar-sulphur* or *Sulphur*, in the same manner. Or *Chinchona* and *Mercurius* may be given, in alternation, once in four hours. *Mercurius* is specially indicated when the disease appears to have arisen from a derangement of the digestive system.

When jaundice arises from a fit of passion, *Nux-vomica* is about the best remedy; it may be given in alternation with *Chamomilla*.

Nux-vomica is specially indicated in those cases which are occasioned by indolence, or sedentary habits; also, when jaundice is accompanied by constipation, or by constipation alternating with diarrhœa.

Sulphur or *Lachesis* will be most suitable to eradicate the disposition to the disease. They may be given alone, or in alternation, one dose of six pills every other night.

Where bilious symptoms appear, with yellowness of the skin, in persons who have taken much calomel, give *Chinchona* or *Hepar-sulphur*. One dose every night and morning.

DIET AND REGIMEN. — The diet should be light and free from stimulants of every kind, either in the form of condiments or drinks. Avoid all indigestible substances or coarse food; live rather upon soups, broth, gruels, plain puddings, and the like.

Baths of lukewarm water are often of great service.

The constipation, which is sometimes very troublesome, may be relieved by cold-water injections, assisted by friction of the abdomen and sides.

The patient should be kept warm and of an even temperature; perspiration should be encouraged as much as possible. In chronic cases the patient should take active out-door exercise.

JAUNDICE OF INFANTS.

DEFINITION. — FREQUENCY. — CAUSE. — It often happens in infants of two or three days old, that the skin assumes a yellowish hue, presenting, as far as appearance goes, a perfect picture of jaundice. Appearances, however, are sometimes deceptive; they are especially so in cases like this. This peculiar tinge of the skin, which appears a few days after birth, we have good reason to believe is *not* jaundice, and has no relation whatever to the biliary organs. In fact, it has no resemblance to the icterus of adults or of older children, except the color of the skin; it causes little or no constitutional disturbance, — the child, in fact, to all appearance, remaining perfectly well.

It is accounted for differently by different authors. Dr. Watson, in speaking of this disorder, says: " The surface of an infant at its birth is frequently of a deep red, from congestion of blood, presenting a condition which falls little short of a mild but universal bruise.   By degrees the redness fades, as bruises fade, through shades of *yellow* into the genuine flesh color." Others, finding fault with this explanation, attribute the discoloration to defective respiration, and the impaired or imperfect performance of the function of the skin, which is no explanation at all.

However, it matters but little as I know of, what causes this peculiar hue of the infant's skin ; it is seldom of any consequence producing no disturbance or sickness in the child, and hardly ever continuing over two or three days.

Of course, true jaundice may, as well as most other diseases, befall children of all ages, — even the infant born only yesterday, — but it is rare indeed.

The treatment of this simple form of jaundice may be embraced in a few words : *Chamomilla* may be given at first, one dose of one or two pills, every four hours, especially should the child be nervous, or irritable, or if the bowels are loose.   Should this fail to remove the disorder, give *China* in the same manner.

If the child is wakeful and extremely cross, and, especially, if, in addition, there is constipation, give *Nux-vomica.*

Now let me tell you what *not* to give, — *saffron tea.*   It has no resemblance whatever to the disease except its color.   It never did, and it never can do any good ; therefore, do not use it.

A species of true jaundice does attack nursing children, from the affect of grief, anxiety, or a fit of passion on the part of the mother or nurse.   The treatment in such cases will be obvious to all.

WORMS.

DESCRIPTION, VARIETIES, SEAT, FREQUENCY. — In commencing this article I would state, that I am an unbeliever in the orthodox faith of worm affections.   Not that I doubt the existence of worms in children, nor that they aggravate existing diseases.   But I do most assuredly doubt that worms were ever yet the sole and originating cause of any disease either in the child or the adult.   That worms do exist in the alimentary canal of all children is a fact beyond dispute ; but all modern writers agree that their significance has been greatly overrated.   Popular opinion, I am aware, attaches great importance to them, and can see proofs of their presence and depredations in almost every case of gastric or intestinal disease. But it is next to impossible to find an educated physician who will point out to you a disease and affirm that it was *caused* by worms. On the contrary, there is no lack of intelligent physicians and physiologists, who not only believe that worms do inhabit children, but that they fulfil some wise and necessary purpose which we know not of. Dr. Rush, an eminent physician, when writing upon this subject, says: " When we consider how universally worms are found in all young animals, and how frequently they exist in the human body, without producing disease of any kind, it is natural to conclude that they serve some useful and necessary purpose in the animal economy."

The origin of these entozoa is a question the answer to which physiologists have never satisfactorily demonstrated. Some assert that their eggs or germs are introduced from the exterior world, while others, not exactly understanding how this can be, particularly when they are found enclosed in shut cavities, fall back upon the theory of spontaneous generation. But undoubtedly they exist, from the child's birth, in numbers sufficient to fulfil the end for which they were designed, and it is only when disease or a hereditary habit of body favorable to their development exists, that worms show themselves in any quantities, or by their presence produce any disturbance.

Worms, as such, are not injurious. They exist in many children without their presence even being suspected. If worms were the *cause* of disease, their mere expulsion would be sufficient to remove the symptoms attributed to their presence. But no such beneficial results follow the administration of vermifuges, although numbers of worms are killed and expelled by their use. On the contrary, if anything, they aggravate the case by driving the worms from the fæces, which they naturally inhabit, to the mucous surface of the intestines, besides exciting a secretion upon which they multiply and flourish.

The proper method of removing what are termed worm-symptoms is to remove that disordered condition of the system which is favorable to the development and support of the worms. The mere *expulsion* of the worms is a matter of no consequence ; keep the children's digestive apparatus in a healthy condition, and you may rest assured that worms will never trouble them.

In treating these cases, always bear this in mind, that it is not the worms merely that you wish to remove, but that habit of body which favors their accumulation in such quantities as we some-

41

times find them. You cannot get rid of them entirely; perhaps it is as well you cannot; but as soon as the system recovers from its diseased condition you will find no more of them; they will return to their unobtrusive quiet again, or, at least, the "worm-symptoms," which were the manifest difficulty, will have subsided, although perhaps it will be impossible to find a vestige of worms in the passages.

There are five different species of worms which infest the alimentary canal, but two of these are peculiar to children.

The first and most troublesome is the common seat-worm, — *Ascaris-vermicularis,* — thread-worm, pin-worm, or maw-worm, as it is variously called. This is the smallest of the intestinal worms, measuring only from two to five-twelfths of an inch in length, and looks about like a small piece of white cotton thread. Their number is extremely variable; there may be three or four, ten or twenty, or they may exist in innumerable hosts, rolled and knotted into balls, making formidable obstructions in the intestinal canal. They are usually found in the large intestines and rectum. They frequently crawl into the urethra and vagina, causing a troublesome itching, and a mucous discharge, and not unfrequently revealing the seat of sensations which may lead to evil consequences.

The next species of worms most frequently found in children is the long, round worm, — *Ascaris lumbricoides,* — which very much resembles the common earth-worm. It is of a yellowish, dirty-white color, and varies from six to twelve inches in length. The small intestines is the favorite locality of this species of worm, but they traverse all parts of the alimentary canal. They are found in the large intestines, from which they are frequently expelled by stool. They are sometimes found in the stomach, and not unfrequently find their way into the throat. It is not at all uncommon for children to eject them by vomiting.

CAUSES. — It is generally believed that a disposition toward worms, which exists principally in children, is favored by the use of coarse farinaceous, or an exclusively vegetable diet. The use of much sugar, fat, cheese, butter, fruit, or any other diet, or circumstance which enfeebles or disarranges the digestive system, strongly predisposes to their production. Children of a lymphatic or scrofulous constitution are more disposed to them than others; and those living in damp, dark, and unclean dwellings, or in marshy regions, are more prone to worm-affections than those otherwise situated.

SYMPTOMS. — There is no single symptom, or group of symptoms, that I know of, by which we can to a certainty, diagnose the presence of worms in children. The expulsion of a number, or their appearance in the stool, is proof that they did exist, but no evidence that more remain ; for, as far as my experience goes, I have seldom known more than eight or ten to be expelled at a time, or rather within a week's time ; though the child may have been all the while under the active effect of vermifuges sufficient to kill any living thing within the system, in fact, coming near killing the patient itself.

Do not jump to a conclusion, and imagine, because your child passed five or six worms, that there necessarily must be more, and that they are the cause of your child's indisposition ; for various disorders of the digestive tube may and do exist, simultaneously with, and yet independent of, the presence of worms. I do not, therefore, think it fair always to conclude that the symptoms under which the child labors are still the result of worms, because there may be no more in the bowels.

The following extract of symptoms is taken from Teste's " Diseases of Children," page 283 : —

" Sudden and frequent changes in the color of the face, which is sometimes red, sometimes pale, sometimes lead-colored ; bluish semi-circles circumscribing the lower eyelid ; increase or diminution of the brilliancy of the eyes ; uncertainty or momentary fixedness of the look ; dilatation of the pupils ; itching of the nostrils ; bleeding of the nose ; headache after meals ; flow of saliva in the mouth, especially during the night ; tongue a little dry, with red dots upon the point and edges ; insipid, acid, or fetid odor of the breath ; thirst on awaking ; capricious appetites, great hunger, or dislike of food ; uneasiness, increased by abstinence ; enlargement of the abdomen ; from time to time a pinching pain, or twisting sensation in the abdomen ; frequent rumbling in the intestines ; sudden vomitings without apparent cause ; slight diarrhœa ; from time to time very abundant and fetid stools ; itching in the anus ; short dry cough, or even violent attacks of cough, like that of a severe cold, with or without glairy expectoration ; sorrowful, unequal, and fantastic humor ; attacks of fainting, which may return a great number of times in the same day ; disinclination for labor ; agitated sleep, talking in the sleep ; nightmare ; aggravation of the symptoms in the morning, especially when fasting ; propensity to onanism, leucorrhœa ; lastly, convulsions, delirium, epileptiform attacks, etc., etc."

It is not necessary that a child suffering from verminous affection should present all the symptoms above enumerated. In fact, I should think it a rarity, indeed, to find a patient in whom all these symptoms were well marked. Many of them represent other disorders, which it is not impossible should exist in conjunction with, yet independent of, vermin.

TREATMENT. — In examining the various symptoms usually attributed to worms, we shall find many of them the same as those which characterize some one of the many forms of gastric or intestinal disorders. In the majority of cases, however, the same remedies will suit both classes of disease, because they both spring from one and the same disturbing influence.

We shall enumerate here only such remedies as may be properly called our worm medicines. And, should the symptoms for which they are prescribed refuse to yield to them, the reader had better examine the case closely, and see if the symptoms do not depend upon some other derangement, such as chronic dyspepsia, diarrhœa, or some inflammatory disease of the intestinal mucous membrane ; and, if such is found to be the case, refer to such diseases under their respective heads.

I generally commence the treatment of these cases with a few doses of *Aconitum;* not that *Aconite* is specially adapted to all cases of verminous affections, but I find it a valuable remedy to quiet the child by allaying the nervous irritability, and thus render the symptoms more tractable to the remedies which follow.

Aconite will also be found of service when there is considerable fever, with colic and distention of the abdomen ; great restlessness at night ; irritability of temper ; constant itching and burning at the anus. A dose may be given once in two hours. If this proves ineffectual, follow it with *Ignatia,* or it may sometimes be alternated advantageously with *Stannum.* Sometimes, though promptly relieved by *Aconite,* the symptoms will return with every new and full moon. When such is the case, *Sulphur* or *Silecia* may be given, one dose every morning for four or five days.

Cina. — This is the chief remedy for all complaints really arising from worms, especially when the following symptoms are present: boring with the fingers in the nose ; picking the lips ; changing of the color of the face, being at times pale and cold, at others, red and hot; tongue coated white ; mouth full of tough mucus ; capricious appetite ; cross and fretful temper ; bloated face, with livid circle around the eyes ; distention and pain in the abdomen ; constipation, or loose evacuations; fever, especially at night, with pain in the head, starting or talking in sleep; grating the teeth; itching at the fundament; the crawling out of thread worms.

Nux-vomica. — When, with worm symptoms, there is constipation; severe itching; burning and pricking sensation at the anus, caused by little worms, which may be discovered by drawing the fundament open.

Spigelia. — In cases of worm colic, when there is abdominal pain, accompanied with fever, urging and scanty discharge of slimy stool.

Silicea. — Especially for scrofulous children, and where worm fever assumes a slow, chronic form, with or without diarrhœa. If this remedy fails to afford permanent relief, give a dose of *Calcarea*, and repeat it every other night for one week.

Lycopodium. — I have accomplished more with this remedy in worm affections than with any other, especially where there is much itching at the fundament.

Sulphur may be given at the end of every case, especially where other remedies have removed the distressing symptoms. One dose may be given every other night for one week.

Teucrium — " Is unquestionably a specific for the irritation, itching, and uneasiness caused by pin-worms. I use it thus : mix from three to four drops of the tincture in a tumbler half full of cold water ; stir it well, and give to a child two, three, or four years old, a large-sized teaspoonful, morning, noon, and at bedtime, for two days." — FRELIGH'S *Materia Medica.*

ADMINISTRATION OF REMEDIES. — In addition to what has been said, we may say that, to make a permanent cure of these cases, the treatment should be continued for some little time, and, further, when you have once selected a remedy, do not be in too great haste to change it for another, even though you do not see any immediate improvement from its administration. If, after a fair trial, however, it fails to afford relief, and the case remains the same, or grows worse, make a new selection, and at the same time pay particular attention to the diet.

DIET. — Avoid all gross, heavy nourishment, such as too much bread and butter, potatoes, or boiled vegetables of any kind ; also, rich puddings, pies, or cakes, and pastry in general. Rather give the patient meat-soups, roasted or broiled meat, plenty of cold water and milk. Exercise in the open air is very essential.

ADJUVANTS. — In ordinary cases of verminous affections, the above mode of treatment will be found amply sufficient to effect a prompt and speedy cure, but we occasionally meet with cases where the itching at the fundament occasions so much annoyance to the patient, that immediate means for prompt relief are absolutely demanded. In such cases I have been in the habit of using an occasional injection of cold water. In cases of severe itching, and when the simple water afforded no relief, I have added a little salt to it, or you can use a weak solution of vinegar and water, or lemon-juice and water ; either will be found at times of great service.

In mild cases, where the itching was troublesome, I have found sweet oil to answer the purpose ; it may be applied both externally and internally.

EPIDEMIC CHOLERA.

GENERAL REMARKS. — This disease, which is said to have destroyed more than sixty millions of human beings, is now raging in Europe, and presenting all the tendencies to travel westward that it heretofore has shown previous to its formerly visiting this country. It can be descried afar off as plainly and as certainly as the rising of a storm. We can trace its progress and direction over the face of the earth as plainly as we can trace the direction and progress of the clouds through the heavens. The probabilities are that it will visit this country with the warm weather of 1866.

It, therefore, becomes the pre-eminent duty of physicians to lay before the public such advice and instruction as to the amount of danger to which we shall all be exposed, and to give each one an opportunity to " set his house in order," and to throw up such barriers against the advance of the disease, as investigation and experience have taught to be of the most avail. For these reasons, this article, though not otherwise within the scope of the present work, has been introduced.

Cholera is, to a great extent, a nervous disorder, and where dread and panic prevail, it will reap its richest harvest. Fear will cause at once the premonitory symptoms. It will be of the first importance, then, to calm the nervous and unreasonable alarm of all classes. This can easily be accomplished, by giving the people to understand that cholera is a disease which can be generally warded off, if they but pay proper attention to known hygienic laws, and, also, that it is a disease easily managed,— at least with homœopathic remedies,— if met early enough, and that we have simple yet efficient remedies, which, if timely administered, will, in nine cases out of every ten, arrest all the premonitory symptoms and cut short the disease.

Cholera is not a disease difficult to manage, or peculiarly dangerous; neither is it a disease that can be communicated from one person to another by the touch or the breath, by clothing or the like. Ordinarily speaking, cholera is not a contagious disease. Therefore, do not fear to assist your friends or neighbors, should any of them be stricken down. Always remember, however, that from the peculiar epidemic constitution of the atmosphere, indiscretions of all kinds, which, under other atmospheric conditions, or which had heretofore occasioned but slight derangements, as headache or cold, are almost certain to produce the prevailing complaint, cholera. Therefore, while assisting others, be careful not to expose yourself by over-taxation, anxiety or long fasting. Care should be taken to avoid all the general causes of disease; such as late hours, late suppers, over-eating, over-drinking, damp apartments or streets, crowded rooms, excitement of all kinds, excessive joy, grief, anger, fear, fatigue, and all exhausting employment of mind and body. Cultivate cheerfulness, and habitual calmness of mind; dispel all fear; keep your person cleanly; live upon a good, generous diet; avoid all acids and high-seasoned dishes; abstain from all unripe fruits and vegetables that are not of the freshest kind. Plainly-cooked meats, potatoes cooked dry and mealy, bread, rice, and a few other wholesome articles of diet, should be strictly adhered to, to the exclusion of all fancy dishes, and indigestible articles.

People who live in clean, well-ventilated houses, have wholesome food, are well-clad, and are regular in all their natural habits, have very little to fear from cholera.

There is not much danger of cholera producing great ravages, except in such localities as are notorious for filth, and absence of general hygienic care. The fever-nests and homes of diphtheria, of typhoid and scarlet fever, are the favorite haunts of cholera.

The districts near the bone-boiling establishments, distilleries, swill-milk stables, slaughter-houses, pig-sties, and malarious pond-holes, suffer most when cholera is abroad.

Open privies, and cesspools, heaps of garbage near the house, or in the cellar, crowds of inmates in dirty rooms and damp basements, and even the moderate use of alcoholic drinks, are what *especially* stimulate the disease.

The gutters of the streets, the privies and cesspools should be frequently cleaned, and daily sprinkled with fresh chloride of lime. Damp basements and rooms should constantly have an *open* fire in them, to burn up the foul air, and cause a more perfect ventilation.

Chloride of lime, or any other disinfectant, should not be used in the house, and especially in the sick-room. There is no evidence that they have the slightest influence in freeing the atmosphere from its infecting property, whatever it may be ; and the fumigation of the house is decidedly prejudicial to good health. Keep your house well ventilated and clean.

An important point to which I wish to call particular attention is, that many people imagine that they fortify themselves against disease, by keeping their bowels in a relaxed condition, and consequently are in the habit of taking an occasional cathartic ; others, with equal wisdom, think, if they have a diarrhœa, it arises from some offending substance within, which must be carried off ; and they take purgative medicines to get rid of it. Both of these are serious and often fatal mistakes.

Diarrhœa, during cholera seasons, has a peculiar tendency to run on, if not checked, into the more perilous form of the disease ; and nothing is more certain to insure and hasten that catastrophe than purgative medicines.

The least looseness of the bowels should receive prompt attention ; and those remedies which will quiet the irritation, and stop the copious discharges, not by their astringent properties, but by their direct specific effect upon the disease, are the best.

The great desideratum for the successful treatment of the disease is to meet it early.

In tropical climates the course of the disease is described as one " *exhausting march*," from the beginning to the termination in death. As it has heretofore appeared in the United States, it has always been preceded by a well-marked premonitory stage, of from one to two or three days' duration ; such as confusion of the head, languor and debility, derangement of the stomach, and a tendency to diarrhœa. As a general thing, these symptoms occasion but little, if any, pain or uneasiness, and, what is singular, excite no alarm. This is unfortunate ; as a little wholesome fear would generate precaution ; and the administration of proper remedies at this stage of the disease, is almost certain to arrest it.

At present the newspapers are teeming with specifics for cholera; and, under the sanction of what the populace may consider high authority, we see recommended a most pernicious course of treatment. It is to place the patient in bed, apply a mustard-plaster to the pit of the stomach, and give him a good dose of opium or laudanum.

In regard to the blood-purifiers, vegetable pills, cholera mixture, and patent nostrums generally, which are thrust before one at every turn, it is scarcely necessary to warn the readers of this book; none but the most ignorant ever think of resorting to them.

The question is frequently asked, " What will the homœopathists do when they are called to treat cholera? Will they rely upon small doses?" In reply, I would state that I think they will, and for the following reasons: that, while allopathia loses 54 cases out of every 100, homœopathia loses but 9 out of the same number. *Homœopathia is the only treatment that has ever proved itself worthy of any confidence in cholera.*

Dr. Lobethal, of Germany, who had charge of a large cholera hospital — *allopathic* — during the epidemic of 1831, and who treated an immense number of cholera patients *homœopathically*, in the summer of 1847 and 1849, observes: " It has been reserved to the 'specific' healing art, generally known under the name of homœopathia, to stand the test of practical observation, and to demonstrate its superiority in combating this fearful disease, — cholera, — the appearance of which, followed by an immense number of well-substantiated cures, has tended in the highest degree to the spread of the new healing art."— LUTZ, *Practice*, p. 128.

Dr. Balfour, of Edinburgh, who is opposed to homœopathists, writes to Dr. Forbes, from Vienna, in 1836, in the following words: " During the first appearance of the cholera here, the practice of homœopathia was first introduced; and cholera, when it came again, renewed the favorable impulse previously given, as it was through Dr. Fleischmann's successful treatment of this disease, that the restrictive laws were removed, and homœopathists obtained leave to practice and dispense medicines in Austria. Since that time, their number has increased more than threefold in Vienna and its provinces. No young physician settling in Austria — excluding government officers — can hope to make his bread, unless

42

at least prepared to treat homœopathically, if requested." — Jos-
LIN, *on Epidemic Cholera*, p. 70.

James Johnson, M. D. — Physician (allopathic) Extraordinary
to the king of Great Britain — says: "When cholera appeared in
Hindostan, the papers so teemed with specifics and cures, that the
government put a stop to their further publication on account of
the mortality they caused." He continues: "For ourselves, what
shall we say? Alas! we must own that we are gloomy, heartless
sceptics, without so much as a grain of faith, or one saving particle of belief. Would that it were otherwise — would we could
only so much as imagine that cholera has been, is, or will be cured
by the thousand and one plans of happy memory, already published, or to be published. In point of fact, *we know no better
mode of treating cholera than when it first* appeared in the island;
and the really severe cases are just as fatal as they ever have
been." — MARCY & HUNT, *Theory and Practice*, vol. i. p. 363.

Probably in no part of America did the cholera appear with
more violence in 1849 than in Cincinnati. Two physicians, Drs.
Pulte and Ehrmann, treated 1116 genuine cholera patients in all
stages of the disease, *and with a loss of only* 35, *or about* 3 *in* 100.

Since homœopathia can produce such results, we feel confident
that it will cure almost every case, provided the physician is called
soon after the attack, and before the patient has been poisoned by
massive doses of calomel and opium; and, while homœopathia
continues to produce such results, we are inclined to think its
adherents will be content with "small doses."

PROPHYLACTICS, OR PREVENTIVE TREATMENT.

During the prevalence of cholera much may be done toward
fortifying the system against its attacks by a judicious administration of homœopathic remedies. I say "judicious administration,"
because I believe the too frequent repetition of large doses, that
is, the crude drugs, or even the first potency, would do more
harm than good. Hahnemann recommended *Cuprum* and *Veratrum*, of the 30th potency, to be taken in rotation every six or
seven days. His advice was, "First, take one dose of *Cuprum*,
30th; then wait one week and take a similar potency of *Veratrum*; after another week has elapsed, take *Cuprum* again, and
so on.

Perhaps this method cannot be improved on. I certainly should
not think of giving a lower potency.

Where a whole family is to be protected, dissolve twelve globules in as many spoonfuls of water, and let each member take a spoonful or two, as above directed.

The *tincture* of *Camphor* in drop-doses once or twice in the twenty-four hours is also recommended. Undoubtedly, *Camphor* is a valuable antidote against the poison of cholera, but its effect is too transient to be of much service as a prophylactic; besides having a tendency to interfere with other medicines, it had better be reserved for the premonitory stage.

Every family should be provided with a well-stoppered phial of Camphor, so that in case of emergency there will be no delay. The ordinary tincture procured at the druggists will answer. Or you can make it yourself, by dissolving one ounce of gum Camphor in ten ounces — *i. e.* 2½ gills — of Alcohol.

Persons who carry Camphor about with them in their pocket, — as some do while travelling, and which is very prudent, — should be particular to have it well corked.

Dr. Hering, of Philadelphia, says: " *The surest preventive is Sulphur.* Put half a teaspoonful of *Flowers* of *Sulphur* into each of your stockings and go about your business; never go out with an empty stomach; eat no fresh bread nor sour food. This is not only a preventive in cholera, but also in many other epidemic diseases. *Not one of the many thousands who have followed this, my advice, has been attacked by cholera.*"

Though Camphor and Sulphur may be good prophylactics, I should rather put my trust in Hahnemann's method.

Cholerine is the name given to the diarrhœa which prevails during cholera seasons, or precedes an attack of cholera. It is, to all appearances, the beginning of cholera. When it is occasioned by fear, give *Chamomilla*. When occasioned by grief, *Phosphoric acid* or *China*. If accompanied by nausea, *Ipecac.* — Consult article on " DIARRHŒA."

SYMPTOMS OF CHOLERA. — Epidemic cholera varies much in its mode of attack. It may seize upon the patient suddenly and without warning, at once prostrating him and almost depriving him of vitality. The expression of the countenance in such cases is sunken and death-like; the pulse is feeble, almost imperceptible; the skin blue, cold, and shrivelled, and covered with a clammy sweat; cramps in the calves of the legs, fingers, and muscles of the abdomen, with stupidity or extreme anguish, vomiting and diarrhœa, with rice-water discharges.

of the legs, fingers, and muscles of the abdomen, with stupidity
or extreme anguish, vomiting and diarrhœa, with rice-water discharges.

This mode of attack, in this country, at least, is rare ; much
more frequently the disease is preceded some little time, even for
a day or two, perhaps, by lassitude ; confusion of the head ; debility ; diarrhœa ; rumbling in the bowels and stomach ; tongue
moist and a little coated, perhaps pasty or gluey.   At this stage
the disease can easily be checked ; but if neglected and allowed
to continue, these symptoms increase in severity and new ones are
rapidly developed.

The diarrhœa, which at first consists of digested food and feculent matter, becomes yellowish or brown, and thin, soon reaching
the characteristic rice-water discharges, accompanied with nausea, vomiting, intense thirst, great anguish and burning in the
stomach.

The amount of liquid matter thrown up from the stomach and
discharged from the bowels is sometimes wonderful.   The discharges resemble water in which rice has been boiled, containing
white flakes or specks floating in it.   This fluid is discharged with
very little effort.   It pours from the bowels almost in a stream,
and is spouted from the mouth as from a pump.   Each evacuation
is preceded by great noise and rumbling in the intestines, like the
rumbling of gas, or the running of water.

These symptoms are soon succeeded by great oppression of the
chest ; the voice becomes husky and faint ; cramps and hard knots
form in the muscles of the legs and abdomen ; intense thirst ; and
great loss of strength.   The urine is suppressed ; the pulse becomes small, intermittent, or imperceptible ; the surface grows cold
and bluish ; the lips purple ; the tongue assumes a leaden hue, and
feels cold to the touch ; the breath also becomes cold ; the eyes
sink deep into the sockets ; the cheeks pale ; the skin, bathed in a
cold sweat, is shrivelled, as though long soaked in water ; the countenance becomes withered and ghastly as that of a corpse.

Attacks of cholera do occur where diarrhœa and vomiting are
absent ; but with all the other symptoms present in an aggravated
degree.   This variety requires the most prompt attention.

Of course, it must not be presumed that all these symptoms are
present in every case ; more or less of them may be absent.   The
picture, more or less vividly drawn, according to the constitutional,
predisposing, and exciting cause which may exist.

Favorable symptoms are cessation of the vomiting, purging, and cramps, restoration of the secretion of urine, and a return of the pulse, warmth of the surface, and natural appearance of the skin.

TREATMENT. — The usual premonitory symptoms are a slight looseness of the bowels, with or without pain; the evacuations may be rather copious and watery, accompanied with more or less weariness. These symptoms, though trivial in themselves, during the prevalence of cholera are forewarnings, — the first alarm, — which should be instantly heeded. The person in whom they appear, if away from home, should return at once, with as little exertion as possible, and remain quiet. Should there be much diarrhœa, it will be as well to lie down; it will not be necessary to go to bed, but it is advisable to keep the recumbent position till the alarm is hushed.

As soon as possible, take a dose of *Camphor*. The ordinary tincture of camphor, which can be obtained from any druggist, is suitable. Put twelve drops in a spoonful of sugar, and then dissolve it in twelve spoonfuls of cold water in a tumbler. Of this solution, take one spoonful at a dose. Should diarrhœa be the principal symptom, take one dose after each evacuation.

If, after five or six doses have been taken, the diarrhœa continues, and especially if the discharges are liquid and light-colored, either painless or attended with colic; the tongue coated, and sticks a little to the finger when applied to it, take *Phosphoric acid*, after each evacuation.

Dissolve twelve globules in six spoonfuls of water, and take of the solution one spoonful at a dose.

By this means the disease will be warded off, or, if these symptoms be not premonitory, but actually the first stage of the pestilence itself, this treatment will immediately arrest it, and shield the patient from the torments of the more advanced stages of the disease.

In whatever manner cholera presents itself, or in whatever stage it may be met with, *Camphor* is the first remedy to be given. It is a true specific for the disease, having the power to destroy the poison or malignant agent.

Should the patient be taken suddenly with burning in the stomach and abdomen, with anguish and tossing about; rapid failure of strength; pulse feeble and slow; heaviness and pressure in the head; bluish color of the face; pressure at the pit of the stomach; inconsolable anguish; dread of suffocation; cramps in the calves of the legs and feet; coldness of the body; little or no diarrhœa or vomiting; cover him warm in bed, and give *Camphor*, prepared as above, every five minutes. Send at once for a homœopathic physician; continue the *Camphor* till he arrives, or until signs of reaction are observed, which will usually take place by the time five or six doses have been given. The size of the dose must then be diminished, and the remedy given at longer intervals, until reaction is fully established.

When the patient gets warm, and begins to perspire, — which is a good symptom, and the effect of the camphor, — great care should be taken that the clothes be not hastily thrown off, because the perspiration, being suddenly checked, would inevitably bring on a relapse. In the course of eight or ten hours, the clothing, though still warm, may be left somewhat to the patient's choice.

If, in addition to the symptoms for which *Camphor* has been recommended, the following symptoms are present, — *frequent violent vomiting and purging;* cramps in the extremities and abdomen; rumbling and griping in the bowels; coldness and blueness of the skin; cold, clammy sweat; thirst and great restlessness; skin withered and wrinkled; watery, flocky stools, and coldness of the breath,— give *Veratrum*. Dissolve twelve globules in twelve spoonfuls of water; give of the solution, thus made, one spoonful every five minutes until decided improvement is manifest, when the interval between the doses should be lengthened.

Arsenicum should be given when there is *intolerable burning in the stomach and bowels*, worse after vomiting, with cutting, cramplike pains in the abdomen; scalding evacuations; excessive anxiety; violent thirst; labored respiration; hoarseness of the voice; pulse weak and irregular; blueness of the face and lips; skin cold and clammy; cramps in the calves of the legs; vomiting and purging, immediately after eating or drinking ever so little; excessive fear of death, with great dread of being alone.

Should the disease, on reaching these shores, present the same characteristics that it now does in Europe, it requires no great foresight to perceive that *Arsenicum* and *Veratrum* will be its specifics. These remedies will prove efficient weapons in the hands of those who know how to use them skilfully. *Arsenicum* may be administered the same as *Veratrum*. Many physicians are in the habit of giving the two, in alternation, when prompt relief is not obtained from either of them singly. I have never found this necessity to exist.

Cuprum. —This remedy should be given when there is vomiting and rice-water discharges; skin cold and livid; pressure in the stomach; spasmodic colic; eyes sunken in their orbits; skin corrugated and withered; spasms of the jaw; cramps in the calves of the legs; diminished secretion of urine; loss of voice; constriction of the chest; cold, clammy sweat; or, should there be bloody evacuations.

Carbo-veg. is called for when the disease has reached the stage of *collapse; pulse imperceptible;* surface cold and bluish; *breath cold,* and *voice extinct.*

Secale cornutum. — Especially for aged persons, and when there is rapid prostration of strength; violent thirst; cold, dry, livid tongue; blueness and withered appearance of the skin.

During the treatment of cholera, the heat of the body should be kept up as much as possible by artificial means; besides having the patient in a warm room and well covered in bed, it is advisable to put hot bricks or bottles of hot water about the abdomen and to the feet. Friction of the limbs with the dry hand, or with a piece of flannel cloth, adds materially in restoring warmth to the extremities, and is also the best remedy for the cramps in the muscles.

To quench the intense thirst which is almost always present, small quantities of ice or ice-water may be given from time to time, provided it does not aggravate the disease. Should the patient prefer warm toast-water, it may be allowed.

Indeed, ice-water itself is a remedy for the colic, vomiting and cold skin, and in most cases may be given to the patient in *small* quantities with marked benefit. Injections of cold water are often serviceable in relieving the colic and cramps in the intestines.

External application of spirits is objectionable; rubbing with the dry hand is advisable; bathing with camphor, patent lotions of every description, plasters of mustard, horseradish, and everything of the kind, are not only discountenanced but positively prohibited.

RECAPITULATION. — *Camphor* should be given in *all* cases, at the commencement. If the diarrhœa does not yield, and the stools are liquid and whitish, and especially if the tongue is covered with a sticky, pasty coating, follow it with *Phosphoric acid.* Should there be great anguish in the chest; immoderate fear of death; lips blue and cold; great thirst; burning, pressure, and anxiety in the pit of the stomach; vomiting after drinking; rice-water evacuations without smell, or dark, putrid evacuations; respiration labored; skin, cold, bluish, and covered with a clammy perspiration, — give *Arsenicum.*

Give *Veratrum* when there is vertigo with nausea; blue face; blue and cold lips; coldness of the tongue; cold sweat on the body; vomiting and purging; vomiting of watery liquid attended with colic and pain in the stomach and abdomen; rumbling in the intestines; anguish in the chest; cramps in the chest and calves of the legs; excessive coldness; skin, withered and wrinkled.

Cuprum — should be given when there are cramps all over the body; cramps in the stomach; coldness and blueness of the skin; diminished secretion of urine; skin withered and corrugated; eyes sunk deep in the sockets; hoarseness of voice.

Secale cornutum. — When the diarrhœa produces great prostration of strength, especially in aged persons.

When complete collapse is present, with coldness of breath, give Carbo-veg.

DIET AND REGIMEN. — As soon as the disease has spent its violence and the patient begins to mend, there will be a demand for nourishment, and it will be necessary that he should early take something to assist nature in regaining the strength that has been so rapidly exhausted. It is quite as essential, however, that only such food should be taken as can easily be digested. A little gruel, at first, perhaps, would be as appropriate as anything; this can soon be followed with toast-bread, afterwards with meat-broth, gradually increasing the diet both in quantity and quality, as the patient regains strength, until he finally gets back to his accustomed mode of living.

CHAPTER VIII.

DISEASES OF THE SKIN. — SCARLET FEVER, OR SCARLATINA.

DEFINITION. — Considerable mystification exists in the minds of many about this complaint, mistaking the latter word scarlatina as representing a modified form of the disease ; you will hear parents speak of one child having had the real scarlet fever, while another only had scarlatina.

Scarlet fever and scarlatina are one and the same disease, and the attempt, which is practised in the profession by those only who wish to appear wise above their brethren, to draw any line of distinction between them, can lead merely to confusion and a disregard of requisite precaution.

Scarlet fever is an epidemic and contagious febrile disease, characterized by a peculiar rash, which appears upon the first or second day, and by inflammation of the tonsils and mucous membrane of the mouth.

The two most important and striking features of the disease are, affection of the throat and affection of the skin; yet either may be entirely absent, or at least be so imperfectly marked as to attract but little attention. And this circumstance has led authors to divide one and the same disease into different varieties. There is no good reason why this malady, any more than any other fever, should be divided into the variety of forms ascribed to it by many.

There are mild and grave cases, and although scarlet fever may present many degrees of severity and apparent differences, as in fact do all diseases, it is scarlet fever still, and as such should be described.

Being contagious, and appearing seldom more than once in the same individual, this is almost exclusively a disease of childhood.

I have never known in my own experience of the second appearance of this fever in any one person; but that it does so appear

43

sometimes, is proved by incontrovertible facts, brought forward by various authors.

Scarlet fever is believed to be a decidedly less prevalent disease than measles. It affects both sexes in about equal proportions, and is most common between the ages of one and five. It prevails at all seasons of the year, but is most frequent in the spring and autumn.

CAUSES. — As stated above, scarlatina is an epidemic and contagious disease. That it is propagated by contagion, no one at the present day doubts; at least, there is no lack of evidence upon this point, and any one who will take the trouble to investigate the subject will soon become convinced of its contagious character; yet all authors agree that it is much less so than small-pox, measles, hooping-cough, or chicken-pox. I am decidedly of the opinion that the majority of cases of scarlet fever are contracted from the epidemic constitution of the atmosphere, and not, as is generally believed, by direct contagion.

I am led to this conclusion by the fact that, in most cases, the mother is unable to tell when or where the child has been exposed to the disease; besides, we often find it attacking young children, infants at the breast, who have not been out of the house for months; again, we find it manifesting itself in a country village, where the disease has not been known for perhaps years before. I certainly see no way in which these cases could have occurred, except through the epidemic influence which was at that time prevalent in the vicinity.

The question is often asked, if the infection can be conveyed from one house to another by a person's clothes. Under ordinary circumstances I apprehend it cannot. By ordinary circumstances I mean casual visits to the sick-room. For instance, I do not think a mother exposes her children by visiting a neighbor's who are suffering from scarlatina, provided she remains but a short time in the sick-room, and especially if she walks a few blocks in the open air before she reaches her own home. Of course, I should not advise any one to visit a house which was infected with the disease merely for the sake of paying a visit; on the contrary, I would advise them by all means to avoid it, unless duty call them and they could be of real service to the suffering ones.

But, whereas you might with impunity visit your neighbor's sick room, it would be quite a different affair, and highly imprudent for one who has been in constant attendance upon a case of scarlet fever, to come in contact with your healthy children. How long the clothes may retain the infection, after being thoroughly impregnated with it, is uncertain; it may be inferred, however, that the clothing which is exposed to a current of fresh air will be less liable to retain the seeds of the disease for a length of time, than that packed away, or shut up closely. Clothing, beddings, or rooms that are frequently and thoroughly aired, soon lose their capability of disseminating disease.

"What is the period of incubation of scarlet fever, or how long after a child is exposed will the malady manifest itself," is a question often asked. Some authors say from nine to fifteen days, others, from two to seven. Dr. Meigs says: "It may be stated to vary between two or three days, and two or three weeks." My own observation would fix it from three to nine days.

How long the patient retains the power of imparting the contagion it is impossible to determine. I certainly do not pretend to know. Some assert that it lasts throughout the period of desquamation, and that during that period it is most active. This may be so.

SYMPTOMS. — Scarlet fever commences, as do all eruptive diseases, with shivering and lassitude, headache frequently severe, sometimes with delirium, and occasionally with nausea and vomiting.

The eruption generally appears on the second day, and simultaneously with the fever, there is in all cases more or less sore throat.

Generally the onset of scarlet fever is sudden, the child going to bed apparently as well as usual, becomes restless, hot, and wakeful in the night, and in the morning presents all the characteristics of the disease; or, as frequently happens, the child goes out to play, or to school, well, is taken sick, perhaps with nausea and vomiting, or with shivering and lassitude, returns home, and in a few hours shows the eruption over face, shoulders, and neck, accompanied with fever and sore throat.

The eruption usually appears first on the face, neck, and breast, and extends rapidly over the whole surface of the body. It first appears in dark-red points, which speedily become so numerous that the surface seems to be universally red. The eruption is not equally diffused, or, at least, not usually so, over the whole body, but is more apparent about the groins, upon the back, and in the flexures of the joints, than elsewhere. On the arms and legs the eruption does not always present the same appearance as upon the trunk; instead of being of a uniform smooth redness, it is more spotty and rough.

In most cases, the fever is attended with a burning irritation of the skin. The redness disappears under slight pressure of the finger, and returns when the pressure is removed.

The eruption reaches its height about the fourth day, remains stationary for about one day, after which it begins to decline, becoming by degrees indistinct, and disappearing altogether in the majority of instances, about the seventh or eighth day. At this time the skin begins to peel off. In some mild cases the whole duration of the eruptive period is not more than two or three days, the skin presenting but a slight blush, and there being but little heat or fever.

Sore throat, I believe, is always present; sometimes, however, it is so slight as to pass unobserved by the patient and nurse, but upon close inspection inflammatory action is plainly visible. The tonsils are swollen and red, the glands of the neck are tumefied and tender to the touch.

In severe cases, the throat symptoms constitute the most important feature of the disease, and should receive early and prompt attention.

There is always fever throughout the whole course of the disease, which does not subside on the appearance of the eruption. The pulse is strong and frequent, running up to one hundred and twenty or even one hundred and sixty.

The appearance of the tongue is characteristic. At the commencement of the disease it is covered over with a thick cream-like fur, sometimes the edges and tip presenting a deep red color. After the first two or three days the tongue clears off and becomes preternaturally red and rough, looking like raw flesh.

Possibly scarlet fever and measles may be confounded by those unfamiliar with eruptive diseases. The distinguishing marks between the two diseases, therefore, are 1st. The eruption of measles is always preceded by catarrhal symptoms, such as coughing, sneezing, and running from the nose, while scarlet fever is not. 2d. Scarlet fever is always accompanied by sore throat; measles is not. 3d. The rash of scarlet fever appears on the second day; that of measles, at least in its most regular form, not until the fourth. Generally the eruption of scarlet fever is smooth and even to the touch, and of a uniform scarlet color; in measles, on the contrary, the eruption consists of minute little pimples which are felt to be slightly elevated, and firm to the touch; besides the eruption is not continuous, but cut up in little clusters by portions of healthy skin. In measles it is said the eruption presents somewhat the tint of a raspberry, and in scarlet fever, that of a boiled lobster.

As before stated, the eruption appearing upon the second day, reaches its height about the fourth, remains stationary for one or two days, and afterwards gradually declines, disappearing altogether about the seventh or eighth day. About this time desquamation begins to take place, or, in plain English, the external or scarf skin begins to peel off. From the face and body it drops off in scurfs or small scales, from the extremities in large flakes, and from the hands and feet, it sometimes separates almost entire.

Scarlet fever does not always present itself in as mild a form as the above account of symptoms might seem to indicate. The mild and severe cases differ so much from each other that a description of the former would give one but a faint idea of the fearful character which the latter sometimes assumes. " In these malignant and terrible cases, the eruption, if it appears at all, is livid, partial, and fades early, and is attended with feeble pulse, cold skin, and typhoid depression; sometimes the patient sinks at once, and irretrievably under the virulence of the poison." Or, where the patient survives the first shock, as the disease progresses, a condition of the throat develops itself, which frequently baffles the skill of the physician and soon destroys the life of the patient.

This malignant form I shall not particularly describe, as it should be treated only by an experienced physician. In fact, scarlet fever in its most simple form is not safe in the hands of a layman; for it not unfrequently happens that for one or two days the case may promise to be mild, and then, suddenly and without any ascertainable cause, assumes the threatening features of the worst form of the disease.

The consequences of scarlet fever are frequently worse than the disease itself. Children who have suffered from an attack of it are liable to fall into a state of permanent ill health, and become a prey to some of the many chronic forms of scrofula, boils, ulcers, diseases of the scalp, sores behind the ears, scrofulous swelling of the glands of the neck, chronic inflammation of the eyes and eyelids. The same results sometimes follow measles and other eruptive diseases.

One of the most frequent and important sequels or results of this fever is dropsy. This dropsical affusion attacks subcutaneous areolar tissues, that is, the structure or tissue just beneath the skin, or any of the cavities of the body. When it affects the head, dropsy of the brain, or water on the brain, is the result, and when the chest becomes the seat of the affusion, we have dropsy of the chest, &c.

The *exciting cause* of dropsy is generally believed to be cold. A child, just recovering from illness, is much more liable to be affected by slight atmospheric changes, than one whose constitution has not been enfeebled by disease. Therefore, we have sometimes very serious results from slight exposures. Simply standing by the window, going down stairs, or into the hall, or changing rooms, is often enough to produce the difficulty in question.

Dropsy appears to have no relation to the violence and danger of the preceding fever; or, at least, if there is any relation whatever, it is an inverse one, for it has been found by experience that dropsy more frequently follows mild cases than severe ones. This may be owing to the fact, that less care and caution are observed in mild cases during the period of convalescence, and especially during the time the cuticle is peeling off. In severe cases, where the recovery is slower and more doubtful, there is apt to be more care; cold is more particularly guarded against, and the child is not permitted to go out until a later period. On the contrary, in mild cases, where the fever and eruption have been but trifling, the mother, considering the child well, or nearly so, and perhaps not having been warned by the attending physician, permits the child to run about the house, stand in the draught of an open door or window, or, what is perhaps more frequent, takes the convalescent into a cold room to sleep, before the new cuticle is formed, or while it is forming; the result is, the child takes cold and dropsy ensues.

Perhaps no period is as dangerous as that of convalescence. At this time the child needs the most watchful care and attendance, and at no stage of the disease is the patient more apt to be neglected. The mother, thinking the child almost well, leaves him to the care of a friend, or older children, while she goes out, and they, not understanding the necessity of great caution, permit him to stand by an open window, or as happened in one case I attended, allow the fire to go out or the room to become chilled. The patient, from this exposure, takes cold, becomes drooping, languid, irritable, peevish, and restless, after which, swelling about the face soon makes its appearance, at first so slight as to be scarcely perceptible. From the face it extends to the hands and feet, and finally to the whole surface of the body.

We are frequently asked at what time it will be safe for the child to be permitted to leave the room. Physicians are somewhat at variance on this point. Thinking it best to err, if err we must, on the safe side, I have been in the habit of directing the mother to keep the patient confined to the room for four full weeks from the commencement of the disease, provided it was in the cool season of the year. Where the house is heated throughout by a furnace, and the halls and other rooms are of the same temperature with the one in which the child has been confined, this precaution is not necessary. Nevertheless, it is not as well to allow free range of the house during convalescence, and especially the lower floor of the house, where the outside doors are being constantly opened.

According to the observations of Dr. Wells, "the dropsical symptoms commonly show themselves on the twenty-second or twenty-third day after the commencement of the disease. They have been known to begin as early as the sixteenth, and as late as the twenty-fifth day."

When no dropsical symptoms appeared before the end of the fourth week, Dr. Wells always ventured to state that it was no longer to be dreaded.

TREATMENT. — Our opponents are very fond of saying, that if the cases of scarlet fever falling into homœopathic hands were as severe as those coming under allopathic treatment, that the mortality of the latter would not so far exceed that of the former as it now does, and that the relative success of the two practices would be about equal. I am inclined to think that there is considerable truth in the assertion, and yet at first sight it seems strange that we should be favored over our brethren. Let us look into the case a little and see how it is. Sydenham, good allopathic authority, has said that "simple scarlet fever is fatal only through the officiousness of the doctor." Now a large portion of all the scarlatina that we meet with is *simple*, and only becomes complicated or serious when the patient is advised to take some mild cathartic, which is invariably the prescription of all allopathic

physicians on first visiting a patient with this fever. A little magnesia, a small dose of castor-oil, or syrup of rhubard, to be followed with antimonial wine and sweet spirits of nitre, is generally
the stereotyped prescription. Why they think it so necessary that
the bowels should be immediately moved, I am sure I do not
know. The intestinal irritation thus produced, certainly seems to
me like adding fuel to the fire. If, with the assistance of the first
prescription, the patient is not better at the next visit, a more
active course of medication is resorted to, such as blood-letting,
emetics, purgatives, etc.; thus, though the case at first presents
itself in as mild a form as possible, it is soon made serious through
" the officiousness of the doctor."

Simple cases need scarcely any treatment at all, least of all a
cathartic. What the sense is in throwing this extra amount of
burthen upon the patient, I never have been able to determine.
Instead of protecting the child from a long and serious illness, this
is the very way to produce it. I know of no means by which a
light case of scarlet fever can be changed to a serious aspect so
quickly as by the action of a sharp cathartic; and I have no hesitation whatever in asserting that this is one of the chief reasons
why allopathic physicians meet with cases so much more severe
than we of the opposite school of practice.

Another reason why our cases are less severe, is, that we invariably make use of *Belladonna* as a prophylactic or preventive.
Experience has confirmed the fact first advanced by Hahnemann,
that *Belladonna*, properly administered, will in the majority of
cases exert a preventive and protecting influence upon the body,
against the contagion of scarlet fever, and where it does not entirely *prevent* the disease, renders it so mild that it seldom proves
fatal. Allopathic physicians are aware of this fact, but, either
from obstinacy or some undiscovered cause, they refuse to administer it, as Hahnemann advised, and prescribe doses so large that
they do not as certainly obtain the desired result, and not unfrequently aggravate the case and really assist in developing the disease in its worst form.

For true scarlet fever, such as Hahnemann describes, *Belladonna*
is *the* specific. When the fever appears free from all complications, *Belladonna* should be prescribed from the onset. It is the
appropriate remedy for all stages of the disease, and is frequently
the only one called for.

It is specially indicated by the following symptoms : —Dry, burning fever; quick pulse; great thirst; dry, red, or whitish coated tongue; soreness and burning of throat; difficulty of swallowing; bright-red appearance of the tongue, mouth, and throat; swelling of the tonsils; stiffness of the jaws and neck; quantities of stringy mucus in the mouth; scarlet eruptions on the face, and over the entire body; starting, and closing the eyes for a few moments; and delirium, or when there is, nausea and vomiting; also when the fever begins with convulsions. For this latter symptom, *Cuprum* is an excellent remedy.

In severe cases, *Belladonna* may be given frequently, as often perhaps, as every hour. The amendment of the symptoms will be a reason for lengthening the intervals between the doses. In mild cases a dose may be given every two or three hours.

If the disease presents a favorable appearance after the second or third repetition, the remedy may be continued at longer intervals for three or four days, or until the cuticle peels off, when it may be withdrawn, and three or four doses of *Sulphur*, one dose every evening will be sufficient to complete the cure.

Unfortunately, scarlet fever does not always present itself in simple uncomplicated form, owing, I am fully persuaded, in not a few cases, to the domestic remedies which are given to work off the cold, for which the first symptoms are sometimes mistaken; or to the hot herb teas which are given to throw out the eruption, and promote perspiration, when the disease is anticipated. According to my experience, gastric, or intestinal irritation, whether induced as above, or by a cathartic, is almost sure to be followed by an aggravation of symptoms; and I think some of the worst cases I have ever seen have been those which at the commencement promised to run a moderate course, but in which hot herb teas were given, full perspiration induced, and then, from some unaccountable cause, or from exposure, the patient would suddenly become chilled, and the case immediately assume a threatening aspect.

As before stated, it is a well-ascertained fact that *Belladonna* will, if not entirely prevent, at least so moderate the disease as to make it easily manageable. Neglect in this particular, may frequently be accounted the cause of the gravity of many cases. To meet the various anomalies which the disease presents, other remedies than the one mentioned are called for.

Aconitum. — is sometimes necessary at the commencement of the attack, before the eruption makes its appearance; when the

44

fever is high, pulse rapid, head hot, extremities cold, and great agitation. After *Aconite* has subdued these inflammatory, febrile symptoms, or after the eruption has appeared with soreness of the throat, *Belladonna* should be given, or *Aconite* and *Belladonna* may be given in alternation every hour, from the onset of the disease, especially if there be violent fever, with dry heat, full, quick pulse, congestion of the head, occasional delirium, or dulness and drowsiness, with starting from sleep when awakened, twirling of the fingers and tossing about. Should the patient be better under the treatment during the day, but at night the symptoms, especially the restlessness and sleeplessness, increase, give an occasional dose of *Coffea*.

Mercurius. — may follow *Belladonna*, or be given in alternation with it, when *Belladonna* alone has failed to produce a favorable change in the symptoms, and there is ulceration of the tonsils, swelling of the glands, increase of mucus in the mouth, and offensive breath.

Arsenicum. — is indicated when there is great prostration of strength ; when the ulcers in the throat present a livid appearance about the edges, and emit an offensive odor. This is a prominent remedy for the malignant form of the disease.

Opium. — is sometimes indicated, especially when the breathing resembles snoring ; starting, or constant delirium ; puffed red face, great restlessness ; burning heat of the skin, with or without perspiration.

When the scarlet eruption strikes in, or assumes a livid bluish hue, give *Bryonia* and *Belladonna* in alternation, every half-hour. If no relief is afforded, give *Ipecacuanha*, or *Camphor*.

Crotalus, *Phosphoric acid*, *Arsenicum*, *Lachesis*, or *Nitric acid* will be found of service in malignant cases.

Sulphur. — is a valuable remedy, and is almost always required . to complete the cure. When the symptoms calling for *Belladonna* do not promptly yield to that remedy, an occasional dose of *Sulphur* will be found beneficial.

Rhus. — For the dropsical swelling which sometimes follows the fever, especially when the swelling affects the inferior extremities. Should this remedy not answer the purpose, and there is lingering fever in the evening, puffiness of the face, swelling of the hands and feet, give an occasional dose of a high attenuation of *Belladonna*.

Digitalis. — When there are indications of dropsy of the chest. When there exists, after taking cold, considerable swelling of the glands of the neck, give *Rhus*, and should this do no good, follow with *Arsenicum*.

When the whole body swells after the patient has taken cold, give *Bryonia*, *Helliborus*, *Rhus*, or *Calcarea*.

Pulsatilla. — For *earache* consequent upon scarlet fever, give one dose every hour, or every two hours, according to the severity of the case. If this does not afford relief after five or six doses have been taken, alternate *Belladonna* and *Hepar-sulphur* every hour.

Lycopodium. — For *running from the ears*, give one dose every six hours ; wait three or four days, and if not better, give *Calcarea*, or *Silicea* in the same manner.

Aurum. — For *running from the nose*, give one dose once in six hours.

ADMINISTRATION OF REMEDIES. — Two drops or twelve globules of the chosen remedy may be dissolved in twelve spoonfuls of water, and one spoonful of the solution given every half hour, hour, or two hours, according to the severity of the symptoms.

In all cases consult a Homœopathic physician if possible.

As adjuvants or assistants to the regular mode of treatment, a variety of articles has been recommended, such as baths, lotions, inunctions, affusions, etc. ; my own experience in regard to the application of these, has not been very extensive. The only external application I have ever recommended, has been simply to bathe the child with a weak solution of saleratus, and apply cold water bandages to the throat, and I am inclined to think that homœopathic physicians generally meet with such good success in the treatment of scarlet fever, that they are seldom tempted to resort to other remedial agents than those which come legitimately under the law of " *similia.*"

Inunction has within the last few years been highly recommended in the treatment of this disease. Inunction means simply smearing the patient all over with oil or fat. It is performed in this manner : — Take a piece of fat bacon about the size of your hand, and rub the patient with it from head to foot, omitting only the face and scalp, every morning and evening. This is to be done as soon as aware of the nature of the disease. It is said that children like this rubbing very much, as soon as they become initiated and learn how pleasant it is. It is recommended especially for those whose skin is very hot and dry, and when the eruption is intense, accompanied by violent itching and irritation. On account of the disagreeable character of bacon, Dr. Meigs recommends the following ointment, which can be obtained from any druggist : —

> Glycerine, one drachm,
> Ointment of rose-water, one ounce.

The two thoroughly mixed together form an unctuous substance of which the doctor says : " In my hands it has had the effect of allaying, in all cases, the violent irritation, caused by the intense heat and inflammation of the skin. This preparation removes, of course, the dryness and hardness of the skin, keeping it, instead, soft and moist. It lessens, or even removes, the burning and itching caused by the eruption." — *Diseases Child.* 3d ed., p. 539.

BATHS. — Though having never myself used baths in the treatment of this disease, I am inclined to the opinion that great benefit may be derived from the judicious external employment of water.

To make a proper application of water, the physician, or at least an intelligent, educated nurse, should be constantly with the patient, because the efficiency of the treatment depends upon the proper adaptation of the temperature of the water to each individual case ; for instance, if ice-cold affusions were made on a scarlet fever patient, when the skin was pale and cool, the pulse rapid and feeble, great injury would be done. Such a case should be treated with hot or warm baths. Some cases require cold water, some temperate, others warm ; again, at some stages of the disease it may be necessary to use cold baths, at others warm ones; in fact, water of various temperatures must be applied according to the state or condition and nature of the case under treatment.

I have always been in the habit of recommending the application of cold bandages to the throat, as soon as the patient complained of soreness, and especially in those cases attended with great heat of skin, high fever, pulse full and strong, violent inflammation of the throat, and the glands about the neck swollen and hard. The bandages should be dipped afresh into the water every few minutes, and the application continued as long as the throat symptoms continue severe.

Sponging the patient off with a weak solution — just sufficiently alkaline to be perceptible to the taste — of saleratus water, produces all the soothing effect claimed so strongly for inunction. In case of repercussion of the eruption, I should highly recommend the course of treatment recommended by Dr. Mundle, an experienced hydropathist, as quoted by Dr. Pulte. "In case the scarlet fever strikes in suddenly, the patient is sponged off in cold water all over ; and if spasms had ensued, cold water is dashed over him in larger quantities until the spasmodic action ceases ; he is then wrapped, without being dried or rubbed, in woollen blankets, if possible, and as much cold water given internally as he can drink ; in most cases a general perspiration will ensue, the eruption reappear, and the patient is saved.

DIET AND REGIMEN. — During the height of the fever, the patient seldom cares for anything to eat. When the mouth is dry and parched, small quantities of thin rice gruel, or gruel made of arrow-root, may be administered, or, if the patient prefer it, he may be allowed rice-water, toast-water, or cold water, to drink. A very pleasant and cooling drink is made by adding a little raspberry or strawberry syrup to pure cold water. The best syrups are those prepared by " Turner Brothers," corner of Franklin and Washington Street, New York. Warm drinks should not be allowed, unless especially craved. When the teeth and lips become covered over with crusts or scabs, they should be carefully cleansed with tepid milk and water. Great care should be taken to keep the mouth as cleanly as possible, and this can only be done by constant attention.

After the fever has abated, there will be a craving for something more substantial, in the shape of food. The return, however, to a more nourishing diet should be gradual, and great care should be taken in order to avoid overtaxing the digestive organs, as neglect in this respect may be productive of the most serious consequences. In mild attacks, the patient may be allowed, during the whole illness, gruel and weak broths ; but in severe cases, and during the raging of the fever, toast-water, barley-water, and perhaps very thin gruel, is about the only nourishment that it is advisable to administer. During the early part of convalescence, gruels, milk-toast, etc., may be allowed as the appetite returns. If this does not interfere, and digestion goes regularly on, recourse may soon be had to broths, soups, digestible meats, etc.

The room in which a scarlet-fever patient is confined, should be as large and airy as possible ; it should be well ventilated, but never fumigated. The bed should be kept sweet and clean, clothes, bandages, in fact, everything about the patient, as soon as done with, should be removed. Of course great care should be taken to guard strictly the patient, lest he should take cold. A room can be kept well ventilated without exposing the patient.

Particular caution is also necessary about allowing a child to go out too early, as often various secondary disturbances are caused by want of proper prudence; but of this we have spoken at length, elsewhere.

PREVENTION. — I believe as a general thing, all schools of medicine now acknowledge the prophylactic properties of *Belladonna* against scarlet fever. During the prevalence of the disease, it is advisable that all who are liable to an attack should take this remedy as a preventative, and then if the disease is not entirely guarded against, it is at least rendered comparatively harmless. One dose of the thirtieth attenuation of *Belladonna* should be taken every other evening, for at least ten days.

SCARLET RASH.

DEFINITION. — This, though frequently mistaken for, is quite a different disease from scarlet fever. Scarlet rash consists of small granular elevations, easily felt on passing the hand over the skin. The eruption is of a dark-red color, sometimes almost purple; the pressure of the finger leaves no white imprint as it does in scarlet fever, and there is seldom any, or at least no great amount of sore throat. Scarlet rash may be easily confounded with measles, as the eruption in the two diseases is very similar.

CAUSES. — This malady is most common in summer and autumn, nevertheless it does occur at all seasons of the year. It attacks children of all ages. It is not a contagious disease, and is said to be occasioned by gastric derangement, also by sudden atmospheric changes, by violent exercise, by the use of cold drinks while the body is heated, and by checked perspiration.

SYMPTOMS. — The eruption is generally preceded by chilliness, alternating with heat, accompanied by loss of strength, heaviness, and fulness of the head, restlessness, sometimes with vertigo, severe pain in the head, and even mild delirium. There is for the first few days in connection with the above symptoms, more or less fever, heat and dryness of the skin, loss of appetite, and perhaps some gastric derangement. After these symptoms have continued for an indefinite length of time, the rash appears; sometimes upon the third or fourth day, and in its regularity and appearance so much resembles measles, that it may be mistaken for that disease. There is this difference, however, between scarlet rash and measles; the latter disease is accompanied by catarrhal symptoms, running at the nose, eyes, &c., the eruption appearing invariably on the fourth day, first on the face, next on the body, and lastly on the extremities. Such regularity is not found in scarlet rash, neither is the rash accompanied by catarrhal symptoms, and the eruption may appear irregularly, or at once over the whole body.

Scarlet rash need not be confounded with scarlet fever, as the rash of the latter is of a bright tint, and generally quite uniform over the whole surface of the body: in scarlet rash the eruption is composed of irregular circular patches, and their color is of a deep rose red, instead of a bright-red or scarlet. In scarlet fever we have a peculiar sore throat; in the simple rash we have none; besides scarlet rash is not contagious.

TREATMENT. — *Aconite* is about the only remedy called for in ordinary cases, perhaps where there is great restlessness it might be advantageously alternated with *Coffea.*

Ipecacuanha and *Pulsatilla* are sometimes called for, especially when the disorder is attended with nausea and vomiting.

If the rash disappears suddenly, alternate *Ipecacuanha* and *Bryonia.*

Belladonna is called for when there is fulness of the head, blood-shot eyes, starting and closing the eyes, and other symptoms of head disturbances.

When there is stupor, give an occasional dose of *Opium.*

ADMINISTRATION OF REMEDIES. — Dissolve twelve globules of the selected remedy in twelve spoonfuls of water, and give one spoonful of the solution at a dose. The repetition of the remedy will depend upon the severity of the symptoms. Generally a dose of *Aconite* every two hours will soon end the difficulty.

DIET AND REGIMEN. — The same as in " MEASLES."

MEASLES. RUBEOLA.

DEFINITION. — This disease is characterized by inflammatory fever ; by catarrhal symptoms ; hoarseness, dry cough ; sneezing, drowsiness, and an eruption. The eruption appears generally on the fourth day, in the shape of small, red dots, like flea-bites, which, as they multiply, unite together into irregular circles or horse-shoe shapes, leaving the intermediate portions of skin of their natural color. These red points are slightly elevated, and can readily be felt by passing the hand over the surface.

In many respects, measles resemble scarlet fever; it has its period of incubation, its introductory fever, its peculiar rash, it occurs but once to the same person, and is contagious.

CAUSES. — The causes of measles are epidemic influences and contagion. Of these two modes of propagation there is scarcely a doubt but that the former is by far the most active.

At precisely what period of the disease its infectious nature is most to be feared is not well ascertained, neither are we certain as to how long it lingers about the patient after convalescence has fairly taken place. The average period of incubation or time required to develop the disease after exposure, is from seven to twenty days.

SYMPTOMS. — As a general thing, the first symptoms complained of are lassitude, irritability, aching in the back and limbs, and shivering, which is soon followed by fever, thirst, and headache, and by irritation of the mucous membrane of the eyes, nose, mouth, and larynx.

The symptoms preceding an attack are similar to those of a catarrh or cold in the head. The eyes assume a peculiar appearance, somewhat as if blood-shot; the eyelids are heavy, turgid, and red. There is generally much sneezing, watering at the eyes, copious defluxion from the nose, soreness of the throat, and a dry, hoarse, peculiar cough. These symptoms are owing to the irritation and inflammation of the mucous membrane lining the throat and nasal passages.

This, the first, or catarrhal stage of the disease, lasts generally about three days; upon the fourth day, seldom earlier, frequently later, the eruption makes its appearance. The eruption itself consists at first of distinct points, not unlike flea-bites, of a more or less rose, or bright red, or crimson color, which, as they multiply, unite into blotches or patches, that are mostly of an irregular, oval, or semi-lunar shape, leaving the intervening portions of skin of their natural color.

The rash is two or three days in coming out; beginning upon the chin or cheeks, or some other portion of the face, it extends to the neck, arms, and trunk of the body, and finally to the lower extremities. This stage lasts from twenty-four to forty-eight hours. The fever does not diminish when the eruption makes its appearance. On the contrary, during this period all the symptoms are at their height, but the moment the eruption passes its highest point of intensity the fever gradually begins to diminish, the catarrhal symptoms subside, the cough loses its hoarseness, becomes looser, and finally dies away.

About the seventh or eighth day of the attack, or third or fourth of the eruption, the disease begins to subside. The rash first fades on those parts where it first made its appearance, and it not unfrequently happens that it has almost entirely disappeared upon the face, while it is still livid upon the lower extremities.

After the eruption passes away, the parts which it recently occupied are left covered with a dry, small scurf, small bran-like scales. The skin does not peel off in large flakes as it sometimes does in scarlet fever, but it crumbles away like dust or fine powder.

This stage of desquamation, as it is called, is more indefinite in its duration than those which precede it; but, as a general thing, it lasts six or seven days, and during this period the patient ought to receive as much care as when the disease was at its height.

The above account of symptoms are those which appear in a regular or usual form of the disease, but the eruption presents various irregularities, which we will now notice.

The severity of measles does not depend upon the amount of the eruption, or rather because the rash appears early and plentiful, it is no sign that the disease will be more severe or more dangerous; on the contrary, the worst cases met with are those where the eruption is but partial, does not come out well, appears late, or irregular. In what is called the black measles, the eruption comes out slowly and imperfectly, and is of a livid, purplish, or even blackish color. This is a very dangerous form of the disease; the patient may die early from exhaustion, or congestion of the brain or lungs.

A retrocession of the eruption is very apt to be followed by unpleasant, if not alarming symptoms.

Sometimes measles are complicated with gastric disarrangements; in such cases the tongue will be found coated; there is some nausea and perhaps sickness at the stomach; the eruption does not stand out as prominent as it should, and the healthy portions of skin between the patches of eruption have a yellowish tinge. Perhaps the most frequent and important complication of measles is inflammation of the lungs. Inflammation of the bowels is also a frequent complication.

As scarlet fever and measles may possibly be confounded one with the other, we have taken pains to point out their distinguishing feature under the head of the former disease, to which the reader is referred.

TREATMENT. — *Aconitum.* — In ordinary cases this is about the only remedy called for ; in fact the simple uncomplicated forms of the disease need scarcely any treatment whatever except hygienic.

In all cases, no matter how mild, the patient should be confined in a large, well-ventilated room. In most cases the patient is quite willing to lie in his bed during the first part of the disease ; but as soon as the eruption begins to disappear and the fever subside, he will want to be dressed, and, when once dressed, he will think it strange that he cannot go out, especially if he feels quite well ; however, he should not leave his room, and certainly not the house, until he has regained his accustomed healthful look.

It has always been the custom to shut a measles patient in a hot room, and allow him nothing but hot drinks. This is a most pernicious habit, and has no doubt led to a great many serious and even fatal results.

The bed, if it be possible, as we before said, should be placed in a *large, well-ventilated* room. The patient should never be allowed *hot drinks,* and especially those which are recommended, to throw out the eruption. If he is thirsty, give him *cold water,* as much as he wants. It is the most palatable, and by far the best drink you can procure. I have never known small quantities — say a wineglass full at a time — of the coldest water do any harm. On the contrary, I have seen the most happy results brought about by its free use. In those cases where the eruption is backward about coming out, give the patient a glass of good cold water and cover him up warm in bed. This is especially advisable where the fever is violent and the heat of the skin very great.

The diet, during the febrile stage, ought to be extremely light. The patient, as a usual thing, will ask for but little ; but that little should consist of thin wheat or rice-flour gruel, barley-water, toast-water, milk and water, tapioca, crackers soaked in water, or some similar food. When the fever begins to abate, the allowance may be increased to plain or toast bread, or bread pudding, or to some light broth, either animal or vegetable, and even to a small quantity of chicken or beef-steak once a day until the strength is fully regained, when the usual diet can be resumed. By observing the above rules strictly, and giving the patient an occasional dose of *Aconitum*, you will have no trouble in managing all ordinary cases.

Owing to complications, which are not unfrequently present, other treatment than the simple one pointed out above will be called for.

Belladonna. — This remedy should be given when there is considerable soreness of the throat, shooting and prickling pain when swallowing, much thirst, and a spasmodic, dry cough; also, in those cases where the eruption is backward about coming out, and when there is congestion to the head, high fever, restlessness, and delirium.

Pulsatilla. — This may be alternated with *Aconite*, especially should the catarrhal symptoms predominate, or should there be gastric derangement. Where the gastric symptoms predominate, it is sometimes necessary to give an occasional dose of *Ipecacuanha*.

Ipecacuanha and *Bryonia* — in alternation, are called for when the eruption does not come out well, or, when the measles strike in suddenly, and look pale, and especially if there be sickness at the stomach, and oppression at the chest.

Bryonia — is called for where there is threatened bronchitis or pneumonia, indicated by shooting pains or stitches in the chest, and violent, dry cough.

Euphrasia — for severe inflammation and watering of the eyes.

Rhus — in alternation with *Bryonia*, when the symptoms assume a typhoid form, the tongue being dry and red, the skin hot and dry, constant or occasional delirium. Should there be, in addition to the above symptoms, great restlessness, intense thirst, brownish or dark diarrhœa, alternate *Arsenicum* and *Rhus*.

SEQUELS. — Disorders consequent upon measles are frequently even more dangerous than the primary affection. Running at the ears, inflammation and swelling of the glands, especially about the neck, are apt to occur. This is frequently the case in scrofulous children.

For running at the ears, and earache, give *Pulsatilla*, or *Sulphur*.

Swelling of the glands of the neck, *Arnica, Rhus*, or *Mercurius*.

For burning and itching of the skin, *Nux-v., Sulphur*, or *Arsenicum*.

For the remaining cough, which often lingers with the patient for some time, give *Bryonia, Drosera, Hyoscyamus, Causticum.* — See article on " COUGH." Comp. *Hepar-s.*

When measles are epidemically prevailing, *Pulsatilla* has been

recommended as a preventive. One dose should be given every
two or three days; this, it is said, will often ward off an attack;
or, if the disease should be taken, render it mild.

DIET AND REGIMEN. — Of this we have already spoken.

ADMINISTRATION OF REMEDIES. — Of the tincture, one drop, or
twelve globules should be dissolved in twelve spoonfuls of water,
and of this solution, one spoonful may be given every two or three
hours, according to the severity of the symptoms.

For the sequels of the disease, the remedy should not be repeated oftener than once in six hours.

NETTLE-RASH. HIVES. URTICARIA.

DEFINITION. — *Urticaria,* or as it is more commonly called, *nettle-
rash,* is a non-contagious eruptive disease, characterized by little,
hard elevations upon the skin, of uncertain size and shape, and
generally of a red color, with a whitish tinge ; sometimes, however, there is little or no redness, and the elevated parts are even
paler than the surface around them ; more frequently, however, I
think the elevated spots are partially red and partially white. The
eruption, on making its appearance is attended with intense heat,
tingling and itching in the spots ; it is much like that produced
by the sting of the nettle, from which it takes its name.

CAUSES. — Some persons have a constitutional predisposition to
this disease, and the slightest error in diet, or the most trivial
functional derangement of the digestive apparatus, is sufficient to
bring on an attack. Children possessing a fine, delicate skin, are
particularly predisposed to attacks of hives ; in such, slight gastric
disturbance, a warm day, excessive clothing, dentition, or almost
any little disturbance will produce an attack.

SYMPTOMS. — As a general thing, the disorder in children manifests itself without any premonitory symptoms. The eruption, as
before stated, consisting of elevated spots, sometimes red, more
frequently partly red and partly white, attended with heat, burning and itching, the blotches are constantly changing from one
position to another, or disappearing in a few hours on one part,
and appearing on another. The most frequent form of the disease which we meet with in small children consists in large
inflamed blotches of an irregular shape, being either round or
oblong, appearing suddenly, and preceded by very slight if any
constitutional symptoms. The blotches are of a bright-red color,
excepting the slightly elevated center, which is white.

This form of the disease is not dangerous, but is very annoying, and occasions great irritability and crying. The eruption most commonly makes its appearance about the face, the upper part of the arms, thighs, and buttocks.

In some cases, especially in the older children, the eruption is preceded by headache, bitter taste in the mouth, coated tongue, nausea, vomiting, and fever. This is particularly the case in that form of the rash which is induced by errors in diet and exposure to cold.

Another form of the disease which is preceded for a few hours or a few days by feverishness, headache, nausea, chilliness, and languor, is where the blotches assume reddish and solid elevations, either round or oblong, often called wheals. They resemble as much as anything the ridges caused by the stroke of a whip-lash. This eruption, like the other forms, is attended with violent itching and burning. During the attack, the patient is usually more or less feverish, and suffers from headache, languor, loss of appetite, and other signs of gastric derangement.

TREATMENT. — *Aconite* should always be given when the eruption is preceded or accompanied by much fever, hot skin, thirst, furred tongue, restlessness, and anxiety.

Pulsatilla. — When the attack has apparently been excited by indigestible food.

Nux-vomica. — When there is considerable gastric derangement with constipation.

Dulcamara. — When caused by taking cold, accompanied by diarrhœa at night; slimy taste in the mouth; coated tongue. This remedy may be given in alternation with *Antimonium-crud.*

Rhus. — For those predisposed to the eruption, and especially when it arises or has been thrown out by some particular article of food.

Ledum Palustre. — This remedy will cure the majority of cases.

External applications, except it be a solution which is being given internally, should be avoided, as their use is liable to cause a sudden disappearance of the eruption, which may have a serious or even fatal consequence.

Should the rash strike in suddenly, and the patient complain of oppression, great weakness, and sickness at the stomach, give *Ipecac.* or *Bryonia;* if not better in a couple of hours, give *Arsenicum.* At the same time endeavor to promote perspiration by covering the patient well and giving him plenty of cold water to drink.

For chronic urticaria give *Calcaria* and *Lycopodium*, in alternation, one dose every fourth day.

ADMINISTRATION OF REMEDIES. — Of the remedy chosen give five globules dry upon the tongue every three hours. Or, in severe cases, dissolve six globules in twelve spoonfuls of water, and give one spoonful of the solution every hour.

DIET AND REGIMEN. — About the same as that recommended in "MEASLES."

ERYSIPELAS. ST. ANTHONY'S FIRE.

DEFINITION. — CAUSES. — Erysipelas is a non-contagious disease, characterized by a deep-red rash, or superficial inflammation of the skin, which has the peculiarity of spreading from place to place, the part first attacked recovering while the neighboring parts are becoming affected.

Erysipelas in childhood is a rare disease; I have seen but a few cases; three are all that I can now call to mind, and they resulted from vaccination; not that the vaccine matter poisoned the child, but rather the local irritation produced in introducing the virus acted as the *exciting* cause or agency, which brought into action a disease, the seeds of which already existed in the system. Any other scratch would have been followed by a like result.

The causes of erysipelas are obscure: slight points of irritation upon the skin may form a nucleus from which the erysipelatous inflammation may spread, but these certainly cannot be the *real cause*. There must be a general epidemic constitution of the air at times, in certain localities or districts, which predisposes to the disease, or else there is a hereditary taint in the system.

SYMPTOMS. — As a general thing, we find but few if any marked premonitory constitutional symptoms; the appearance of the eruption being the first sign of the disease, after which we soon have fever, heat, dryness of the skin, and thirst.

The inflamed surface is at first of a bright-red and shining appearance, but it soon assumes a purplish hue, and, as this change takes place, the parts become tense, hard to the touch, and more or less swollen and painful. The color disappears under pressure of the finger, but returns as soon as the pressure is removed.

When the inflammation once begins, if not soon arrested there is no knowing where it will end. When it starts upon the face it may extend to the scalp and cover its whole surface, or, when commencing upon the arm, it may extend down to the fingers, or up to the shoulder, and from there over the whole trunk of the body.

TREATMENT. — *Aconite.* — This remedy is called for when there is high inflammatory fever; hot, dry skin; thirst; etc.

Belladonna. — This is the principal remedy for this disease, at least it has proved itself so in my hands. It is especially valuable for erysipelas of the face, with swollen eyes, great thirst, dry skin, and delirium.

Lachesis. — When there is considerable swelling; also, swelling of the adjoining glands, and when little blisters appear upon the surface and turn yellow or blue.

Arsenicum. — When the eruption assumes a dark hue; also, when there is great prostration of strength.

Pulsatilla. — Especially in that form of the disease where the eruption disappears in one place to reappear in another. *Graphites* is also useful in this form of the disease. *Pulsatilla* should be given when the disorder follows some particular article of diet; also, when it affects the ear. It may be followed by *Rhus* or *Bryonia.*

Mercurius and *Hepar-s.* — are called for when erysipelas terminates in abscesses.

For this disease it is always best, when possible, to consult a homœopathic physician.

ADMINISTRATION OF REMEDIES. — Dissolve six globules in as many spoonfuls of water, and give of the solution one spoonful every two or three hours. Should the first remedy, after waiting a reasonable length of time, fail to afford relief, select another and administer it in the same way.

DIET AND REGIMEN. — The same as for any other febrile disease, measles or scarlet fever.

To allay the itching which is sometimes almost intolerable, dust the parts over with powdered starch, or burnt rye flour.

Wet or greasy applications of every description should be abjured, as they *always* aggravate disease.

ITCH. PSORA. SCABIES.

DEFINITION. — Psora, scabies, or *itch*, which, if not the most elegant, is certainly the most expressive name, is a contagious eruptive disease, characterized by more or less numerous distinct pointed vesicles, transparent at the summit, and filled with a viscid serous fluid, while, from the base of each vesicle, as a general thing, run off small red lines.

At present, itch is a rare disease; at least as far as my observation goes, it is so in this city. By foreign writers it is said to be very common among the poor in the large cities of Europe.

CAUSES. — Itch is a contagious disease, and in all probability is contracted only by actual contact.

The little vesicles which rise upon the skin are caused by the presence of a small insect, called *Acarus Scabiei*. The zigzag track, which the mite makes in burrowing beneath the scarf-skin to lay its eggs, can readily be seen; not so, however, with the mite itself, for it is very small, measuring, according to Wilson, between $\frac{1}{147}$ and $\frac{1}{77}$ of an inch in length, and between $\frac{1}{303}$ and $\frac{1}{83}$ of an inch in breadth. — WILSON's *Diseases of the Skin.*

SYMPTOMS. — As a general thing, the eruption first appears upon the wrists and between the fingers, and extends more or less rapidly over the whole body except the face. It is frequently, however, confined to the hands, fingers, and bend of the joints.

As above stated, the eruption appears in the form of small, pointed, transparent vesicles. Their number is variable; in some cases they are very abundant, while in others they are but few, and confined for the most part to the flexures of the joints. At first, the vesicles are of a pinkish color, and contain a drop of viscid or sticky transparent serum; these soon become broken by the clothes or fingers, or burst spontaneously and form thin scabs.

The disease is always attended by severe itching; in fact this is the most prominent symptom and distressing feature of the disease. The itching is most troublesome at night, being increased by the warmth of the bedclothes.

TREATMENT. — For true itch, as we have described, *Sulphur ointment* is the specific and only remedy called for.

Take of the finest Powder of Sulphur — sold by all druggists under the name of "Milk of Sulphur" — one part to two parts of lard; mix thoroughly and rub it well into the skin, before a fire, night and morning, for two days.

During this treatment the patient should wear a flannel gown and keep his bed. On the third day the skin should be washed off with soap and water. Should the first attempt not succeed in removing the trouble, repeat it.

The disease scarcely requires any constitutional treatment, but should the *Sulphur ointment* do no good, simply because there are no mites for it to destroy, give the patient an occasional dose of *Mercurius*, say three globules once in four hours.

When the vesicles are dry and small, or when from neglect the vesicles have spread so as to acquire the appearance of ulcers that have discharged their contents, give *Hepar-s.* the same as *Mercurius*.

Sulphur may be used when the above remedies fail. *Sulphur* may be followed by *Causticum*.

ITCHING OF THE SKIN.

Simple itching of the skin is scarcely a disease, of itself, but rather a symptom of some disease, and, indefinite though it is, it may direct us in the selection of a remedy for the morbid condition which gives rise to the irritation.

Should the itching always commence after the patient gets warm in bed, give *Pulsatilla, Mercurius*, or *Cocculus*.

When the itching moves from one place to another on being scratched, give *Ignatia*.

If it always commences when undressing, as is sometimes the case, *Nux-v., Pulsatilla*, or *Arsenicum*, sometimes *Cocculus*.

When the itching is accompanied by intense burning, give *Rhus, Ledum, Apium-v., Nux-v.*, or *Bryonia*.

Stinging in the skin, *Drosera, Staphysagria, Thuja*, or *Byronia*.

When the skin bleeds readily from scratching, *Mercurius*, or *Sulphur*.

When the itching comes on in the daytime, and arises from over-heating, *Lycopodium*.

Should either of the above remedies fail to produce the desired relief, give one dose of *Sulphur*, night and morning, when better results may be obtained.

In administering the remedies, give from four to eight globules, morning, noon, and night. When parts of the body itch so intolerably that the child will scratch them till they bleed, besides giving the remedies above recommended, rub the parts with sweet oil. Sometimes powdered starch, sprinkled thickly over the skin, will allay the irritation. Again, washing with camphor-soap will stop it, or rubbing with spirits of camphor.

For itching, produced by mosquito bites, *Camphor* is a specific,—applied externally.

HERPES, OR TETTER. ZOSTER, OR SHINGLES. CIRCINATUS, OR RINGWORM.

DEFINITION. — By the term herpes, physicians understand a peculiar, non-contagious eruptive disease, characterized by an

46

assemblage of numerous little vesicles or watery pimples, in clusters. These patches of vesicles are surrounded by more or less inflammation, or rather the vesicles are situated on an inflamed surface, and are separated from each other by portions of perfectly healthy skin.

The fluid which fills the apex of each little vesicle is at first transparent and colorless, but soon becomes milky and opaque, and in the course of eight or ten days is entirely absorbed, or concretes into furfuraceous, bran-like scales.

There are several varieties of herpes; those that are most common in children are *zoster* or *shingles*, and *circinatus* or *ringworm*. The other varieties are so seldom met with, that we deem it useless to mention them.

Herpes is quite a frequent disease in childhood, especially that variety known as " Ringworm."

CAUSES. — The causes of skin diseases are obscure and uncertain. Perhaps the most appreciable cause, the one most clearly ascertained, and certainly the most frequent, is some disturbance of the digestive function. Bilious disorders of all kinds, sudden transitions of temperature, suppressed perspiration, irregularity in diet, and local irritants may be set down as the exciting causes.

RINGWORM, or " Herpes Circinatus." — Almost every one is more or less familiar with this form of the disease. Its characteristic feature is the peculiar arrangement of the vesicles, which are very small, and disposed in circular rings. The vesicles present the same appearance, except that they are smaller, dry up and exfoliate, as they do in the other varieties of the disease.

The first indication we have of the presence of the disease, or of its threatened appearance, is the more or less vivid redness of the skin at the point affected. This redness or inflammation is soon studded over, more or less thickly, with vesicles. The circular patches — as a general thing they are exactly circular — vary a good deal in size, being in some instances not larger than a ten-cent piece; in others it may present a diameter of two or three inches.

When small the whole surface of the patch is inflamed, the centre being of a lighter shade than the circumference. When large, the circumference alone is red, the centre retaining the natural color of the skin.

These eruptive patches or rings may appear upon any part of the body, but are most frequent perhaps upon the upper extremities and neck.

Ringworm of the scalp, is considered by many authors, as a separate and distinct disease, differing from herpes circinatus, as it appears upon other parts of the body. For my own part, I can see no good reason for considering it anything but the simple herpes which we have just described.

If, indeed, there be a difference in the two diseases, we not unfrequently have them both existing in the same patient at the same time.

The latter is said to be contagious; the former, not. This may be so. I have not sufficient data upon this point to give an intelligent opinion.

HERPES ZOSTER, or, as it is sometimes called, " Shingles," is quite an uncommon variety of the disease, at least, I think so, from the fact that I have never met with but one case. The word " zoster " signifies a girdle or belt, and is applied to this disease, from the fact that eruption appears in the form of a half zone or belt surrounding the body.

Old ladies will tell you, that, if the two ends of this vesicular zone meet,—that is, if the belt extends clear around the body, —the child will die ; but, as this *never* happens, it need give you no alarm. The most frequent seat of shingles is at the waist, the belt seldom extending more than half-way around the body, commencing, as a general thing, at the mesial line in front, and extending to some point behind.

From most descriptions of the disease, one would think that the eruption formed a perfect half zone, but this is not the case ; it is rather made up of distinct patches arranged in a line, but separated from each other by portions of healthy skin.

Shingles is generally preceded by constitutional symptoms more or less severe, such as languor, loss of appetite, rigors, headache, sickness, and fever. The local symptoms are pungent and burning pain at the points where the eruption makes its appearance.

The duration of herpes is variable ; it is an acute disease, and seldom lasts over eight or ten days; some cases, however, last longer, especially of the variety called *circinatus.* Sometimes the rings appear, and in a short time fade away, but to re-appear upon some other part of the body ; and thus, by the formation of successive rings or patches, the disease is continued for three or four weeks.

TREATMENT. — As a general thing, ringworm yields readily. under the action of *Sepia.*

Give three globules of the 200th potency every night, for three days; then omit four days, and repeat in the same manner.

Should this prove insufficient, or should the rings reappear after having once vanished, give *Rhus* and *Sulphur* in alternation, one dose every third day. As soon as amendment takes place, discontinue the remedies; the cure will steadily progress to completion.

Occasionally, it will be found necessary to give either *Natrummur.*, *Calcarea*, *Graphites*, *Silicea*, *Nitric-ac.*, *Phosphorus*, or *Mercurius*.

Should there be violent itching, give *Nitric-ac.*, or *Graphites*.

Should the surface be scaly, give *Sepia*, *Silicea*, or *Sulphur*.

Should it be moist or running, give *Calcarea*, *Graphites*, or *Rhus*.

For *ringworm* of the scalp, first give one dose of *Rhus*,—200th— every evening for three days; then omit three, and repeat as before. Should the case improve, continue the remedy at long intervals, until the cure is completed.

Should the eruption be moist and offensive, give *Staphysagria*, as above directed for *Rhus*. *Rhus* may be advantageously followed by *Staphysagria*. Should these remedies fail, give one or two doses of *Arsenicum*, when better results may be expected.

Should the eruption affect the scalp and face at the same time, give *Hepar-s.* or *Calcarea*.

When the glands of the neck become painful and swollen, give *Mercurius* or *Bryonia*.

The only external application that is called for is a solution of the remedy which you are giving internally. This may be made by dissolving eight or ten pills in a teacupful of tepid soft water.

For *Zona*, or *Shingles*, give *Aconite*, especially when, on the first breaking out of the eruption, there is languor, headache, pain in the chest, and fever. When these symptoms have somewhat subsided, *Rhus* will be the appropriate remedy, and may be given once in four hours.

Should there be nausea and vomiting, give *Tartar-emetic*, one dose every two hours until relief is obtained.

When the fluid in the vesicles becomes dark and maturated, give *Hepar-s.* the same as *Rhus*.

When there is great thirst, dry skin, with burning and uncomfortable restlessness, give *Arsenicum*.

Sometimes, after resisting all other remedies, it will yield kindly to a few doses of *Sulphur*.

Should the eruptive patches degenerate into ulceration, give one of the following remedies: *Mercurius*, *Lycopodium*, *Sulphur*, or *Sepia*.

DIET AND REGIMEN. — As the complaint often arises from gastric derangement, particular care should be taken as to what the patient is allowed to eat. Avoid all high-seasoned food, all rich dishes, all irritating substances; in one word, place the child upon a plain, farinaceous diet.

The skin should be kept clean; avoid all irritating and all scented soaps, and be a little careful to have the clothes so adjusted that they will not rub and irritate the eruptive patches.

PRICKLY HEAT.

DEFINITION AND CAUSES. — During the heats of summer, infants and young children are frequently much annoyed with an eruption consisting of small papulæ or pimples, few of them being larger than a pin's head, scattered more or less thickly over the affected surface. The pimples are about the size of a pin's head, and are of a red color, more or less bright, according to the intensity of the eruption. As a general thing, the skin between the papulæ retains its natural appearance. The eruption is most abundant on those parts covered by the dress; its development is undoubtedly favored by warm rooms and excess of clothing. As a general thing, we find it more copious about the neck, the upper part of the chest, and on the arms and legs.

More or less fever usually accompanies the disorder; and the intolerable itching of the parts affected causes much fretfulness and a constant desire to scratch. In the infant there is great restlessness, worrying, and disturbance of the sleep.

TREATMENT. — In the majority of these cases, scarcely any treatment whatever is called for. Only when the eruption is very abundant, and the child is much annoyed by the heat and itching it occasions, is it necessary to pay any attention to it. It is regarded as salutary, rather beneficial than otherwise, to the child; and it is therefore considered bad practice to apply anything which has a tendency to repel it.

Aconite and *Chamomilla* — will generally afford relief, especially when the eruption is attended with fever and restlessness.

Rhus, *Arsenicum*, and *Sulphur* are sometimes called for.

ADMINISTRATION OF REMEDIES. — Dissolve six pills in as many spoonfuls of water, and of this solution give one spoonful, every two hours.

Great comfort and benefit will be obtained by frequent bathing. Sponging the child off two or three times a day with bran-water, or slippery-elm water, or any other mucilaginous substance, will often allay the irritation, and afford considerable relief.

STROPHULUS. — RED GUM. WHITE GUM. TOOTH-RASH.

DEFINITION. — Under the general head of Strophulus, we have included, *red gum, white gum*, and *tooth-rash*. These eruptions are most common during dentition. Their causes are, various disturbances of the digestive apparatus. They are never attended with any danger, and as they are about the only *pimply* eruption to which young infants are subject, there is no difficulty in distinguishing them. We shall describe them separately.

RED GUM. — The papulæ or pimples, characterizing this variety of the disease, rise sensibly above the level of the skin ; they are of a vivid red color, and scattered here and there over different parts of the whole body : we more commonly find them, however, on the cheeks, fore-arm, and the back of the hands. Red gum occurs chiefly within the first two months of lactation. The eruption remains upon the skin some time, one or two weeks, perhaps ; the pimples disappearing and reappearing in successive crops. It usually terminates in exfoliation of the scarf-skin.

WHITE GUM. — This variety of strophulus is characterized by white instead of red papulæ.

TOOTH-RASH. — In this variety the pimples or papulæ are much smaller, more numerous, and set more closely together than in the others ; their color is not so vivid, but they are generally more prominent, and constitute a more severe disorder. The eruption, for the most part appearing only during dentition, has on this account received the name of tooth-rash.

TREATMENT. — As a general rule, it is hardly worth while to make a prescription for either of these complaints. If the bowels be out of order, as they sometimes are, a corrective should be administered. Cleanliness and attention to the dress, however, are usually all that is necessary.

Either *Coffea, Chamomilla, Aconite*, or *Belladonna* may be given when there is great restlessness. Dose, the same as in " PRICKLY HEAT."

CHICKEN-POX.

DEFINITION. — Chicken-pox, or varicella, as it is technically called, is a contagious, eruptive, febrile disease, characterized by more or less numerous transparent vesicles or little bladders, which appear first as a small red dot or stigma, and gradually change into vesicles about the size of a small pea, containing a watery, sometimes a milky fluid.

At one time many considered varicella as but one of the varieties, or modified forms, of small-pox. Observation and experience, however, have proved that it is an independent and specific disease, and not in the remotest way related to variola or varioloid. It is propagated by contagion, and by epidemic influences.

SYMPTOMS. — The constitutional symptoms of varicella are, as a general thing, trifling: occasionally we have, as precursory symptoms, chills followed by heat, hurried pulse, loss of appetite, nausea, and sometimes vomiting. After which, the eruption makes its appearance, but without that regularity which marks variola. Sometimes it is first observed upon the back or face; perhaps more frequently upon these parts than elsewhere; however, it may appear upon any part of the body. As a general thing, the premonitory symptoms are entirely absent, or, at least, so imperfectly developed as to pass unnoticed, and the appearance of the eruption is the first declaration of the presence of the malady.

The eruption appears in the form of small red stigmata or pimply spots, which, in the course of a few hours, exhibit small vesicles in their centres. About the second day the papulæ are converted into globular vesicles, the size of a small pea, and filled with a transparent fluid, which is either entirely colorless, or of a faint orange tinge. Generally the vesicles are not very numerous, and are scattered over the body. Sometimes we find them crowded together, even running into each other. On the fourth day, they begin to shrink; those that have not been broken by accident, or scratched open by the child, in its effort to allay the itching which they give rise to, assuming a shrivelled appearance at the margins, and soon turning into a thin, brownish, horny scurf, which falls off in two or three days, leaving behind no scar, but a faint-red spot, which soon disappears.

The eruption is usually accompanied by a sensation of heat and itching, which is the occasion of a great deal of uneasiness. The child rubs and scratches those vesicles that are within reach, breaking and thus preventing them from running the regular course above described.

TREATMENT. — Unless complicated, this disease requires but little treatment beyond attention to diet and the avoidance of cold during convalescence.

Poor people let their children, during the whole course of the disease, run about the streets the same as ever, and they recover.

Should the constitutional effects be marked, the fever and headache being considerable, an occasional dose of *Aconite* or *Belladonna* may be given.

The disease may be very much shortened by giving an occasional dose of *Pulsatilla*. This remedy has also been recommended as a preventive.

Coffea. — When there is restlessness and considerable nervous excitement; disturbed sleep, with dreams and moaning.

Tartar-emetic. — When the eruption is very severe. Mercurius, Cantharides, and Ignatia are useful in some cases.

ADMINISTRATION OF REMEDIES. — Dissolve six globules in six spoonfuls of water, and give one spoonful of the solution every three hours. Should headache and fever be present, a dose may be given every hour.

DIET AND REGIMEN. — The same as in "MEASLES."

VARIOLA AND VARIOLOID.

DEFINITION. — Small-pox is an epidemic and contagious eruptive, febrile disease, characterized by an initial fever, which upon the third or fourth day is followed by an eruption of red pimples. In the course of two or three days these pimples are gradually changed into small vesicles, which contain a drop of transparent fluid. From the fourth to the sixth day these are changed into pustules, for the suppurative process now commences, converting the serum or transparent fluid, contained in the vesicle, into pus or matter, after which the pustules dry up, and are converted into scabs, which fall off between the fifteenth and twentieth day.

Owing to the attention now everywhere given to vaccination, small-pox is comparatively a rare disease in children, especially among the middle and upper classes of society. I have never seen a case in a person under twenty years of age, and so deem it almost superfluous to write this description of the disease. But, as among the careless and the poorer classes vaccination is sometimes neglected, the disease will occasionally break out, and one case is enough to alarm the whole neighborhood, it is as well that all should understand its nature and appropriate treatment.

CAUSES. — Variola is a contagious and epidemic disease. The principal cause of the disease is universally acknowledged to be contagion. That it is propagated by epidemic influence, however is doubted. At what period of its course the disease acquires its power of infection is not clearly ascertained, and as it is always best to err, if err we must, on the safe side, it is advisable to avoid the patient and his house, from the moment the real nature of the disease becomes apparent.

The period after one is exposed to the disease, — that of incubation, — which elapses before the first symptoms manifest themselves, varies from nine to twelve or fourteen days.

Like scarlet fever, one attack protects the constitution, in the majority of cases, against subsequent contagion.

SYMPTOMS AND TREATMENT. — The disease has been divided into four stages, which we will proceed to describe, and give the treatment appropriate to each as we go along.

FIRST OR FEBRILE STAGE. — This commences, as we have before observed, from nine to twelve or fourteen days after exposure to the contagion. The patient first complains of pains in the bones and loins, similar to, and indeed they are often mistaken for those of a common cold, or he may be taken with a chill, more or less severe, accompanied with headache, and soon followed by fever; dry, hot skin, and great thirst. Nausea and vomiting often exist from the beginning of the attack; there are at the same time loss of appetite, oppression in the stomach, and constipation, more or less obstinate; tongue red and dry. The principal symptom, during this stage of the disease, is the pain in the loins, which, though varying much in degree, is always severe.

In some cases the head symptoms are severe, consisting of restlessness, and irritability; light hurts the eyes; there is swimming in the head; the mind wanders; the patient is flighty, and occasionally there are convulsions.

These symptoms continue up to the time the eruption makes its appearance, which is usually from forty-eight to seventy-two hours.

As appropriate remedies for this stage of the disease we have recommended *Aconite, Belladonna, Bryonia, Rhus,* and *Tartaremetic.*

Aconite. — Is especially called for during the chill and first few hours of the fever, and when there is severe pain in the head; full bounding pulse; thirst; intolerance of light, and delirium.

47

Belladonna — may follow *Aconite*, especially when there is severe headache and delirium; also, when there is intolerance of
noise.

Bryonia. — For the severe backache, pains in the bones, soreness
of the chest, and constipation. Should this not relieve, alternate
it with *Rhus*.

Tartar-emetic. — For nausea and vomiting with clamminess of
the skin.

DR. TESTE, in writing upon this disease, after recommending
Causticum and *Mercurius-cor.* (each of the 30th potency), to be
given in alternation, once in four hours, says: " and we shall
see, in an immense majority of cases, that, under the influence of
this medication, the exanthema and all its concomitant symptoms will be extinguished as if by magic." How true this assertion is, I cannot say, as I have never applied these remedies in
this disease. I can say, however, that *many* of Dr. Teste's *positive*
assertions cannot be substantiated in practice.

Another remedy, which is gaining great repute, as a complete
annihilator of the disease, is *Thuja-occidentalis ;* it is said to
" throttle " the disease, in whatever stage it is administered.

SECOND OR ERUPTIVE STAGE. — Some time in the course of the
third day, after the patient is first stricken with fever, the eruption
begins to make its appearance in the shape of small red pimples,
of the size of pin-heads ; as the eruption comes out, the fever
subsides. This pimply eruption first shows itself upon the face,
and then extends to the neck, trunk, limbs, hands, and feet. This
stage of the disease lasts about three days, during which time the
papulæ or pimples gradually increase in size, and are changed into
vesicles, or little pouches, filled with a transparent fluid.

At the same time the eruption appears upon the skin, we have
something corresponding to it, affecting the mucous membrane of
the mouth, throat, and nose. Sometimes there is severe inflammation of the throat, with tenderness and swelling of the glands
about the neck.

The *treatment,* called for in this stage of the disease, depends,
in a great measure, upon the effect of that instituted for the
preceding one. If *Thuja*, or *Mercurius-cor.*, and *Causticum*
should have arrested the disease, other remedies will be uncalled
for.

Tartar-emetic, and *Thuja*, to a certain extent, undoubtedly have
the power of arresting or mitigating the eruption. We frequently

have cases reported in our medical journals, where, by these remedies, the disease in full progress has been arrested, and the pock dwindled away without ever arriving to maturity.

If, during the eruptive stage, the delirium for which *Belladonna* has been given still remains, *Stramonium* should be administered, and especially if the eruption is at all backward about coming out.

Bryonia. — When the eruption does not make its appearance as promptly as it should; also, when there are constipation, headache, and pain in the back.

Should there be, during the eruptive stage, much or any chest difficulty, with a hoarse, rattling cough, give *Tartar-emetic* or *Ipecac.*

THIRD, OR SUPPURATIVE STAGE. — At this point the eruption changes from vesicular to pustular; that is to say, the transparent fluid, which we have spoken of as being contained in the vesicles, gradually becomes opaque, whitish, and finally yellow, having changed from serum to pus. This change takes place from the fourth to the sixth day of the eruption, or about the eighth or ninth day of the disease. During this stage the pustule completes its development; the pock becomes distended, and as large as a split bean. During the filling up of the pock the face swells, often to such a degree that the eyes are completely closed.

As the eruption occupies about three days in coming out, those pustules which appeared first upon the face are quite in advance of those which appeared last upon the extremities. In fact, while those upon the face are in the third — the suppurative — stage, those upon the breast are in the second — the vesicular; while those upon the extremities are in the first, or just making their appearance. Without this division of the burthen the disease certainly would be unbearable.

The *treatment* adapted to this stage depends in a great measure upon the general feature of disease at its arrival. If there are no alarming symptoms; if the fever which is reproduced during this time is not severe; if the color of the skin between the pustules is not of a livid hue, the remedies which the patient is already taking may be continued.

Mercurius. — This remedy is called for when there is sore throat and considerable fever.

Arsenicum — should be given when the skin between the pustules becomes dark, livid, or brown. Should the pox itself become black, and typhoid symptoms set in, *Muriatic acid* should be given. For stupor during this stage, give *Opium.* For diarrhœa, *China.*

FOURTH, OR STAGE OF DESICCATION. — This is the stage of decline. At about the eighth day of the eruption a small, dark spot makes its appearance on the top of each distended pustule. At this point the pock bursts; a portion of the matter oozes out, and the pustule dries up into a scab. This, however, is not always the case; sometimes the dark point formed upon the apex extends itself until the whole pustule is converted into a hard crust.

The formation of crusts begins upon the face, and extends thence to the trunk and extremities. When at length these crusts fall off, the appearance of the skin beneath is peculiar; there is left a purplish-red stain, which gradually fades away, or else, in severe cases, where there has been true ulceration of the skin beneath, there is a depressed scar, or, as it is said, the patient is "*pitted.*" Desquamation, or the falling off of the crusts does not reach the limbs until about three or four days after it has commenced on the face.

The above description corresponds to a regular and favorable course of the disease, where the pustules are not so numerous as to run together; it is called the *distinct,* in contradistinction to the *confluent,* or that severe form where the pustules are numerous, come in contact, and, running together, form one immense scab covering the whole surface; the latter being necessarily more severe and dangerous than the former.

The *treatment* for this stage is very simple; scarcely anything is called for except cleanliness. Simple ablution with tepid water will generally be all that is required.

At the beginning of this stage it is as well to give an occasional dose of *Sulphur;* say, one dose of three globules every morning until four doses are taken.

DIET AND REGIMEN. — The room in which the patient is-confined should be as large and airy as possible; it should be kept at a moderate temperature, well ventilated, and almost dark. A straw bed or mattress is preferable to a feather bed.

The diet should be cooling, such, for instance, as water, ice-cream, lemonade, oranges, roasted apples, stewed prunes, strawberries, gruels, toast, etc. Avoid the fruit and acids if diarrhœa should be present.

Animal food should not be used until convalescence is pretty well established.

VARIOLOID

This is simply a modified form of small-pox. The treatment which has been given for that applies equally well for this disease.

VACCINATION.

As a preventive against small-pox, vaccination is favorably known and practised by all civilized nations. Many persons object to vaccination for fear that, by this means, some other disease may be introduced into the system. To avoid this, employ a physician whose ability and integrity are above suspicion.

Vaccination and revaccination, from time to time, are considered by every physician as an imperative duty, and the only safeguard against the encroachment of one of the most loathsome and fearful of all diseases.

INTERTRIGO. — EXCORIATIONS.

DEFINITION AND CAUSES. — By intertrigo is understood those superficial sores, excoriations, or gallings, which sometimes appear behind the ears, between the thighs, in the folds of the neck, or other parts of the body where the skin folds back upon itself. This troublesome disorder, as a general rule, is peculiar to fat children. It is said to be caused by the mother or nurse's indulging in high-seasoned or acrid food, and particularly pork. Fat children are predisposed to the disease; and, no doubt, anything which irritates the skin will act as an exciting cause; a want of cleanliness, or, on the contrary, too frequent washing, especially with coarse soap, acrid perspiration, especially when combined with some of the various " baby powders " that are sold by druggists, frequently assist, I have no doubt, in the development of the disease.

TREATMENT. — For this disease, we have recommended *Chamomilla*, that is, when the disease has not originated from the nurse or child's drinking " Chamomile tea," in which case we should rather make use of *Ignatia* or *Pulsatilla*. Should these remedies fail to remove the unpleasant effect of the drug, give one or two doses of *Sulphur*, — three pills, night and morning.

Sometimes the disease, if not caused, is kept in existence by *Lycopodium*, which forms the chief ingredient of most of the " baby powder." When this is the case, discontinue the powder, and give an occasional dose of *Camphor*, *Pulsatilla*, or a little weak *Coffee*, such as we use upon the table, and follow this up with *Graphites*, — one dose of the 30th potency every night. ·.

When the eruption covers a large surface and keeps spreading, give *Mercurius*, 30th, three globules, night and morning. Should this not bring about a favorable appearance of things, discontinue the *Mercurius*, and give one or two doses of *Sulphur*, when better results may be looked for from a readministration of *Mercurius*. Should there at any time during the course of the disease be fever, with hot skin, a little *Aconite* will be called for. Obstinate cases call for *Carbo-v.*, *Graphites*, *Sulph-acid*, *Silicea*, or *Sepia*.

PIMPLES ON THE FACE. — ACNE. PUNCTATA. COMEDONES.

We not unfrequently find upon the face of children and young persons, small, black-headed pimples, from which, by pressing upon their sides, we can squeeze out a small vermiform, or worm-like cylinder, about one line in length. The disease received the name "*comedones*," from the fact that they were for a long time believed to be small animals; investigation, however, has proved that the white cylinder which we squeeze out is nothing more nor less than an accumulation of fatty matter in the follicles of the skin, and the black head is caused by the dust which adheres to it.

The causes of comedones are, anything which obstructs the excretory ducts of the cutaneous follicles; or, indeed, the secretion itself may be of a morbid character, which is frequently the case in persons with a torpid skin; the contents of the oil tubes become too thick and dry to escape in the usual manner. The obstructed and distended tube sometimes inflames, even suppurates, and the pimples become very sore.

TREATMENT. — The remedies recommended for this disease are, *Belladonna*, *Calcarca*, *Carbo-v.*, *Sulphur*, *Nux-vomica*.

Of the one chosen, dissolve six globules in twelve spoonfuls of water, and give of the solution one spoonful every six hours.

ABSCESSES.

DEFINITION. — By the term abscess is understood what in popular language is called a "gathering." A collection of pus or matter in any part of the body, resulting from inflammation, which may be either acute or chronic.

Abscesses are of various kinds : a gathered breast, in technical language, is called a mammary abscess ; a boil is an abscess ; but of these we have spoken, or will speak elsewhere, and shall in the present article confine ourselves to what are frequently called lymphatic tumors and superficial gatherings, such as we so often meet with in children, especially about the head and neck.

CAUSES. — An abscess is not an original disease, but is always the result or termination of inflammatory action. Inflammation and suppuration of the cervical glands of the neck are frequently concomitants of other diseases. Scald-head, scarlet fever, measles, and many other diseases, are frequently followed by glandular inflammation, which terminates in the formation of pus, — true abscesses.

There is about some children a hereditary dyscrasia, or constitutional taint, — scrofula, or some kindred disease, for instance, which predisposes to the disorder.

SYMPTOMS. — Abscesses, as above stated, may be either acute or chronic. The *acute* species is preceded and accompanied by sensible and inflammatory action in the affected part ; it is hot, tumefied, throbbing, and painful. The commencement of the suppurative process, that is, when the formation of matter takes place, is to be known, or, at least suspected, by the change in the character of the pain which takes place at this time, and by the appearance of the skin. The pain which has previously been acute, loses its intensity, becomes dull and throbbing ; the skin changes from a red to a livid color. The tumor presents a somewhat conical shape, and the skin over its apex becomes thin and of a dark livid color. At this point, if left alone, the abscess will burst, and allow its contents to escape.

In abscesses of any magnitude, during the suppurative process, we usually have more or less definitely marked rigors or chills, succeeded in turn by increase of fever. After an abscess is fully formed, provided it is not too deeply seated, fluctuation in the tumor is always perceptible.

CHRONIC ABSCESSES. — Although all abscesses are the result. of inflammation, the inflammatory action in chronic abscesses is sometimes of so low a grade as to be almost imperceptible ; indeed, during the first stage of the disease, it is entirely so ; and were it not for the swelling, which always becomes apparent before it reaches any great magnitude, we would scarcely know that anything ailed the child.

The entire absence of all local and constitutional symptoms renders the disease obscure, until it begins to approach the surface, and form an external swelling.

An acute abscess readily heals, as soon as the pus is freely evacuated. Not so with a chronic abscess; the latter, instead of contracting and filling up with healthy granulations, that is, portions of new flesh, remains open, and discharges copiously of thin, acrid matter; and this state, if continued any great length a time, results in the production of hectic fever, or, in other words, the patient goes into a decline.

TREATMENT. — As abscesses do not always end in suppuration, but sometimes in resolution, as we say, — that is, the inflammation and swelling subside, without the formation of pus, the tumor not gathering, — it is not always advisable to apply poultices, as this may cause it to gather, when it otherwise would not. Should a swelling appear anywhere upon the surface of the body, which we apprehend may terminate in an abscess, our first endeavor should be to bring about resolution, that is, to cut short the inflammation before it reaches the point of suppuration. This can best be done by the internal administration of *Aconite* and *Belladonna*, in alternation, and by the external application of cold water. This treatment is especially recommended when there is considerable constitutional disturbance, with intense pain and extensive inflammation of the parts.

Should this treatment fail to arrest the disease, the next best thing to be done is to hasten suppuration; and this we endeavor to do by the internal administration of *Hepar-sulph.*, one dose of the third potency every four hours, and the external application of hot fomentations and poultices. Ground flaxseed makes the best, because it retains its heat and moisture — the chief properties of all poultices, — longer than almost any other substance.

As soon as the abscess points or comes to a head, the skin becoming livid and thin, and there is distinct fluctuation, it is advisable to make a free incision into the tumor, and evacuate the matter. Some do not advise this treatment, but wait until it bursts of itself. The sooner, after suppuration takes place, the matter is discharged, the sooner the abscess will heal. I see nothing gained, therefore, by waiting: it is but prolonging the patient's suffering, and retarding the cure.

After the abscess has opened, and the matter has been freely discharged, the poultices should be discontinued, and simple dressing substituted. During this stage the patient can take *Calcarea-carb.* or *Silicea*, one dose every evening.

The remedies for chronic abscesses are, *Mercurius, Hepar-s., Silicea, Calcarea, Lachesis, Phosphorus,* and *Sulphur*.

Silicea or *Sulphur* are especially indicated in severe, prostrated cases, in which the suppuration is long continued and seems to exhaust the system.

For hard and swelled glands on the neck, under the ears or chin, *Mercurius* and *Calcarea* are the principal remedies.

Of either of the above remedies, when selected, you can dissolve six globules in twelve spoonfuls of water, and of this solution give one dose every four hours.

DIET. — In acute abscesses, where there is considerable fever, the diet should be about the same as in fevers. During the long and tedious course of some exhausting chronic abscesses, it will be found necessary to select such a diet as will nourish and strengthen the patient. The food should be nutritious and of easy digestion. Broiled steak, mutton chop, meat broths, rice and barley gruel, etc., may be allowed.

BOILS.

DEFINITION. — A boil consists of a round, or rather cone-shaped, inflammatory, and very painful swelling, immediately under the skin. It varies in size from a pin's head to a pigeon's egg. It always has a central "core," as it is called, and is mostly found in strong and vigorous children. A boil always suppurates, and sooner or later discharges its contents, the matter being at first mixed with blood, and afterward composed of pus. A boil never discharges freely, and never heals until the core comes away.

CAUSES. — I do not know why it is that children, who, to all appearance, are otherwise perfectly healthy, should be troubled with boils. It is very easy to say, that some children possess a constitutional predisposition to them, or inherit them, or that they frequently follow eruptive diseases and fevers ; but this is not a satisfactory explanation of their cause, yet nevertheless it is all that we can say. You may attribute them to bad blood, — which is a very common way some people have of accounting for many diseases, — but what does that amount to ?

TREATMENT. — The treatment is about the same as that for abscesses. Apply a poultice early, and bring the tumor to a head as soon as possible, for the sooner the matter is discharged, the sooner the process of reparation will begin. — See "ABSCESSES."

Arnica given internally is said to lessen the pain and inflammation ; one dose every four hours. Should the boil be very red, hot and painful, and attended with fever, thirst, and headache, give *Belladonna* once in three hours, either alone or in alternation with *Mercurius*. When it is slow in coming to a head, give *Hepar-sulph*. After the tumor has broken, if the suppuration is excessive, give *Mercurius*, and especially should there be much swelling and hardness about the base of the tumor.

After the matter has been discharged, discontinue the poultice, wash the parts clean, and dress with lint and simple salve. The lint should be placed *next* to the sore, and the salve *over* the lint.

To eradicate the predisposition to boils, one dose of *Sulphur* of a high potency — 200 — twice a week.

ADMINISTRATION OF REMEDIES. — Of the remedy selected, dissolve twelve globules in twelve spoonfuls of water, and of this solution give one spoonful at a dose as above directed.

SCALD HEAD. — TINEA CAPITIS. FAVUS.

DEFINITION. — Tinea capitis is a contagious eruptive disease of the scalp. It is characterized at first by small yellow pustules, situated on an inflamed ground. The pustules are of a peculiar shape, depressed in the centre, and scarcely raised above the level of the skin. Each pustule, as a general thing, surrounds a hair. Perhaps the whole disease, as some authors assert, consists in an inflammation of the hair follicles.

The disease is not a common one, much less so than ringworm of the scalp, or milk-crust. Among the upper and middle classes of society it is seldom if ever met with. I have never seen but one case in this city, except in dispensary practice.

CAUSES. — There is very little doubt but that this disease is contagious, although some authors think otherwise. This difference of opinion may arise from the fact that the disease is not unfrequently confounded with other similar disorders of the scalp. My experience in this disease has been quite limited ; still, I think I can say, with a good degree of certainty, that it may be propagated by direct contact of the diseased with a healthy skin, or by means of combs, brushes, towels, etc.

Although chiefly found in children, it is by no means exclusively confined to them.

Children living in low, damp, and ill-ventilated dwellings, and those subjected to an unwholesome or an insufficient diet are most prone to it.

SYMPTOMS. — The feature which distinguishes this disease from other eruptions of the scalp is the peculiar shape of the scabs or crusts. Commencing as a small yellow pustule, scarcely raised above the level of the skin, it gradually increases to perhaps an inch in circumference. As it spreads, the watery portions of the pustule dry up, leaving a large yellow crust with inverted edges and a depressed centre. This cup-formed yellow crust, *pierced by a hair*, is peculiar to this disease, and distinguishes it from all other eruptions of the scalp.

At first, when the pustules are small, they are usually isolated; but, as they increase in diameter, their edges come in contact, and thus a number of pustules blended together form irregular patches of larger or smaller size.

When the crusts have been removed, the surface beneath is seen to be red and moist, having the appearance of ulceration.

By no other eruptive disease of the scalp with which I am acquainted is there a permanent loss of the hair. In this disease the hair falls out, and the scalp is left shining and uneven. The hair seldom ever reappears; if it does, it is short, woolly, and unhealthful.

TREATMENT. — Until you can secure the services of a good Homœopathic physician, follow the treatment given for "milk-crust."

The first essential step is to remove the hair. This may be done sufficiently well with a pair of scissors; shaving the head is hardly practicable. No attempt whatever should be made to remove the crusts. Strict attention should be paid to cleanliness. A good and soothing wash for the head is bran-water.

Calcarea-carb., *Sulphur*, *Lycopodium*, *Sepia*, *Arsenicum*, and *Rhus* are among the prominent remedies for this disease.

CRUSTA LACTEA. — MILK-CRUST. IMPETIGO.

DEFINITION. — This is almost exclusively a disease of infancy. It is characterized by an eruption of small, round, yellow, flattened pustules, which are crowded together upon a red surface.

The pustules end by the drying up of their contents into thick, rough, and yellow scabs.

The eruption may appear upon the forehead, cheeks, or scalp, the latter place being the more frequent seat of the disease.

CAUSES. — Like most other varieties of infantile eruptive diseases, the real cause is very imperfectly understood.

As it occurs for the most part during first dentition, many suppose the evolution of the teeth to be the cause of the disease. Among other speculations it is asserted, that it arises from exposures to unhealthy hygenic conditions; as, for instance, improper or unwholesome food, want of cleanliness, damp or ill-ventilated apartments. Not a few think it arises from some constitutional taint existing within the child, as scrofula, or some kindred disease, and that it more frequently manifests itself in fair, fat children.

My own opinion is, that children so affected possess a constitutional tendency to the disease, and that the exciting cause, in nine cases out of ten, is some gastric derangement.

SYMPTOMS. — In some cases the eruption is confined entirely to the face; in others, entirely to the scalp; or, again, it may implicate both, extending up the side of the face, affecting the ear, neck, and portions of the scalp.

The disease may either be acute or chronic in its nature. When acute, it is not unfrequently attended with severe inflammation of the skin. It appears in all grades of severity; in some cases it is very light, extending over a small surface, remaining stationary, or quickly drying up and disappearing; or, when severe, the whole scalp may become completely scabbed over, presenting an offensive and disgusting appearance.

As a general thing, it attacks but a small spot at first, and then gradually spreads to the surrounding parts. When they first appear, the pustules are numerous, small in size, of a light yellowish or straw color, and not unfrequently attended with severe burning or itching. These soon break or get broken, and discharge a sticky fluid, which glues the hair together and forms into thick, uneven crusts. The successive discharges from the surface beneath constantly add to the thickness of the crust, and as the fluid escapes from under the crusts, it irritates or inoculates the parts with which it comes in contact, and thus extends the disease until in some cases the whole scalp is covered with a thick, rough, brownish-yellow crust. In warm weather, or from the warmth of the head and exposure to the air, these crusts sometimes undergo partial decomposition, and exhale a sickening and most offensive odor.

When the crusts are removed, the surface beneath is found inflamed and wet; the secretion which oozes from them plainly visible; little excoriated points soon form new crust similar to the one that has been removed.

The disease, as it appears upon the face, passes through about the same course, as when appearing upon the scalp, except that the large crusts are seldom allowed to form. The severe itching attending the disease causes the child constantly to scratch or rub the part, sometimes to such an extent as not only to prevent the scabs from forming, but to cause the surface to bleed quite freely.

In most cases, the general health of the patient remains good; sometimes, when the inflammation and itching are severe, it makes the child cross and peevish, disturbs its sleep and makes it feverish. The glands, situated upon the neck and especially behind the ears, not unfrequently inflame, become hard and painful, and finally gather and break.

The duration of the disease depends upon the severity of the case, and the treatment which is instituted for its removal. Some cases yield in a few weeks; others, more stubborn, may continue for months, and, if improperly treated, even for years. The whole course of the disease may not be of the same severity; it not unfrequently subsides to such an extent that the mother is already congratulating herself upon the speedy return of her child's health, when a fit of indigestion, or the cutting of a new tooth, or even some change in the weather, may bring it back with renewed violence.

TREATMENT. — I frequently meet with children, who have had the disease for months, their parents refusing to do anything for its removal, under the impression that an attempt to cure it would be attended with serious risk to the health, and even the life of the patient. Now, the idea that the disease is useful, and beneficial to the future health of the child, is preposterous. Perhaps it originated from the fact that a sudden suppression of the disease, by active, external means, has been followed by dangerous and even fatal symptoms. But then it should be remembered that suppressing is not *curing* a disease. I believe it to be unsafe to procure, by the employment of external means, the suppression of any eruptive diseases. We are all aware that alarming and dangerous symptoms frequently follow the " striking in," as it is called, of measles and scarlet fever. Every physician can call to mind cases of acute disease of the brain, resulting from the sudden drying up of this very disease by the application of some one of the numerous specific ointments. This is a literal " striking in " of the disease, a metastasis or translation from the scalp to the brain.

The idea that the disease is in any way beneficial to the health of the child cannot be entertained by any one who has had much acquaintance with the suffering it produces. The dreadful itching produces restlessness, crying and sleeplessness; in fact, it keeps the infant in a constant state of actual suffering, and which, I am certain, cannot continue, as the disease sometimes does, for months, or even years, without seriously injuring the constitution of the child.

I consider it therefore, an entirely mistaken act of kindness which permits the disease to continue a single day without an endeavor to arrest it, under the impression that the child is thereby being permanently benefited.

Active treatment should be instituted as soon as the first symptom of the disease is observed. In the first stage, when the itching is severe, and seems to be particularly aggravated at night, *Aconite* is the best remedy, either alone, once in two hours, or in alternation with *Chamomilla*.

Rhus. — Should *Aconite* fail to relieve the itching and inflammation.

If, after the acute symptoms have subsided, the incrustations still continue to form, or if they remain stationary, we should give *Viola-tricolor*. Frequently this remedy will be sufficient to complete the cure.

In obstinate cases, when the above remedies fail to produce a favorable change, or when the improvement is slow, *Sulphur* is a good remedy. *Sulphur* and *Rhus* alternate well together.

Calcaria-carb. — When the eruption is dry and scurvy.

Lycopodium. — When the eruption is moist, the discharge profuse, and of a fetid smell.

Mezereum, Antim.-crud., Sepia, Hepar, Arsenicum, Graphites, and *Nitric acid,* are sometimes called for.

Should the eruption, which had previously been moist, suddenly dry up, and the child become drowsy, unnaturally sleepy, and sleep with its eyes half open, give *Hellebore* or *Bryonia.* Consult a Homœopathic physician, in all cases, when it is convenient.

ADMINISTRATION OF REMEDIES. — During the acute stage of the disease, the remedy may be repeated as often as once in three hours. Afterward, one dose, night and morning will be sufficient. Of the remedy chosen, dissolve twelve globules in twelve spoonfuls of water; and, for a dose, give two spoonfuls of the solution, or, if preferred, give three globules dry upon the tongue.

Apply nothing externally but a little *Glycerine*. Keep the head clean by washing with weak soapsuds of *Castile soap.*

CHAPTER IX.

DISEASES OF THE BRAIN AND NERVOUS SYSTEM.

REMARKS. — The study of diseases of the brain and nervous system, being surrounded with many and peculiar difficulties, our knowledge in regard to them is less definite perhaps than that of most other diseases which we are called upon to treat.

We can ascertain the state of many disorders by our sense of sight, and in others we depend upon that of touch or of hearing. But the brain and spinal cord we can neither see, nor hear, nor handle. We can make no physical examination of either; so we are compelled to base our conclusions entirely upon the array of objective symptoms which each particular case presents. .

The functions of the brain and nerves are, sensation, thought, volition, and the power of originating motion. Now the multiform disturbances to which these functions are often subjected present us with a host of motley symptoms which it is our duty to arrange and interpret. When it is remembered, also, that many of these symptoms *may* be sympathetic with diseases of other parts than the brain, it will readily be seen that the language they speak is often *very* hard to construe, and that we frequently fail to ascertain, by these outward signs, what inward things they denote. — WATSON.

I shall not enter into any minute or lengthy description of the various disorders to which the brain is exposed, for several reasons. In the first place, if it be the difficult matter that I have stated for those who are familiar with all the morbid pitfalls into which frail humanity is continually stumbling, to diagnose a disease of this important organ, of what practical use would be the little instruction which I could impart to a layman within the short space allotted to each article in a work like this. Besides, a class of diseases, always serious in their nature, involving a vital organ, can at best be combated with any reasonable hope of success, only by an experienced physician. It would not be prudent for a lay-man to take the responsibility upon himself of treating them, neither would he wish to do so, unless a confirmed egotist. I shall, therefore, content myself with simply enumerating some of the more important and prominent symptoms which precede and accompany the first stage of the disease, and especially if it be one of an inflammatory nature, recommending that, in all cases where it is possible, the patient, without loss of time, be placed under the care of an intelligent Homœopathic physician.

INFLAMMATION OF THE BRAIN.

DEFINITION. — Under the general head of inflammation of the brain, I shall enumerate some of the prominent symptoms which characterize all the diseases of an inflammatory nature, which attack the interior of the skull, without regard to the particular tissue affected.

Inflammation of the brain itself is called *Encephalitis ;* inflammation of the membranes which invest the brain is called *Meningitis*.

SYMPTOMS AND CAUSES. — Inflammation of the brain and of its investing membranes has no fixed and uniform train of symptoms by which it declares itself; perhaps the most common and striking phenemenon is a sudden and long-continued attack of general convulsions. Still, convulsions, especially in children, frequently arise from various other causes; for instance, from teething, from overloading the stomach, or from worms.

The attack may come on with but slight pain in the head, with vomiting and impatience to light. More commonly, however, there are severe pains over the entire head, throbbing of the arteries of the neck and temples, fits of shivering, vertigo, sleeplessness, or restless sleep, disturbed dreams, unsteady gait, quick pulse.

In those cases which are occasioned by blows or falls upon the head, the patient may recover entirely from the shock and external wound, if there be one, and remain for a certain period, to all appearances, perfectly well. But, after some days, or even weeks, he begins to complain ; may come in from his play with headache and chilliness ; the skin soon becomes hot and dry ; he is restless ; cannot sleep ; his countenance becomes flushed ; his eyes red and fiery ; the pulse is hard and frequent ; nausea and vomiting su-

49

pervene ; the substance thrown up is, generally, a greenish or yellowish fluid ; and, as the case draws to a close, delirium, convulsions, or profound stupor takes place.

Inflammation of the substance of the brain, either when it invades the whole organ at once, or begins in one part of either or of all the membranes, and extends rapidly to all the rest, is always attended with high excitement, much fever, and great delirium. The face is red and bloated; the eyes are bloodshot and brilliant ; the pupils contracted ; great sensitiveness to light and noise.

The deeper the interior of the brain is affected, the more the senses are stupefied, until the patient becomes entirely unconscious ; he can neither see nor hear ; the pulse is small, frequent, and tremulous.

Owing to the fact that the organization of a child's brain is much more delicate, and, therefore, much more sensitive, than that of adults, its disease must necessarily be more frequent and much more dangerous. It is well, therefore, promptly to heed, and notice critically every symptom, no matter how trivial, which points a finger of suspicion toward the brain.

Generally speaking, these diseases are uncommon as a *primary* disorder ; but they are frequently met with, as a *secondary* affection, and in the following manner : — a child, suffering from a discharge, either acute or chronic, from the ear, takes cold ; the discharge stops, or from any other cause is suddenly suppressed ; the inflammation, so to speak, travels inward, or, in other words, is propagated to the membranes covering the brain. The patient becomes dull and drowsy ; sometimes, when there is high fever, he is delirious ; he puts his hands up to his head, or bores his head into the pillow, and by degrees sinks into a complete stupor, from which he may never recover.

There is no doubt that catarrhal difficulties of the head and throat are frequently transmitted to the brain. Or the inflammation creeps along from one membrane to another, until it finally reaches the brain, and produces fatal results. This is especially the case when astringent lotions and injections are made use of for the cure of such complaints.

In eruptive fevers, especially when the eruption does not come out well upon the surface, or, after being well out, it suddenly strikes in ; also, during difficult dentition, or even in some forms of severe colds, the child will complain of chilliness, with alternate flashes of heat; also, of heaviness of the head; or, if too small to complain, it will be noticed that the head tends to gravitate backwards. There will be pain in the head, manifested by the child's constantly putting its little hands up to its head; or it frequently moans, and finally becomes drowsy and stupid, or is restless; starts during sleep; rolls its head, and screams upon the slightest provocation. Now, almost any old lady, standing by and observing these symptoms, will remark, that " the disease is going to the head;" and she will be right. These symptoms are seldom present except when the brain has become involved. Any general irritation may bring on the disease. It sometimes supervenes upon the drying up of eruptions, such as scald-head, or sores behind the ears, especially in scrofulous children. In fact, anything which tends to bring scrofulous disease into action may be set down as an *exciting* cause of the disease.

It has been observed, by all writers upon diseases of children, that diseases of the brain, during infancy, are much more frequent among those born of parents who are either suffering from some tubercular disease themselves, or in whose families such complaints have existed to a greater or less extent.

Perhaps the most frequent form of brain-disease, in childhood, is that known to the profession as *Tubercular Meningitis*, or *Acute Hydrocephalus*, which is, strictly speaking, a localization of tubercular matter in the highly vascular membrane which directly invests the brain, — the *pia mater*, — and, in the majority of cases, coincident inflammation of that membrane with an effusion of serum into the cavities and substance of the brain. It is commonly known as " *Water on the Brain.*"

In all probability, if those children who are carried off in infancy with this disease, had reached adult life, they would eventually die of consumption; or, in other words, of a localization of tubercular matter upon the lungs.

The predisposing cause of acute hydrocephalus is, undoubtedly, a hereditary scrofulous or consumptive taint, which forms a constitutional tendency toward tubercular diseases. The *exciting* cause may be a fall, difficult dentition, or gastric irritation. These, in children otherwise constituted, would produce scarcely any trouble.

Tubercular meningitis is very insidious in its attack; it sometimes steals upon the patient before the attendant is fairly aware of its approach.

It is of great importance to recognize this disease in its *earliest* stages. Dr. Watson gives the following precursory symptoms: — " The child loses his appetite, or his appetite becomes capricious; he sometimes appears to dislike his food, and sometimes devours it voraciously; his tongue is foul, his breath offensive, his belly enlarges, and sometimes is tender; his bowels are torpid, and the evacuations from them unnatural; the stools are pale, and contain but little bile, or are dark, with vitiated bile, fetid, sour-smelling, slimy, or hard and lumpy, and the child loses his former healthy aspect. Even already there are obscure indications of derangement in the cerebral functions; the child is heavy, languid, and dejected; his customary spirit and activity are gone; he gets fretful and irritable, and is manifestly uneasy, and sometimes he shows a little unsteadiness and tottering in his gait." . . . . " A frequent, sudden cry, or scream, a clenching of the little fists, and a turning in of the thumb toward the palm of the hand, give warning also of the approaching malady."

When these symptoms are observed in a child who has any hereditary title to scrofula, or in a child who is precocious or particularly clever, there will be much reason to apprehend that mischief is brewing within his head.

After the above symptoms, which are characteristic of this stage of the disease, have continued during a variable length of time, the disease assumes a more active form; the pain in the head, which is nearly always present, becomes extremely severe; it is usually located just over the brows, but may extend all over the head; the infant frequently puts its hand up to its head, and presses against it, and rolls its head from side to side. At the beginning of an attack, there is often pain and stiffness at the back of the neck; sometimes extreme tenderness of the scalp; the child cries and shrieks when taken up. Vomiting is nearly a constant symptom, and is often excited by raising the child to an erect position; the headache and vomiting are both aggravated by motion; there is a total loss of appetite; the tongue is coated white; the breath is offensive. Constipation is almost always a prominent symptom. Occasionally we meet with diarrhœa, but it is rare; the constipation is generally obstinate for the first week or ten days of the disease, and then, toward the termination of the case, gives way to a diarrhœa with involuntary stools.

The head is usually hot; the pulse is variable; it is sometimes frequent through the whole course of the disease, and sometimes perfectly natural. The senses of sight and hearing become painfully acute; the patient shuts his eyes and contracts his brows, whenever the light is thrown upon his face; the slightest noise or jar, even walking across the room, irritates and distresses him.

As the disease passes from this stage to the one following, as it does at times with great rapidity, a marked change in the case is plainly manifest. To quote again from Dr. Watson: "Noises do not now disturb or irritate the child, who lies on his back, with his eyes half closed, in a state of drowsiness or stupor, which is occasionally interrupted by some cry or exclamation expressive of pain. Convulsions frequently occur, but not uniformly; slight and partial spasmodic twitchings, or general and long-continued convulsions, sometimes paralysis. The urine and stools are passed unconsciously. Sometimes the child, with feeble and tremulous hands, is incessantly picking his lips, or boring his finger into his ears or nostrils. This stage may last a week or two, and, what is remarkable, is often attended with remissions, sometimes sudden and sometimes gradual; deceitful appearances of amendment, and even of convalescence. The child regains the use of its senses; recognizes those about him; appears to his anxious parents to be recovering; but, in a day or two, it relapses into a state of deeper coma than before. And these fallacious symptoms may occur more than once."

During the last stage of the disease, "The child rolls his head perpetually from side to side, waves his hands in the air, or one hand, the other frequently being palsied; sometimes there is paralysis of one side, and convulsive twitchings of the other. The circulation is very unequal; one part of the body will be found hot and dry, and another covered with a cold sweat; the checks are alternately pale and flushed; the child is raving or insensible; the rapid pulse gets more and more weak; and, at length, the patient expires. In many instances death takes place in the midst of a strong convulsion."

TREATMENT. — The remedies, most to be relied upon in this disease are, *Aconitum, Belladonna, Helleborus, Hyoscyamus, Stramonium* and *Opium.*

Aconitum. — If, at the commencement of an attack, there be violent inflammatory fever, delirium, burning pains in the head, red face and eyes. It is often given in alternation with *Belladonna.*

Belladonna. — This is one of the most important remedies in

this affection, and is frequently called for at the very commencement of an attack. It is indicated, when there is great heat in
the head; red and bloated face; violent pulsations of the arteries
of the neck and temples; red, sparkling eyes, with a wild expression; boring the head into the pillow; great sensitiveness to the
slightest noise and light; contracted or dilated pupils; violent and
furious delirium; shooting pains through the head; which are
aggravated by the least movement, even of the eyes; loss of consciousness; low muttering; convulsions; spasmodic constriction of
the throat; vomiting; and involuntary evacuations of fæces and
urine. In alternation with *Aconite*, or, should the symptoms call
for it, with *Stramonium* or *Hyoscyamus*.

If, at any time during the course of an inflammation of the
brain or other head difficulty, there should be great restlessness,
with inability to sleep, an occasional dose of *Belladonna* or *Coffea*
may be given, with almost a certainty of producing the desired
result, and that, too, without in the least interfering with other
remedies, which the patient may be taking.

Helleborus. — Painful heaviness of the head, with heat in the
head, and cold extremities. The pain is less violent when the
patient lies still and quiet. The pain in the head is of a stupefying nature, as if the brain were bruised. There is a predisposition
to bury the head in the pillow. The child sleeps with its eyes
half-open. The child shrieks and cries when roused up or taken
from its cradle. *Helleborus* may be given alone, or in alternation
with *Bryonia*.

In all cases where head-symptoms supervene upon any other
derangement, as teething, or intestinal irritations, or where from
suppression of eruptive disease, or from suppression of discharges
from the ear, the head becomes affected, *Helleborus* and *Bryonia*
should be given without delay, in alternation, from one to two
hours apart, according to the urgency of the symptoms.

Should these remedies prove insufficient, and the case grow
worse, with the addition of the following symptoms: loss of consciousness; eyes motionless; with insensible pupil; skin of a bluish
color; coldness of the surface, and pulse imperceptible; give *Belladonna* and *Zincum*, in place of *Helleborus* and *Bryonia*.

Hyoscyamus. — Drowsiness; loss of consciousness; delirium;
sudden starting; red face; constant muttering; desire to escape;
picking at the bedclothes; low muttering; — to be given either
alone, or in alternation with *Belladonna*.

Stramonium. — When there is starting or jerking in the limbs, staring look; red face; sleep almost natural, but followed by absence of mind after waking. Sometimes the patient has frightful dreams or visions; moans and tosses about during sleep.

Bryonia. — Constant inclination to sleep; sudden starting from sleep, with delirium; starts, cries, burning and shooting pains through the head; cold sweat on the forehead. *Bryonia* is often of great service in those cases where *Belladonna* and *Aconite* seem to be indicated, but still fail to afford permanent relief. It may be followed by *Helleborus.*

Opium — is called for, when there is lethargic sleep, with heavy breathing; eyes half-open; dizziness on waking; great listlessness and dulness of sense; apathy, and indifference to everything.

When the head is hot, cold water may be applied externally; care should be taken that the cloths are frequently removed, and not allowed to become warm. The better method is, to apply a bladder filled with pounded ice; by this means, the cold is made continuous.

Other remedies, as *Zincum, Cuprum-met., Apis-mel., Rhus-tox., Lachesis,* and *Sulphur,* are sometimes called for.

ADMINISTRATION OF REMEDIES. — Dissolve twelve globules of the selected remedy in twelve spoonfuls of cold water, and give either alone, or in alternation with another remedy, similarly prepared, every hour or two hours, one spoonful, until amelioration, or till another remedy is called for. Have a separate spoon for each remedy.

DIET AND REGIMEN. — The same as in fevers.

CHRONIC HYDROCEPHALUS.

DEFINITION. — SYMPTOMS. — Chronic hydrocephalus is an actual dropsy of the brain. The disease is frequently congenital, the child being born with a head out of all proportion to the rest of its body. From some cause not well understood, a watery fluid collects within the brain, and the skull, being but imperfectly developed, yields to the inward pressure; and the head is augmented in some cases to an enormous size. When this accumulation of water takes place, as it frequently does, while the child is yet in the womb, it is sometimes impossible for it to pass through the natural outlets into the world. In such cases, the head must be emptied by the accoucheur, in order to preserve the life of the mother. In many cases, however, the uterine contractions are able to expel the enlarged head entire and unhurt, and the child lives, for a longer or shorter period.

Sometimes, however, the accumulation of water does not take place till after birth ; but when it does take place, which may be in a few days, weeks, or even months, it will be perceived that the head enlarges with great rapidity, is quite disproportionate to other parts of the body, and, of course, gives the child a very strange aspect.

The greater part of those who are afflicted with this form of dropsy recover or die during infancy ; they seldom grow up in this condition; there are cases, however, on record, where they have reached adult life possessing fair intellects.

A consideration, however, of these cases, is of but little practical value in a work like this. We shall not, therefore, consider them further.

CONVULSIONS, SPASMS, OR FITS.

DEFINITION. — By the term fits, spasms, or convulsions, — and they are used indifferently,— is meant a violent and involuntary agitation of a part or of the whole body. These agitations consist in alternate contractions and relaxations of the muscles of the part affected. Convulsions may be either general or partial. When general, or universal, the muscles of the face and body, as well as those of the extremities, are affected. When partial, the spasmodic action is confined to one particular part.

All convulsive diseases consist in affections of the true spinal system of nerves.

CAUSES. — Among the predisposing causes of the disease, may be mentioned a highly susceptible irritable or nervous temperament. It has been stated that convulsions are more common in girls than in boys. Whether there is any truth in this or not, I am unable to say. It is also said, that delicate children are more subject to them than robust ones. This may be so.

We frequently meet with families, in which all the children, during infancy, are afflicted more or less with spasms. This may be owing to a similarity of nervous temperament inherited from the parents.

Convulsions occur most frequently in children under seven years of age, and, particularly, during first dentition.

The *exciting* causes of convulsions are as numerous and dissimilar as one could well imagine. Perhaps the most common are, irritation of the bowels, difficult dentition, and worms.

A dangerous form of convulsions is often produced by overloading the stomach, or by eating heavy or indigestible substances. The most alarming variety of convulsions, however, — alarming because frequently terminating unfavorably, — are those occasioned by heavy blows or falls upon the head.

Spasms in children are frequently occasioned by the inordinate use of drugs. I have seen many a child thrown into severe convulsions by the disturbance produced by the action of some simple cathartic. I have in my mind now a case of this kind, produced by giving the child a dose of sulphur and molasses. When we come to remember how delicate the organization of an infant's nervous system must be, it is not to be wondered at that the enormous quantities of patent and domestic remedies, which children are compelled to take, frequently derange the equilibrium. The only wonder is, that they ever recover from convulsions thus produced.

Among the causes of convulsions, we would also mention, excessive crying, either from anger or pain, excessive joy and fear, exposure, with the head uncovered, to a hot sun; severe pains, as earache or colic, and repelled eruptions.

Eruptive fevers, — such as measles, scarlet fever, small-pox, etc., — are frequently ushered in by convulsions; but the spasms generally soon disappear as the eruption shows itself upon the surface. In cases like this, it seldom proves fatal; on the contrary, when convulsions appear during or at the termination of such diseases, or where it appears in connection with hooping-cough or pneumonia, it not unfrequently produces serious, and often fatal, results.

It is a well-attested fact, that severe mental emotions, as fright or anger, may change the milk of the mother in such a way as to make it almost a deadly poison to the child. Cases are by no means rare where such a change, brought about by such causes, has produced convulsions in the nursing infant. Evil results have also been produced by the mother's nursing the child while she was overheated.

SYMPTOMS. — As far as my experience goes, convulsions are not preceded by any symptoms which, to the experienced eye, denote their near approach. This, however, I would say, is contrary to the experience of many writers, who have set down, as premonitory symptoms, drowsiness, excessive irritability, a peculiar expression of the face, and drawing of the thumbs into the palms of the hands. If, from a light and easily disturbed sleep, the child starts and moans, or appears to have frightful dreams, or wakens with a start and screams, seems bewildered and terrified, rolls its eyes about, or has a fixed and staring look, with nervous agitation, these certainly might be regarded as symptoms indicating the approach of a convulsive seizure. But such symptoms have not preceded those attacks which it has been my lot to treat.

However, whether preceded by premonitory symptoms or not, the attack usually commences in the eyes, which are, at first, fixed in one position, staring; but as the case advances, they become agitated, and are turned up beneath the upper lid, leaving only the whites visible; the eyelids are sometimes open, at others shut; the eyes are frequently crossed; the pupils may be either contracted or dilated. The muscles of the face next become affected, and the contractions produce at times most horrid contortions. There are sometimes only slight twitchings of the muscles of the face, with alternate contractions and relaxations. The mouth is distorted into various shapes; the corners of the mouth are drawn down and fixed in this position, or the muscles of one side may contract, while the others relax, and so keep the parts in a constant state of agitation. The tongue, when it can be seen, will be observed to be in constant motion. It not unfrequently gets between the teeth, and severely bitten.

Sometimes the jaws are firmly set; again they are in violent motion. At times, but rarely, there is foaming at the mouth. There is usually more or less congestion of the head; the whole surface often becomes slightly violet-colored; there is always a blue shade around the eyes and mouth.

In slight cases, the muscular twitchings may be confined to the face, or rapid shocks may affect a few parts, the face, neck, and arms, or half of the body. But, in severe cases, when the spasm becomes general, the whole body is violently convulsed; the head is drawn backward, or to either side; the body may become stiff and rigid, or variously contorted; the fingers are drawn into the palms of the hands; the arms are thrown backward and forward, or jerked and drawn into all conceivable positions; the lower extremities are likewise affected, but not generally in so violent a manner.

In mild attacks, though consciousness is entirely destroyed, the child is sensible to external impressions, will open his mouth when a spoon is put against his lips. The child is always able to swallow, I believe, even in the worst forms of the disease.

The duration of a fit is exceedingly uncertain. In some cases, it is but a flash, or of a very few moments' duration, while in others it continues for hours. As a general rule, I think it may be stated that a spasm lasts from one or two minutes to half an hour. When convulsions continue for a long time, they are almost always interrupted by remission. Sometimes the child will have several fits through the course of the day; or, perhaps there will be several fits with intervals of half an hour between each of them; during these intervals the child may be conscious, or partly so, or he may be delirious, or dull and drowsy, or stupid, according to the nature of the attack.

The duration and recurrence of an attack depends entirely upon the *cause* of the disorder. As long as this continues in action of course we can expect no permanent improvement. Our first effort, then, is to remove it. This, however, is not always easily accomplished, because in many cases it cannot be detected.

TREATMENT. — In the treatment of convulsions, our first and chief attention should be given, first, to the *cause* of the disorder, and second, to the *disordered condition* of the nervous system, which this cause has produced.

Before recording the treatment applicable to *special* and *particular* cases, we will speak first of the *general* treatment, which applies to *all cases*, no matter what may be the nature or cause of the attack. The first thing, on being called to a case of convulsions, order the child immediately to be put in a warm bath, — 96° F. While this is being done, the particulars of the attack, its history, mode of seizure, whether general or partial, etc., can be inquired into. I have frequently seen children come out of the bath wonderfully improved, -and I have never seen or heard of its doing harm. I order it in almost every case that I have to treat. In some slight forms of the disease it is scarcely necessary to make the bath general; a foot-bath with a little mustard in it will be sufficient. The patient should be kept in the water from ten to twenty minutes, or until the convulsion ceases. And, when taken out, do not stop to dress him, but just wrap him up in a warm flannel or woollen blanket; do not even stop to wipe him off.

Cold water to the head has also proved an efficient adjuvant, according to my experience.

During the application of the foot-bath, cold applications can be made, and should be continued until the head feels quite cool. The best way to apply it is, to pour cold water from the nozzle of a teapot, held two or three feet above the child's head. This process must be repeated as often as the head gets hot, or rather as often as the head begins to get warm again; it should not be allowed to get hot.

If possible, the child should be placed in a large, well-ventilated room, where the air is pure. When this cannot be done, the next best thing is to expose the patient to fresh air at an open window.

Some persons are very fond of applying mustard draughts to the ankles and wrists. I cannot say that I have ever known them to be of much service, still I never object to their use, because they satisfy bystanders even if they do not help the patient; besides, I am not one of those who imagine the action of homœopathic medicine can be impeded by such a proceeding. Onion draughts, however, are quite a different thing; they certainly are of no earthly use to the sick one, and can just as well be dispensed with as not. The fact is, I think they are abominable, disgusting. If it were possible to arrest or prevent the action of remedies by odors, I should certainly put a great deal of confidence in that emitted by onion draughts.

SPECIAL TREATMENT. — INDIGESTION. — If the convulsion has been occasioned by the child's overloading its stomach, or from partaking of some indigestible substance, evacuate the stomach at once by an emetic. Lukewarm water is as good as anything for this purpose; its operation may be hastened by tickling the throat with the feathered end of a quill. The feet, and legs up to the knees, should be immediately placed in hot water, as hot as it can be borne.

Give, internally, *Nux-vomica, Ipecac., Pulsatilla,* or *Veratrum.*

Nux-vomica — should have the preference when there is constipation; colic; eructations; violent spasms attended by shrieks; bending of the body backward; jerking backward of the head; eyes set in the head; trembling of the limbs.

Pulsatilla — when the trouble can be traced directly to an overloaded stomach, may be given in alternation with *Nux.*

Veratrum. — Especially if the patient is cold and pale, with cold perspiration on the forehead.

Ipecac. — If the vomiting continues after the convulsion has ceased.

Should this treatment fail to break up the fit, place the child, all over, in a warm bath; at the same time pour a small stream of *cold* water upon the head from a height of two or three feet. If the bowels are constipated, or if you have failed to produce vomiting, give an injection of simple water, or one of equal parts of sweet oil and warm milk.

For the symptoms, which remain after the spasm is broken, give *Nux-vomica, Chamomilla,* or *Aconite.*

Nux-vomica. — For griping and distention of the abdomen.

Chamomilla. — If the child is cross and fretful, and, especially, if it is teething.

Aconite. — Should there be any fever.

TEETHING. — When the spasms arise from difficult dentition, as they very frequently do, and when the gums are red and swollen, take a sharp penknife, and make a free incision down upon each tooth that appears to be causing irritation. The idea advanced by some that, should the cut heal up, the scar formed over the tooth would add to the difficulty of its working its way through the gum, has no foundation in truth. A warm bath is, also, advisable. The remedies, in such cases, are *Aconite, Belladonna, Chamomilla, Coffea, Ignatia.*

Belladonna and *Coffea* may be given in alternation, at the commencement of an attack, as often as every ten or fifteen minutes. *Belladonna* is especially called for when the following symptoms are present, and, especially, in fat children. The child starts suddenly from sleep; stares about wildly; the pupils of the eyes are dilated; the body is rigid; the forehead and hands hot and dry; clinching of the hands; great nervousness; the least touch being sufficient to excite an attack; convulsions attended by smiles and laughter. Should *Belladonna* afford no relief, or but partial relief, it may be followed by *Chamomilla,* or *Belladonna* and *Chamomilla* may be given in alternation.

Chamomilla — is especially applicable to nervous children; those that are extremely sensitive, cross, and fretful. It may be given when there are convulsive jerkings of the limbs; twitching of the muscles of the face and eyelids; eyes drawn up beneath the upper lids; constant rolling of the head from side to side; clinched thumbs; great restlessness; disposition to drowsiness between the fits; jerks and convulsions of the arms and legs; one cheek red, and the other pale. When *Chamomilla* fails to suppress the convulsion, give *Belladonna,* or give the two in alternation.

Aconite. — When there is high fever; congestion of blood to the head, with red face; great restlessness, crying, and starting.

Ignatia. — Convulsive starts; tremors of the whole body, attended with violent crying and shrieks. Convulsive movements of single limbs, or of single muscles, in different parts of the body, the fits returning at regular hours each day, and followed by fever and perspiration.

When *Belladonna* and *Chamomilla* fail, try *Ignatia.*

WORMS. — When convulsions are caused by worms, — and this is a difficult matter to decide, — give *Cina, Mercurius, Cicuta, Ignatia,* or *Hyoscyamus.*

Ignatia. — When there are sudden and violent startings from sleep, with the symptoms already recorded.

Cicuta. — When there are violent gripings in the abdomen, sudden shocks and jerkings of the limbs.

Cina. — When there are spasms of the chest, with stiffness of the entire body; especially suitable for scrofulous children, for those who are troubled with cough, and for convulsions during a siege of hooping-cough; may follow *Mercurius,* or may be given in alternation with it, or with *Hyoscyamus.*

Mercurius. — Tossing and stiffness of the limbs; distention and hardness of the abdomen; painful eructations; drooling of water from the mouth; fever, and moist skin. After a paroxysm there is great weakness, the child lying a long while in an exhausted condition. This remedy may precede or follow *Cina.*

Hyoscyamus. — For sudden attacks after eating, — the child gives a shriek, and becomes insensible; nausea and vomiting; convulsive movements of the limbs and whole body; especially twitchings of the muscles of the face; pale-blue ring around the eyes and mouth; foaming at the mouth, and great wildness.

REPELLED ERUPTIONS. — If the eruption has been a chronic one upon the head, and its suppression has caused spasms, give *Tartaremetic.* If this makes no impression, give *Sulphur.* If not better, in a couple of hours, give *Stramonium;* and, especially, if the following symptoms are present: sudden flashes of heat; trembling of the limbs; the eyes fixed; pupils dilated; rigid stiffness of the body; respiration labored; tossing of extremities, and involuntary evacuations of fæces and urine.

If the spasms arise from suppressed *scarlet fever* give *Belladonna;* if from *measles,* give *Bryonia.* Should these not give relief, give *Cuprum*, either alone or in alternation with *Belladonna*. If not better within three hours, give *Stramonium* or *Opium*.

Suppressed *Catarrh*, or cold on the chest, is, occasionally, the cause of convulsions; the disease is transmitted to the head, or spinal nerves, and produces spasms which are exceedingly obstinate.

Should the head be hot, apply cold water, and persevere in its use till the trouble leaves the brain. Also, put the child in a warm bath, and give immediately *Belladonna* and *Cuprum;* afterward *Opium* and *Camphor*.

FRIGHT. — Convulsions from fright require *Opium*, especially if there is trembling over the whole body, or when the child lies as if half-stunned, breathes heavily like snoring, with blue face, loud screaming during the fits. *Ignatia* is also useful. *Stramonium*, should *Opium* fail.

Hyoscyamus. — When there is foaming at the mouth, twitching of the muscles of the face.

Where the child is over-excited, and presents symptoms bordering on convulsions, caused by fright, fear, or joy, give *Aconite;* if not better in an hour, give *Coffea*.

MECHANICAL INJURIES. — When convulsions arise from blows or falls upon the head, give *Arnica*. In such cases it may be necessary to have surgical assistance.

UNKNOWN CAUSES. — There are many cases for which we can find no adequate cause. In all such cases use the warm bath, and apply cold water to the head, as heretofore directed, and give internally, *Ignatia:* should this remedy fail to prove beneficial, give either of the preceding remedies, whose symptoms coincide with the convulsive phenomena.

ADMINISTRATION OF REMEDIES. — Of the selected remedy, dissolve six globules in six teaspoonfuls of water; of the solution thus made, give one spoonful at a dose, and repeat it every ten, fifteen, or twenty minutes, according to the urgency of the case. As the child improves, lengthen the interval between the doses.

After the child has, to all appearance, recovered from the attack, it is as well to give two or three doses of *Sulphur*, at intervals of six hours.

DIET. — After a convulsive seizure, it is best that the child should be kept upon a low, or, rather, non-stimulating diet, for a few days. Especial care, in this respect, should be taken if the spasm has been caused by indigestion.

CHOREA.—ST. VITUS' DANCE.

DEFINITION. — Chorea is essentially a disease of the nervous system, of a spasmodic nature. It is characterized by tremulous, irregular, and, in some cases, most ludicrous motions of all or any or the voluntary muscles. These contractions are, to a certain extent, involuntary; they are more marked upon one side, usually the left, than on the other; they are without pain, and affect females more frequently than males, and occur chiefly with persons between six and fifteen years of age.

CAUSES. — It is said to occur most frequently in children of a nervous, delicate, excitable temperament; some suppose it to be hereditary. No doubt but that a disordered condition of the digestive system, as well as uterine diseases, predispose to the disease.

Among the exciting causes, may be mentioned anything which makes a forcible impression upon the nervous system; strong mental emotion, of which fright, perhaps, is the most common. "Injuries affecting some part of the nervous system, as falls upon the head and back; the improper employment of lead, mercury, etc.; suppressed eruptions; discharges, particularly scald-head, itch, herpes, perspiration of the feet, etc.; metastasis, or extension of rheumatism to the membrane of the spinal cord; second dentition; anxiety; concealed mental impressions and moral emotions; excited jealousy and envy; onanism; retained or difficult or suppressed menstruation."

SYMPTOMS. — The invasion of chorea may be either sudden or gradual. In very nearly every case that has come under my observation, I have been able to trace back a variable state of the patient's health, for several weeks previous to the first noticeable convulsive movement. I have usually found imperfect digestion, constipation of the bowels, loss of appetite, and general derangement of the digestive function; and, in not a few cases, various derangements of the menstrual function. These symptoms of disordered health are followed by, at first, slight, irregular twitchings of the muscles of the face, or, perhaps, of one of the extremities; and, by degrees, the spasmodic action becomes more decided and general.

This, unlike many other convulsive diseases, does not render the subject unconscious, neither does it affect volition. The patient knows perfectly well what he is about, and what he wants to do; but he cannot always do exactly as he wishes. The ordinary movements of the body, to some extent, can be performed; but there is some other power besides the will at work, and this power is constantly interfering with all the movements; misdirecting the hand that is put out to seize something, or jerking it back and giving it a new direction, rendering unsteady and imperfect every act, bringing into play muscles that should be quiet, and arresting those which the will has set at work. The muscles of the face are jerked about with an agility that is truly surprising, drawing the face into all sorts of shapes and grimaces. The hands and arms are twisted and jerked into every conceivable position. The inferior extremities are affected in the same way. In fact, it seems sometimes as though the whole muscular system had gone crazy. If you ask the patient to put out her tongue, she will have to make several attempts before she can accomplish it; when she does succeed, it is out and in again, as quick as a flash: you have to look quick, or you will not see it at all. She cannot keep her foot or hand in one position half a minute at a time; if you take hold of a limb that happens to be agitated, and, by main force keep it still, some other limb, or part, will take on the convulsive movement. When the patient is at her meals, and desires to take up something and carry it to her mouth, it is with difficulty that she can seize hold of what she intends to; and when, after several efforts, she finally succeeds, she cannot do with it what she wishes; the hand will be arrested midway, or jerked back, or pushed off in some other direction.

Walking is always more or less difficult. When the foot is lifted up, it is difficult to say how long it will be kept there, which way it will be moved, or where it will be put down. The patient progresses in a zig-zag direction; totters from side to side; goes by fits and starts. In some cases standing, even, is almost impossible, — the knees suddenly-giving way and letting the body down.

The organs of speech, chewing, and swallowing, are alike affected. In consequence of the irregular motions of the lips and tongue, the patient stutters, or speaks slowly and with great difficulty; in some severe cases articulation is almost impossible.

Mastication is interrupted by the irregular contraction of the muscles which move the jaw; the mouth opens and shuts without any regard to the patient's wishes; and often so violently as to injure the tongue or teeth.

51

As before stated, the disease is unaccompanied by pain ; and, what is a little singular, these violent and constant muscular contractions do not seem to occasion fatigue.

As a general thing, you will observe the convulsive movements affect one side more than the other ; occasionally they are entirely confined to the muscles of one side. They are always increased, I believe, when the patient is in the company of strangers, or when she is being observed. Terror, anger, or excitement of any kind increases them.

How long an attack of chorea endures, depends upon its cause, and upon the treatment which is instituted for its removal. I have seen not a few cases recover very quickly, after all treatment was suspended. From this and other reasons, I am inclined to think that it is frequently prolonged rather than cut short by some, no doubt, very learned medical prescriptions. The books say that the average duration of *acute* cases is from one to two months; while that of chronic cases is from months to years; and that relapses are not unfrequent. I am satified that the generality of physicians meet with better success than that ; Homœopathic physicians, I am certain, do. Chorea is not a dangerous disease.

TREATMENT. — We strongly advise our readers to apply early in a case of this kind to a Homœopathic physician. Though the disease seldom has a fatal termination, it will undoubtedly, if allowed to continue for a length of time, affect the intellect.

In all cases the temper is rendered irritable and capricious. It is very important and necessary that the patient should be under the care of some one in whom she has great confidence.

Where the patient's general health has been injured either from gastric or uterine derangement, it will be necessary to take this into consideration and direct the treatment accordingly.

The remedies are, *Belladonna, Ignatia, Cocculus, Colchicum, Hyoscyamus, Pulsatilla, Nux-vomica,* and *Sulphur.*

I have used with good success, in a number of acute cases, *Colchicum,* one dose every day for one week, after which I gave *Cocculus* in the same way.

When the muscles of the face are chiefly affected, give *Belladonna.*

Hyoscyamus. — When the convulsive movements are confined to the jaw or tongue ; or, when they affect single parts only ; also, when they interfere with the patient's swallowing ; also, when they are occasioned by worms.

Nux-vomica, or *Pulsatilla* — especially when the extremities are principally affected, and when the difficulty arises from derangement of the stomach or womb.

Ignatia. — When it is impossible to say what has occasioned the derangement, and when there are convulsive movements of the extremities and twitching at the corners of the mouth; also, twitchings of single parts, or when there are occasional spasms of single muscles here and there, in different parts of the system.

Sulphur. — Should the other remedies fail, or afford but partial relief.

I have used in some obstinate cases the galvanic battery, with very satisfactory results.

ADMINISTRATION OF REMEDIES. — After selecting a remedy, dissolve six globules in twelve teaspoonfuls of water, and of this solution give one spoonful three times a day. The repetition of a remedy depends a good deal upon the nature of the case, and the susceptibilities of the patient.

DIET AND REGIMEN. — The diet should be perfectly plain and nutritious; all articles of pastry and all rich and high-seasoned dishes should be avoided. Coffee and tea are decidedly injurious. Out-door exercise, plenty of it, is decidedly beneficial.

HEADACHE.

DEFINITION. — By the general term headache is understood a pain of any description in the head. Headaches are usually attended with intolerance of noise and light, and always with incapability of mental exertion.

As headache appears in many forms, and arises from many causes, it has been divided into several varieties, and each variety has received a specific name; hence we have rheumatic, nervous, congestive, neuralgic, bilious, and sympathetic headache. All such divisions, however, are, in the main, quite unsatisfactory, owing to the great difficulty of ascertaining the true nature of the disorder; nevertheless, to facilitate description, and the selection of appropriate remedies, some such arrangement though imperfect, must be accepted.

CAUSES. — A careful inquiry into the *cause* of the pain, in every case of headache, is very essential. In fact, a definite idea of the disordered condition upon which it depends, and a clear conception of the *kind* of pain from which the patient suffers, forms the only basis from which a successful indication of cure can be drawn.

Many females, those, for instance, possessing a highly sensitive nervous organization, as well as those who suffer from any of the various uterine derangements or nervous affections which exhaust the vital energy, are predisposed to what are termed nervous headaches. In them a cold, exposure to dampness, want of sleep, alarm, grief, anxiety, low diet, fasting, decayed teeth, impure air, over-exertion, too copious menstruation, long-continued or excessive nursing, is sufficient to bring an attack. This form of headache is very frequently attended with a degree of gastric disturbance, acidity, and flatulence, or constipation.

Headaches are not unfrequently occasioned by the sudden suppression of discharges and eruptions; suppression of the menses from cold or other causes is not unfrequently followed by a determination of blood to the head, causing a congestive headache. The same result follows when leucorrhœa is suddenly checked by astringent injections; so, also, when eruptive diseases, as hives, prickly heat, etc., strike in, as it is termed.

Many forms of headache are attended with nausea and vomiting, and hence are called *sick-headache*. It is an error, however, to suppose, as many people do, that the gastric disturbance is always the *cause* of the disease; for, beyond a doubt, in many cases, perhaps in the majority, the nausea and vomiting as well as the pain in the head are occasioned, not by an overflow of bile, but by an affection of the brain. As a general thing, this form of headache affects persons who consider themselves of a bilious habit; those who suffer from indigestion and other dyspeptic symptoms.

There is a variety of headache which, for want of a better name, has been called *bilious-headache*. It does not arise, however, as one might infer from its name, either from an exuberance or a deficiency, or any derangement of the bile. Bilious headache occurs chiefly in those persons who are subject to mental and brain excitement, and this very excitement, perhaps, has more to do with causing the disease than anything else. It is not unfrequently occasioned by intense mental application, or anything, indeed, which keeps the mind in a long-continued state of excitement. This variety of headache frequently attacks school-girls, or those who, in connection with leading a sedentary life, are devoted to studious habits. Such persons are predisposed to the complaint, and anything which deranges their general health will induce an attack. It is not unfrequently excited by errors in diet, especially too great variety or quantity; indigestible, greasy, or rich articles; also, by long fasting.

It may follow a hearty meal and cease in two or three hours, or it may not occur for several hours after a meal, and then continue till the offending substance is thrown up. Sometimes the ejection consists of an undigested meal; at others, of merely an insipid water mixed with frothy mucus. If the vomiting continues for a length of time, bile is frequently discharged. As the pain is frequently excited by long-continued fasting, so it is frequently abated by the patient's partaking of some stimulating food or drink. Usually, if the patient can lie down and have a quiet nap of a few hours, the pain will disappear.

Rheumatic headache is occasioned by anything that would induce an attack of rheumatism, exposure to cold, getting the feet damp, getting chilled while in a perspiration, sleeping in damp rooms, etc.

Neuralgic headache is of very common occurrence. It differs from all other forms of the disease by the intensity of the pain and the suddenness with which it seizes the patient. The causes of Neuralgia will be found upon page   .

Headache is very often a sympathetic affection; so much so, that there are very few diseases in which it does not take place. It is a prominent symptom in all fevers and inflammations, and in many nervous diseases. Affections of the spine seldom exist without occasioning pain in the head in some of its forms. Irregularity of menstruation, or when the discharge is painful or scanty, is almost always attended by sympathetic headache; so may be leucorrhœa, or any irritation of the uterine organs.

SYMPTOMS. — The symptoms of headache are too well-known to need any lengthy description. I shall therefore simply give a few of the characteristic symptoms of some of the particular forms of the disease.

BILIOUS HEADACHE is attended with a dull, aching, or racking pain, which moves from one point to another; there is, also, often tenderness of the scalp; the digestive organs are deranged; the tongue is coated, breath foul, and bowels are constipated; there are sometimes nausea and vomiting; flatulency; coldness of the extremities; pain in the eyes; the mental faculties are weakened, and exertion of the mind is irksome.

NERVOUS HEADACHE is characterized by acute, excruciating, lancinating, or darting pains; at times, the pains are constrictive, or attended by a sensation, as though the temples were being pressed together. It is sometimes attended with dizziness, or a

feeling of sinking down; also, with great nervous agitation or
restlessness; exertion, either physical or mental, is almost frequently quite impossible; vision is more or less affected, the patient frequently seeing motes or small dark spots flickering before
the eyes. Generally speaking, the head is cool and the face pale.
The pain, which is frequently confined to a narrow space, is generally worse in the morning than during the day.

CONGESTIVE HEADACHE is usually confined to one part of the
head; the pain is of a dull, heavy nature; there is a sense of
weight in the head; the pain is worse when the patient looks upward or downward, and so he is inclined to carry his head very
straight and steady. There is stupor and heaviness, or a feeling
of giddiness, with dimness of sight and noises in the ears. Sleep
is heavy, and frequently disturbed by dreams or convulsive movements; usually there is great depression of spirits. As a general
thing, the pulse is languid and weak; sometimes there is violent
throbbing of the arteries, with fulness in the forehead and temples; with a sensation as if the head would burst; with a burning
pain through the whole brain.

Congestive headache frequently presents symptoms of an inflammatory nature; a high excitement of the circulatory system;
countenance flushed; pulse full and quick; eyes bright and glassy,
or suffused and heavy; pain pulsative or throbbing, occasionally
with beating noises in the ears; great sensitiveness to light and
noise; pain aggravated by moving, sitting up, or talking. There
are also, at times, nausea and vomiting, with coldness of the extremities.

The neuralgic headache differs from all the other forms, both
by the severity of the pain and the manner of its approach. While
ordinary forms of the disease come on gradually, after some excitement or indiscretion, this one suddenly attacks the patient
without a moment's warning. The pain is usually confined to a
single spot, or a single nerve, and is, at times, so severe as to drive
the sufferer almost distracted. As a general thing, it does not last
long, and sometimes disappears as suddenly as it came. It is frequently intermittent. There is frequently increased sensibility of
the scalp over the region of the pain. This, as well as most other
forms of headache, is often attended with gastric disturbance.—
COPLAND'S *Med. Dict.*

CATARRHAL HEADACHE usually consists of a severe, dull pain,
and sense of weight in the forehead, and pain in the eyes; it is

usually accompanied by a sense of chilliness and lassitude. In its severe forms, the symptoms correspond to those given under the head of CONGESTIVE HEADACHE.

RHEUMATIC HEADACHE is usually complicated with similar pains in other parts of the body, — down the neck, for instance, and into the shoulders. The pain in the head is heavy, distracting, or aching, and attended with a sense of coldness; sometimes it is accompanied by congestive symptoms.

TREATMENT. — As headache is a very common complaint, and one, too, for which almost any intelligent person can prescribe, we shall endeavor to make our treatment as explicit and comprehensible as possible, so that the necessity of calling a physician for such cases need never occur.

For convenience' sake, we shall treat each variety of the disease separately.

NERVOUS HEADACHE OR NEURALGIC HEADACHE.

For the Symptoms and Causes see page 405.

Aconite. — This remedy may be given when the following symptoms are present: pain-like cramps through the forehead, or above the root of the nose, with a sensation as though one would lose his senses; headache as though the brain were raised or moved up, especially during motion; the least noise or motion is intolerable; the pain is also aggravated by reading or speaking; headache from cold, with catarrh; buzzing in the ears, and oppressive weight in the forehead, with a sensation as though the contents of the skull would burst through; a feeling on the top of the head as though the hairs were being pulled; also an undulating sensation, as of water in the head; when there are throbbing or shooting pains, and the patient is easily excited and apprehensive of death. *Aconite* may either be given alone, or in alternation with *Belladonna.* Should these two remedies, though seemingly indicated, fail to afford relief, and, especially, if there is a pain in the head which gradually increases and then as gradually decreases, — a sort of undulating pain; or, if there be a sensation of coldness in the ears, with roaring in the head, as of water, give *Platinum.* — See this remedy further on.

Belladonna. — This is an important remedy in some cases, and, especially, when the pains come on in the afternoon, and last till the next morning; also, when the pain commences gently, and afterwards increases to a fearful intensity.—For further indications requiring this remedy, see " SICK HEADACHE."

Coffea. — Pain, as if a nail were driven into the brain, or head-ache, as if the brain would be torn, shattered, or crushed; pain which seems to be intolerable, driving the patient almost dis-tracted; she is fretful, weeps, screams, tosses about, and is in a constant state of agitation; the pain is excited or aggravated by the least thing,—the slightest noise, a footstep, or even music, being sufficient. For headache arising from cold, vexation, close-think-ing, or excitement. It may be followed by *Hepar-sulph.*, *Nux-vomica*, or *Cina*.

Nux-vomica. — Headache in the morning, after breakfast, worse after every meal, increased by motion or stooping forward, with nausea and sour vomiting; when there are shooting, drawing, tearing, or stitching jerks in the head, worse upon one side, espe-cially after the excessive use of coffee, or wine; pain, as though a nail were driven into the brain, or as though the head would fly to pieces, from a pressive pain from within outwards; heaviness of the head, with a buzzing noise and giddiness when walking, especially early in the morning, or when moving the eyes; shaking of the brain at every step; headache, when reflecting, as though the brain would be crushed; face pale and dejected; when the pain, beginning in the morning, grows worse and worse as the day wears on, until the patient is well-nigh distracted; headache from loss of sleep or long watching. For headache caused by dissipa-tion, by the excessive use of wines or coffee, by a sedentary life, fits of anger, or constipation.

Ignatia. — Aching or pressing pain above the nose, with nausea, pain lessened by bending the head forward; pressing in the head from within outwards; boring, stitching, throbbing, tearing, or shooting pain deep into the brain; sensation as though a nail had been driven into the head, with nausea, dimness of sight, and pale face; the pains relieved, for a moment, by change of position, but frequently returning after eating, or on waking in the morning, the patient is full of fear, inclined to start, despairs of her health, is irresolute, impatient, and wants to be alone. For headache caused by grief or mortified feelings.

Pulsatilla. — Pain of a jerking, stitching, or tearing nature, especially when confined to one side of the head, and accompanied by giddiness, sickness at the stomach, dimness of sight, humming in the ears, or jerking, tearing pain in the ears; countenance pale or yellowish, and haggard ; loss of appetite ; no thirst ; chilliness ; palpitation of the heart, and great nervous agitation ; pain as if the brain would be torn, or as if the head were in a vice, or as if the skull would fly to pieces, especially when moving the eyes ; cracking or snapping in the head when walking; headache after lying down in the evening, or early in the morning in bed ; headache worse in the evening, and continuing all night ; the headache is increased by rest, or sitting still, but relieved by being out in the open air, and by pressure of a tight bandage. It answers best for mild, cold, dull, or sluggish persons, those not easily excited into action or passion.

For headache occasioned by suppression of the menses.

Bryonia. — Great fulness and heaviness of the head, with pressive or burning pain in the forehead ; worse when walking ; with a sensation, when stooping, as though everything would issue through the forehead ; pressing in the brain, either from within outward, or from without inward ; external, tearing pains, which extend to the face and temples ; jerking, throbbing headache, or rending in particular spots ; heat in the head and face, with red cheeks and thirst ; dizziness and headache ; tongue coated yellow ; insipid or foul taste in the mouth ; everything tastes bitter ; nausea and vomiting. The headache is when moving about, or walking about, or when first opening the eyes ; especially indicated in warm and damp weather ; also, for persons suffering from rheumatism.

Nux-moschata. — For compressive or distensive headache, with a sense of looseness of the brain when shaking the head ; beating headache, especially above the left eye ; or for pains that go from left to right ; pressure increases the pain ; stitching and tearing pains, extending to the ears and temples.

Chamomilla. — Stitching headache, affecting half the head, arising from checked perspiration, or from cold ; pains extend to the jaw ; shooting, sharp pains in the temples ; rush of blood to the head, with beating in the brain ; pain in the eyes ; sore throat ; cold on the chest, especially if one cheek is red and the other pale.

For headaches caused by ill-humor, or vexation ; it is suitable for persons who are unable to bear the least pain, and for children whose nervous system is very irritable and sensitive.

Platinum. — For headache gradually increasing and decreasing ; numb feeling in the head ; pain in the sides of the head, as from a plug ; compressive pain in the forehead and temples ; roaring in head, as of water ; sensation of coldness in the ears, eyes ; twitch-

52

ing of the eyelids; objects appear smaller than they are; buzzing
in the ears; burning heat and redness of the face, with violent,
cramp-like pains over the root of the nose; stupefying pressure in
the cheek-bones; hysteria, with lowness of spirits; restlessness,
and whining mood.   Is useful after *Belladonna*.

Mercurius. — Headache, as if the head would fly to pieces, with
fulness in the brain; tearing headache, especially upon the left
side, and when the pain shoots down into the teeth; tearing headache, worse at night; stitching pain in the ears.   The pain is
relieved by pressing the head with both hands.

Mercurius may follow Belladonna; and where both these remedies fail to afford relief, especially if there is a boring pain at the
root of the nose, or when there is a pain as if a nail were driven
into the head, give *Hepar-sulph.;* also, when there are violent,
rending pains during the night, as though the head would fly to
pieces; or when sore spots appear upon the head.

China. — Headache from suppressed catarrh; pain in the forehead, when opening the eyes; heaviness in the head, pressing from
within outward; the brain feels bruised; headache, especially at
night, with sleeplessness; tearing pain in the temples, as if the
head were bursting; rush of blood to the head, with great heat
and fulness in the head; a sensation as though the brain jolted
about and hit against the skull; the scalp is tender to the touch;
the roots of the hair are sensitive to contact; a sensation as though
the scalp were contracted.   The pains are aggravated by contact,
by motion, by stepping; also, by a draught of air, or walking in
the wind.   It often suits after Coffea.

Colocynth. — Excruciating, tearing pain on one side of the head;
pressing in the fore part of the forehead, worse when stooping or
lying upon the back; tearing, oppressive, squeezing pains, with
nausea and vomiting; violent headache, not allowing the patient
to lie quiet, but obliging her to scream and moan; headache which
comes on every afternoon or toward evening, and is attended
with a copious flow of urine, very offensive; also, profuse perspiration, smelling like urine.

Arsenicum. — Beating pain in the forehead, with inclination to
vomit; buzzing in the ears; stupefying pain, with heaviness in
the forehead; weeping and moaning; tenderness of the scalp;
cold applications relieve the pain for a time; the pain is worse
within doors, and relieved on going into the open air.   May follow
Pulsatilla.

Antimonium-crudum. — For nervous headache, excited by a disordered stomach; headache after bathing, or headache from suppressed eruptions; boring pain in the forehead and temples; painful rush of blood to the head, with bleeding at the nose. Answers well after Pulsatilla.

Veratrum. — Oppressive headache on the top of or on one side of the head, accompanied by pains in the stomach, or diarrhœa; also, by nausea and vomiting; painful sensitiveness of the hair; headache in paroxysms, as though the brain were crushed; pain so severe as almost to deprive the patient of his reason; stiffness of the nape of the neck; pain is worse when lying in bed; cold perspiration, chill, and thirst. Suits well after Arsenicum.

Silicea. — Headache ascending from the nape of the neck to the crown of the head; headache from getting heated; pressure in the head, with ill-humor; tearing pain every forenoon; everything has a tendency to the forehead and eyes; stitches in the head, especially in the temples; throbbing headache, pain extends to the nose and face; determination of blood to the head; scalp painful to contact; hair falls out; perspiration of the head in the evening. Good in stubborn and chronic cases.

Sulphur. — Headache with nausea; nightly headache; feeling of fulness and weight in the head, especially at the crown; pain as from a hoop around the head; pressing in the temples from within outward; throbbing, tearing pains, with heat, principally in the morning; headache every day, as if it would fly to pieces; periodical headache every week, consisting of pressure, tearing and stupefaction; bubbling or throbbing headache, mostly with heat in the head, from a rush of blood, especially in the morning and in the evening; humming in the head; coldness of the scalp; hair painful to the touch, or falls out. The pains are worse in the open air, abate while in a room.

May follow any of the other remedies, especially applicable to chronic cases.

When water affords relief, it may be applied either hot or cold, according as either is most agreeable.

For ADMINISTRATION OF REMEDIES, see " SICK-HEADACHE."

SICK-HEADACHE. BILIOUS HEADACHE.

For the symptoms and causes, see page 405.

Belladonna. — For periodical headache, pains returning every afternoon and continuing till after midnight, aggravated by the warmth of the bed, or by lying down; stupefying headache, mostly in the forehead, sometimes with loss of consciousness, often with a sensation as though the head would split; pains of a violent, burning, rending, or shooting character, commencing sometimes· gently, but afterward increasing to a fearful intensity, at times the pain so severe as to force from the sufferer tears and agonizing screams; headache after taking cold in the head, when there is a jolting sensation in the head and forehead at every step, or on going upstairs, or when there is a sensation of a painful, waving motion in the head, buzzing in the ears and dimness of the eyes; also, a sensation of distention or pressing outward. The pains, which often extend to the eyes and nose, are mostly confined to one side, usually the right, and aggravated by every motion, by turning the eyes, by a bright light, by leaning the head backward, by the warmth of the bed, by every noise, by concussions, and particularly by stooping forward. The external parts of the head are very sore; there is alternate chilliness and heat, coated tongue, nausea with loathing of food. Headache connected with uterine complaints, especially falling of the womb, accompanied with severe pressing down.

Sanguinaria. — Severe pain affecting but half of the head and attended with bitter vomiting, coming on early in the morning and lasting till night, with fulness of the head as if it would burst; pain worse upon the right side, and aggravated by motion, or by stooping forward; pressing pain in the eyes, from within outward; shooting, beating pain throughout the head; headache attended with chills; tongue coated; nausea, with empty eructations; inclination to lie down.

Ipecac. — Stitching headache, with great heaviness; giddiness when walking; pressure in the head, especially the forehead; pain affecting all the bones of the skull, through even to the root of of the tongue; coldness of the hands and feet; aversion to every kind of food; tongue coated white or yellow; nausea and vomiting. In all cases of headache which commence with nausea and vomiting, and are accompanied with a bruised sensation about the head.

It may be given in alternation with *Nux-vomica,* when there are shooting pains in the side of the head, which are worse in the open air, and when stooping.

Aconite. — Symptoms calling for this remedy will be found under the head of " NERVOUS HEADACHE."

Antimonium crudum. — For headache excited by a disordered stomach, and when there are dull, boring pains, especially in the bones; boring in the temples and forehead, from within outward; the pains are relieved in the open air; painful rush of blood to the head, followed by bleeding of the nose; nausea and vomiting of bile and mucus. For headaches which cause the hair to fall out.

Pulsatilla. — For its indications, see "NERVOUS HEADACHE." In some obstinate cases where *Pulsatilla* does not afford permanent relief, it may be followed by *Antimonium-crud.*

Hyoscyamus. — is appropriate for nervous and highly excitable persons, and those whose symptoms appear for the most part in the evening; stupefying headache in the forehead, especially after eating; constrictive feeling in the head. The pain is aggravated by noise, as shutting a door, by the warmth of the bed, by lying down, or a draught of air.

Spigelia. — Pain in the left side of the head, with beating in the temple, which is aggravated by the least exercise or noise, or even opening the mouth; pain in the whole left side of the head, in the face and teeth, also, deep in the orbits of the eyes; headache appearing at a regular time each morning, increasing in severity as the day wears on; shocks in the head when walking; pain in the eyes, especially when moving them.

May be given after or in alternation with *Belladonna.*

Should there be great sensitiveness to *light*, give *Belladonna;* to *noise*, *Spigelia;* to all kinds of *odors*, *Sulphur* or *Aconite.*

The remedy which best corresponds to all the symptoms in a given case, and especially to its characteristic or distinguishing features should be administered, no matter whether such remedy be found under the "bilious," "nervous," or any other form or division of the disease.

ADMINISTRATION OF REMEDIES.— The remedies may be given dry or dissolved in water. When dry, let three or four globules dissolve upon the tongue. When dissolved, put twelve pills in as many spoonfuls of water, and take one spoonful at a dose. The repetition of the dose depends in a great measure upon the nature of the diseases; in some forms of nervous or neuralgic headache, one dose will banish the pain in an instant; again, it will be necessary to repeat it often, perhaps as often as every half hour; in other cases it may be given every hour, or every two, three, or six hours, according to the necessity of the case. Marked improvement should be manifest by the time the fifth or sixth dose is taken ; if such is not the case, take another remedy.

To eradicate a *disposition* to certain kinds of headache, the remedy should. be taken at long intervals, say once in three days, for four or six weeks.

CONGESTIVE HEADACHE.

For symptoms and causes, see page 406.

Aconite and *Belladonna* may be given in alternation, when there is violent throbbing, heaviness, fulness in the forehead and temples, with a sensation as if the head would burst asunder ; burning sensation over the whole head ; congestion of blood to the head, with redness of the face and eyes ; burning and undulating sensation as though the head were full of hot water ; also, when the pain is accompanied with incoherent talking and raving ; worse on moving, talking, drinking, or rising up.

Belladonna. — Especially when there is a violent aching pain, deep-seated in the brain, as if the head must split ; pressing in the head from within outward ; undulating feeling as of water in the forehead ; headache during menstruation ; violent throbbing of the arteries of the neck and temples ; great sensitiveness to light, noise, and touch ; face pale and haggard ; drowsiness, low muttering, and delirium. — See " BILIOUS HEADACHE."

Bryonia. — Great fulness of the head, with distending pressure from within outward ; stitches in the head, or only through one side ; desire to lie down ; jerking, throbbing headache, with dulness of the eyes ; vertigo and headache, followed by shuddering and chilliness, with heat in the head, red face, and thirst ; constipation of the bowels. The pain is worse when moving about fast, or when first opening the eyes ; headache from suppressed menstruation.

Pulsatilla. — For pain which is relieved by binding something tight around the head ; also, when the pain is dull, oppressive, and affects only one side of the head ; pain extending from the back part of the head to the root of the nose ; heaviness of the head ; pale face ; dizziness ; inclination to weep ; great agitation ; pain is worse toward evening, when sitting or looking upward.

Nux-vomica. — Headache worse in the morning, and in the open air ; constipation, with congestion of the head ; headache, when reflecting, as if the head would fly to pieces; humming in the head; shaking of the brain at every step. Especially for persons of sedentary habits or who use much coffee.

Rhus. — When there is a sensation as though everything in the head were loose; pain as if the brain were torn; tingling in the head; stitching headache, day and night, extending to the ears, root of the nose, and cheek-bones; burning, throbbing pain, with fulness of the head.

Glanoine. — For throbbing headache, which is increased by the slightest movement; fulness in the head, with severe pain, almost causing delirium. The pain is temporarily relieved by the application of cold water.

Congestive headache from *joy* requires *Coffea;* from a *blow,* *Arnica;* from *fright, Opium;* from *chagrin, Chamomilla;* from *grief,* or *mortified feelings, Ignatia;* from *constipation, Nux-v.* or *Opium ;* from *suppressed menstruation, Belladonna, Pulsatilla, Bryonia;* from *sedentary habits, Nux-v.* or *Sulphur ;* from *suppressed eruptions, Antimonium crudum, Helleborus,* or *Sulphur.*

The application of cold water, especially in this form of headache, is often of great service. It may be applied by putting upon the head cloths wrung in ice-water. Should there be cold extremities and hot head, put the feet in hot water and the cold to the head. Some persons prefer alcohol, or vinegar and water; there is no objection to either.

For ADMINISTRATION OF REMEDIES, see " SICK HEADACHE."

DIET. — During a siege of this headache the diet should be low; all articles of a stimulating nature should be strictly avoided, as also should tea, coffee, and all spirituous drinks.

RHEUMATIC HEADACHE.

For SYMPTOMS and CAUSES, see page 407.

Chamomilla. — Rheumatic pains in the head, which change their seat frequently, and have been excited by suppressed perspiration; tearing pains in one side of the head, which extend down into the jaw; tearing and stinging earache. When *Chamomilla* fails, give *Pulsatilla* or *Nux-v.,* the latter, especially, when there are shooting pains in the side of the head; head painful, externally, or when moved.

Mercury. — Burning, rending, or shooting pains, which extend into the face, teeth, and ears; worse in bed and at night.

China, Ignatia, and *Ipecacuanha* are also, at times, of service. *Colocynth* will sometimes relieve when all other remedies fail.

CATARRHAL HEADACHE.

For SYMPTOMS and CAUSES, see pp. 406, 7.

Nux-vomica and *Mercurius.* — These two remedies may often be given, with good results, for those headaches which accompany epidemic catarrh or influenza; and when there is pressing pain in the forehead, and over the roots of the nose, with constant sneezing and running at the nose; heaviness of the head, especially in the forehead.

Hepar-sulph. — Boring headache, especially at the root of the nose; headache, as if a nail were driven into the head; pain in the forehead, above the eyes; catarrh, especially upon one side.

Cina. — Drawing, tearing headache, which is aggravated by reading and meditation; pressing downwards in the head as from a load; dull headache early in the morning, with irritation of the eyes; dimness of sight when reading.

Arsenicum. — Throbbing pain in the forehead, with excessive discharge of an acrid, burning water from the nose; also, with hoarseness and restlessness; buzzing in the ears, with hard hearing, as though the ears were stopped; pains are aggravated by other people's talking, and are relieved by external warmth.

Aconite. — Pressing, dull feeling in the forehead, better in the open air; fever intermixed with chills; running at the eyes and nose.

Euphrasia. — Oppressive headache with intolerance of light and heat, especially in the forehead; stitching in the temples and forehead; profuse discharge from the head; confused feeling in the head; pain in the eyes, especially when looking at the light; pressure in the eyes; profuse discharge of smarting tears.

I think, of this disease, I cure nine out of every ten with *Arsenicum* or *Nux-vomica,* and I never use anything below the 200th potency.

For ADMINISTRATION OF REMEDIES, see " SICK HEADACHE."

HEADACHE FROM CONSTIPATION.

For SYMPTOMS and CAUSES, see p. 405.

Nux-vomica, Pulsatilla, Bryonia, Opium, Ipecacuanha, and *Lyco-*

podium, are the remedies upon which chief reliance is to be placed.

For particulars, see article upon " CONSTIPATION."

NEURALGIA.

DEFINITION. — The meaning of the term *neuralgia* is, pain in the nerve. When occurring in particular parts of the system, it is frequently confounded with rheumatism; yet it differs quite materially from that disease. Rheumatism is a specific kind of inflammation, affecting particular tissues of the body, while neuralgia is quite independent of inflammation, as it is simply a pain in the nerve, unaccompanied by fever, and unattended by any noticeable change of structure in the affected part. The pain of neuralgia is often severe, sometimes excruciating; it occurs in paroxysms of irregular duration, and after either regular or irregular intervals. It affects various parts of the body. It attacks males as well as females. I introduce it in this work from the fact of its frequent occurrence.

CAUSES. — In not a few cases, the causes of neuralgia are very obscure. One great difficulty in making out the causes and origin of these pains is, that they are so frequently occasioned by some source of irritation situated in a part distant from where the pain is felt. For instance, you strike your elbow in a certain way, and you produce a peculiar tingling sensation — not in the part struck, but at a distance — in your little finger. The same thing is constantly happening in disease. Something taken into the stomach, which arrests digestion, may cause pain in a remote part; some affection of the brain or spinal cord may cause it. The sensation is apparently reflected from the real seat of the disease to the extremity of some nerve, which is not the subject of the irritation.

Among the most frequent and better understood causes of the disease, we may mention exposure to *damp* and *cold*, in any form; sitting upon cold, damp seats; standing on cold floors, or on the damp ground; getting the feet wet; exposure of the face to cold and wet, while walking; sitting in a current of air; sudden changes in weather, especially prolonged wet seasons, etc., etc. Also, sleeping in damp rooms, especially basements, or residing in low, damp, miasmatic localities, no doubt assists in producing neuralgic complaints.

Facial neuralgia is often occasioned by decayed teeth, while the tooth may be perfectly free from pain.

53

SYMPTOMS. — The symptoms of neuralgia are so familiar that it is hardly necessary for me to record them in detail. Sometimes the pain commences as a slight uneasiness or obtuse pain in a certain part, and gradually increases in violence, with more or less rapidity, becoming sharp, darting, lancinating, or twinging, until it reaches a height almost unbearable. Sometimes the pain shoots like an electric shock through the part.

As a general thing, I am inclined to think it is sudden in its invasion, reaching a degree of great intensity in a short time, and not unfrequently disappearing in an instant.

The paroxysms are extremely variable in their duration, as they may last a minute or an hour, a day or a week. Generally speaking, the more intense the pain, the shorter its duration.

These attacks may recur at regular intervals of a few moments, or they may take place daily, or they may be separated by a much longer period. Sometimes the pain, in a moderate degree, is continuous for days, but wonderfully exalted or aggravated at spells.

Neuralgia being purely an affection of the nerves, the pain may become most intense and agonizing, throwing the patient into convulsions and delirium.

TREATMENT. — For convenience' sake, we shall treat neuralgia as it affects particular parts, and commence with that variety known as

TIC-DOULOUREUX, *Prosopalgia* or *Face-ache*. — The remedies are,

Aconite. — When there are throbbing; burning, shooting pains, worse at night, appearing in paroxysms, and preceded by slight itching or crawling pains; great sensibility of the whole nervous system; redness of the face, with heat and swelling.

Belladonna. — Paroxysms commencing gradually, with creeping or itching in the affected part, which by degrees changes into violent, darting, lancinating, drawing pains in the cheek-bones, nose and jaws; stiffness of the neck; spasmodic twitching of the facial muscles; twitching of the eyelids; shooting, tearing pains through the eyes; violent, cutting pain, especially upon one side, generally the right, almost insupportable, and, when the pain is most violent, just below or in the orbits of the eye.

The pain is aggravated by the slightest noise or movement, also by the warmth of a bed.

Bryonia. — Especially for persons affected with or subject to rheumatics. Red, burning face, attended with pressing, drawing, lacerating, piercing pains; pain relieved by pressure; also, by movement; pains in the limbs; chilliness, followed by fever.

Chamomilla. — Especially for females with extreme sensibility, with twitching of the musles of the face; tearing, pulsative pain, with a sensation of numbness in the affected part; redness of one cheek.

China. — Especially when the attacks are periodical; excessive pains, with great sensibility of the skin; aggravation by the slightest touch; severe pain through the check-bones.

Colocynth. — Violent rending and darting pains which extend to the ears, nose, temples, teeth, and all parts of the head, principally upon the left side; and pains aggravated by the slightest touch.

Arsenicum. — When the pains return periodically, and are of a burning, pricking character, are worse at night or during repose, and are relieved by the application of extreme heat; great anguish and prostration; severe pain in and around the eyes; also in the temples, so severe as to almost drive the patient distracted.

Platinum. — Tensive, stupefying pressure in the check-bones, with a feeling of coldness, pain worse at night.

Coffea. — Morbid exaltation of sensibility; excessive painfulness of the affected part; great irritability of mind and body.

Spigelia. — Jerking, tearing, burning, and pressure round the cheek-bones, aggravated by the slightest touch and motion, the pain coming on at a certain time of day.

Staphisagria. — Violent aching and throbbing in the face, from the teeth to the eyes; heat, with cutting, tearing pains in the cheek-bones, pain worse upon the left side.

Nux-vomica. — Tearing pains in the check-bones, aggravated in bed, and in the cold air; tingling and beating in the muscles of the face; also, when there is gastric derangement.

Pulsatilla. — For females, especially when connected with uterine derangements; painful sensitiveness of one side of the face, with shivering.

Mercurius. — Tearing or shooting pain on one side, from the teeth to the temples, pain worse at night, and in the warmth of the bed; nightly perspiration; sensation of coldness in the part affected.

Kalmia. — Violent lacerating or throbbing pains. In some of the worst cases of tic, when other remedies fail, this one will afford relief.

ADMINISTRATION OF REMEDIES. — Of the remedy chosen, dissolve twelve globules in twelve spoonfuls of water; of this solu-

tion, take one spoonful at a dose, and repeat it every fifteen minutes, half-hour, hour, or two hours, according to the urgency of the
case.   As soon as the symptoms ameliorate, lengthen the interval
between the doses to four, six, eight, or ten hours.

As an external application, perhaps nothing is better than cold
water; however, in some rare cases, cold aggravates the pain;
when this happens, try warm water.   I have known considerable
relief to be obtained by bathing the affected part with a mixture,
composed of six drops of the *Tincture* of *Aconite*, and six table
spoonfuls of water, either warm or cold, as the patient prefers.

Other applications, according to my experience, are of no avail.
I have a friend, however, who thinks the very first thing to be
done, in all cases of neuralgia, is to bind draughts of *ground horse-
radish* upon the patient's wrists.

DIET AND REGIMEN. — All that can be said about diet, is, that it
should be plain and nutritious.   *Coffee* and *green tea* should be
avoided.   As gastric disturbance frequently is the exciting cause
of neuralgia, care should be taken to avoid all those articles which
are known to disagree, as well as all articles which are known to
be of an indigestible nature.

Frequent bathing in cold water and plenty of out-door exercise
are always beneficial.

ANGINA PECTORIS. — This is a neuralgic affection of the heart.
It is sometimes called *breast-pang*, or *suffocative breast pang*.

"Those who are afflicted with it are seized whilst they are
walking, and more particularly when walking soon after eating,
with a painful and most disagreeable sensation in the breast, — in
the region of the heart, — which seems as if it would take their
life away, if it were to increase or to continue.   The moment that
they stand still, all this uneasiness vanishes.   In all other respects,
the patients are, at the beginning of this disorder, perfectly well,
and, in particular, have no shortness of breath." — DR. HEBERDEN.

The remedies for this complaint are *Arsenicum, Digitalis, Belladonna, Rhus, Spigelia*, and *Phosphorus*.

During the paroxysm, the patient should remain perfectly quiet,
in an erect position, with all the clothing loosened.

SCIATICA. — This is a neuralgic affection of the great sciatic
nerve.   The pain starts in the region of the hip-joint, and extends
to the knee, or even to the foot, accurately following the course of
the great sciatic nerve.   The pain is sometimes so severe, as not
only to impede the motion of the foot, but to deprive the patient

of all rest. It frequently produces stiffness and contraction of the limb.

As diseases of the hip and knee joint not unfrequently result in serious deformity, it is *always* best, when these parts are threatened, to place the child, *at once,* under the care of an intelligent Homœopathic physician.

HYSTERIA. HYSTERICS.

DEFINITION. — Hysteria is purely a nervous disorder occasioned by some morbid modification of the reproductive apparatus. It manifests itself mostly in unmarried ladies between the ages of fifteen and thirty-five, and occurs oftener at, or about, the period of menstruation, than at any other time. There are instances recorded where men of very irritable nervous disposition have suffered from hysterics. I have never met with such a case.

The disease attacks in paroxysms or fits. These are usually preceded by dejection of spirits ; difficulty of breathing, or a sensation of suffocation ; palpitation of the heart ; pain in the side, or the peculiar sensation of a ball rising from the left side, or from the stomach, into the throat.

Females, possessing great susceptibility of the nervous system and of mental emotion, are more liable to the disease than those otherwise constituted.

Hysteria manifests itself in such a variety of forms, that it is quite impossible, in an article so short as this must necessarily be, to give a complete account of the disease, embracing all its modifications consequent upon age, habit of body, temperament, physical and moral education, nervous susceptibility, association, etc., etc.

CAUSES. — The predisposing causes of hysteria are to be found in a variety of conditions ; among them we may mention a highly nervous, irritable, and sanguine temperament, a lax or delicate, impressible habit of body, or a full habit with deficient tone. Children of weak and exhausted or aged parents, as well as those who possess a feeble or impaired constitution, whether this delicate state of body be hereditary or occasioned by mismanagement during early education and development, are most likely to be subject to this disorder. These diseased propensities are fostered and accelerated by a sedentary and luxurious mode of living ; by high-seasoned or stimulating food; the use of wines, green tea, and coffee; by an over-refined education, and the excitement of the imagination and emotions by the perusal of exciting novels, and the various ways by which the feelings are affected and acute sensibility is promoted.

The immediate exciting cause of an attack of hysteria may be some trifle, which would scarcely discompose an ordinary female; some word or gesture, or the failure of some word or gesture gives the occasion, "and off she goes." An individual, whose nervous system is habitually in this abnormal condition, is like a surcharged thunder-cloud, requiring but the smallest point to draw the flash. Prominent among the exciting causes may be mentioned sudden mental emotions; anger; jealousy; longing after objects of desire; frights; the sight of objects distressing or disgusting; sudden intelligence of an exciting nature; disappointments in love; spurned affections; protracted expectation; "hope deferred." Immoderate fits of laughter produced by humorous occurrences, or fits of crying caused by vexation, may pass into hysterical spasms. There is no doubt, but that these paroxysms can be brought on at pleasure, and frequently are, by recalling various feelings, emotions, or circumstances, just as tears can be shed by some at will by the same means.

SYMPTOMS. — A detailed account of the symptoms of hysteria would fill a volume; therefore a synopsis of them here will have to suffice. As before stated, the disease attacks in paroxysms or fits, and commonly begins with pain in the left side, anxiety, palpitation, difficulty of breathing, and, less frequently, with nausea and vomiting.

A peculiar symptom, complained of by most hysterical patients, is the sensation of a ball rising from the stomach or bowels into the throat. This ball, or *aura*, as it is called, starting from some definite point, passes either slowly or rapidly along an easily defined track, rises to the throat, and explodes, so to speak, in a paroxysm of hysteria. The sensation produced, when this ball reaches the throat is one of suffocation; this feeling of suffocation or strangulation, is not always plainly marked, but the moment the ball reaches the throat, the patient goes off into a fit, and catches at her throat, or pounds and rubs her chest to relieve the spasmodic action there in progress, or else she bursts forth into fits of laughing, or crying, or both, in alternation, accompanied with reproaches, sobs, tears, wringing of the hands, tearing of the hair, spasmodic clinching of the teeth, twisting of the body and general convulsions; these symptoms last for a longer or shorter time, when there is a gradual return to a state of gentle composure.

The lighter forms of the disease may not proceed farther than a slight sense of suffocation and general uneasiness, followed by depression of spirits and weeping. Or the patient may have what her friends will term a " fainting fit;" some little thing goes wrong, or perplexes her, when she falls over, has perhaps a few spasmodic twitchings, and remains perfectly insensible for a short time.

The truly spasmodic form of the disease presents alarming symptoms. As soon as the ball reaches the throat the patient is thrown into convulsions; the trunk of the body is writhed to and fro, and the limbs are variously agitated; the patient beats her breast with her fist, or beats her head against the bed; tears her hair, screams, shrieks, laughs, sobs, and cries immoderately; she talks incoherently, and foams at the mouth. The breathing is much affected, and rendered laborious by the spasms about the throat; she frequently rubs and pounds her abdomen, and in her struggles she sometimes bites her arms and hands, or even those who are near her.

The duration of a fit of hysteria varies from a few minutes to two or three hours. The attack usually passes off with eructations, sighing, sobbing, with a flow of tears, or by a fit of laughing, or by an exclamation. The recovery is generally rapid and complete. After an attack, the patient sometimes complains of headache and a general soreness of the whole body; and not unfrequently there is a copious discharge of limpid urine.

Although the patient is very desirous to be thought entirely unconscious of all that has transpired, it is generally conceded that she retains more or less consciousness of what has taken place during the fit.

The general health of a person suffering from hysteria is seldom good; in the intervals between the paroxysms, it will be found that she suffers from some functional derangement; menstruation is seldom regular as to quantity or time of appearance. It may be either delayed, retained, suppressed, too frequent, excessive, or it may be painful and difficult. The patient may, as in fact she usually does, suffer from dyspepsia. Digestion is impaired; there is a morbid appetite; a craving for indigestible or hurtful articles; there is derangement of the bowels; the patient either suffers from constipation or diarrhœa.

Leucorrhœa is usually present. As a general thing, an impairing of the patient's general health is a sure precursor of the hysterical annoyance.

Hysteria may simulate various other diseases and lead the unwary prescriber into many errors; it is, therefore, always best in severe cases at least, to place such patients under the care of an intelligent Homœopathic physician.

TREATMENT. — During the hysteric fit or spasm it is only necessary to place the patient in an easy position; admit all the fresh air possible; remove everything tight from around the waist, chest, and neck, and sprinkle the face with cold water. This, in mild cases, will be sufficient; when, however, the paroxysms are attended by severe convulsions, care should be taken to prevent the patient from injuring herself. As a general thing, though her struggles are severe, she retains consciousness enough to avoid danger, and therefore but little restraint is necessary. There is no necessity in trying to keep the patient quiet; it is just as well that she should agitate herself as long as she does herself no harm; in fact, it is rather beneficial, as it equalizes the circulation. If, however, the patient is entirely unconscious, as they sometimes are, some restraint will be necessary; a folded napkin should be placed between the teeth to keep the tongue and lips from being bitten.

If the extremities are cold and the head hot, it will be advisable to keep the head wet with cold water; cloths wet with cold water, or with vinegar and water, should be placed around the head, and the feet put into warm water to which a little mustard has been added.

Sometimes a fit will be instantly stopped by the patient's drinking a tumblerful of *cold* water; or in cases where the patient cannot swallow, from spasms of the face and throat, the same result may be obtained from an injection of *cold* water.

There is no doubt but that a determined resolution on the part of the patient would often prove successful in suppressing an attack. Most females, however, subject to this disease, give way to the current of their feelings until a paroxysm is fully developed, or until they become, as it were, surcharged, and must give way. They sit like Tam O'Shanter's wife, —

> " Gathering her brows, like gathering storm,
> Nursing her wrath, to keep it warm."

till some little thing gives the occasion, when off they go. Hysterical persons should nip all gloomy and exciting feelings in the very bud, and never allow themselves to be carried to the very verge of a paroxysm, over which some trifle may precipitate them.

As hysteria is entirely devoid of danger, the simple means above mentioned will be all that is necessary while the fit lasts. The time for active treatment is during the intervals, and this can only be conducted by an experienced physician, because the predisposing causes, which are often abstruse, must first be entirely eradicated before permanent relief can be reasonably looked for.

The remedies are, *Aconitum, Cocculus, Cuprum, Coffea, Ignatia, Lachesis, Platina, Pulsatilla, Nux-vomica, Sulphur, Conium, Calc.-c., Natrum-m., Veratrum.*

Perhaps there is no better remedy for hysteria than cold water. Frequent bathing and washing are the principal modes of application. Active exercise in the open air and change of scenery will undoubtedly contribute much to invigorate the nervous system, and the consequent abatement of the hysterical attacks.

Dr. Copland says: "Where a female is liable to hysterical attacks, she should be confined to a light and nourishing diet; take much exercise in the open air; use cold sponging, or the shower-bath in the morning; avoid tight-lacing, tea and coffee, hot rooms and late hours, strong moral emotions, and novel reading; sleep on a hair mattress, in a large, well-ventilated apartment; and — what is of equal importance — the mind should be strengthened by being employed in healthful and interesting pursuits, with frequent indulgence in innocent and rational amusements. The objects aimed at are, to restore the nervous system to the requisite degree of stability, and to correct the disordered functions of the uterine system.

54

CHAPTER X.

DISEASES OF THE EYES, EARS, AND NOSE.—DISEASES OF THE
EYES.

GENERAL REMARKS. — Of all the senses with which we are endowed, perhaps none adds so much to our real comfort and happiness as the sense of vision. By it we are made acquainted
with the form, color, and position of objects that surround us.
Through it, flows into the soul of man, as he goes forth amid
the beauties of nature with which a bounteous Creator has
surrounded us, an ever-constant stream of happiness and joy.
Through it, we realize how great and good, how kind and thoughtful is He who thus mingles pleasure with our existence.

Without the eye, all the beauties of field and forest, in fact the
very face of nature, would be to us a perfect blank. A picture of
blindness need not be drawn to illustrate and enforce an argument to prove that no greater misfortune could befall one than the
deprivation of his sight. The distance from sight to blindness,
from perfect day to total darkness, though great, indeed, is spanned
by one of Nature's most delicate, yet most wonderful pieces of
mechanism which man has ever been granted the privilege of
comprehending. ·

The eye is the most perfect and beautiful optical instrument in
the world ; but, like all other organs and parts of the body, this
organ, upon the perfect working of which so much of our comfort
and happiness depends, is subject to disease and decay. Its delicate structure renders it extremely liable to accidents of various
kinds, and diseases of various forms ; and, what is indeed unfortunate, for all these diseases and accidents, every one whom the
patient meets has a certain cure, — one that has never been known
to fail. Washes, salves, ointments, and eye-waters are advised
and prescribed, with recommendations and assurances that really
astonish one who is familiar with the nature and course of the disease.

It is not asserting too much to say, that, without a doubt, more permanent injury has been done to the eyes by the use of local applications, than has ever been done by natural disease. Slight ailments, which would have been but trifles under rational treatment, have been aggravated into serious diseases by irritant washes and lotions. Allopathic physicians seldom treat a case of opthalmia, or diseases of the margin of the eyelids, by other means than applications of lunar caustic or nitrate of silver, and frequently to the permanent injury, sometimes to the total destruction, of the eye. We do not have to look far for melancholy instances, as the result of this rash treatment. I would advise every one to abjure all eye-waters, lotions, salves, ointments and the like, that are so highly lauded and perseveringly recommended by patent-medicine dealers and officious friends.

The only wholesome eye-water is that which nature has supplied us, pure and cold. Pure cold water is an excellent remedy when the eyes are red, burning, and painful; also, in chronic affections of the eye. It may be used, by washing the eye frequently, or by applying linen cloths which have been wet with it. In cases where the patient cannot bear cold water, or when it proves useless, lukewarm water may be used instead.

When Erysipelas affects the eyes,—which will be known by the surrounding redness,—nothing wet should be applied; but, instead, dry and warm applications should be used.

SORE EYES OF YOUNG INFANTS.

This affection is very common among young infants,—setting in, frequently, when the child is but a few days old. Generally, I think, the eyelids are first affected, but the eye proper soon becomes implicated, if the disease continues long, or is neglected. It is occasioned either by some irritating substance getting into the eye — soap, for instance — when the child is being washed, or by cold.

The following course of treatment will generally remove this disorder: —

Bathe the eyes in lukewarm water, and give the child *Aconite* and *Belladonna* in alternation, one globule three times a day, for several days. Should this afford no relief, then give one dose of *Æthusa*, — one globule every three hours.

Chamomilla. — When the eyes are swollen, and glued together in the morning with yellowish matter.

Mercurius and *Pulsatilla*. — When there are small, yellowish ulcers along the margin of the lids; a profuse discharge of yellow matter from the eyes, with redness of the whole interior of the eye.

When there are symptoms indicating a scrofulous tendency, *Calcaria* or *Sulphur* should be given. When there is a great accumulation of matter, which remains in the eye, *Euphrasia* or *Rhus* should be given, one dose of one globule, every three hours.

STY ON THE EYELID.

A sty is simply a little boil upon the margin of the eyelid. Sties usually appear near the great angle of the eye. They are quite painful, suppurate slowly, and have no tendency to burst. The specific remedy in the commencement of this complaint is *Silicea* — the 200th potency; one dose will generally be sufficient.

When the remedy does not check the advance of the disease, and suppuration is about to take place, or when there is considerable redness, with throbbing pain, a warm poultice should be applied, to facilitate its breaking.

Staphysagria. — When there is a strong tendency to these tormenting visitors, and when, without suppuration, they disappear, leaving hard spots behind. Dose, from three to four globules, according to size, every four hours.

SQUINTING — STRABISMUS.

Squinting, as Strabismus is commonly called, is an affection of the eyes, in which the axes of the two do not retain their natural relation. Squinting may be spasmodic, caused by some cerebral difficulty, or it may be occasioned and confirmed by a permanent shortening of one of the lateral straight muscles of the eyeball. In this case the only cure is an operation to divide the muscle. Children are very prone to squint from habit; and not unfrequently the cause in young children may be attributed to their being so placed in their cradle that the light constantly falls upon the same side. The habit, growing upon children, becomes after a time, a permanent defect.

I have never been able to effect much towards the cure of strabismus by the administration of medicines; still, *Hyoscyamus* and *Belladonna* are highly recommended by some physicians; and, no doubt, in conjunction with the ordinary mechanical means, which are used in these cases, may be of great service.

In giving these remedies, dissolve six globules in twelve spoonfuls of water, and give one spoonful, every three hours, for two or three days; then discontinue for the same length of time; and then proceed with another remedy, if necessary.

INFLAMMATION OF THE EAR.

This is a very painful disease. The inflammation affects the passage or tube of the ear, sometimes causing it to swell to such an extent as to entirely close it; and at times occasioning such severe pain as to scarcely allow the affected member to be touched.

The symptoms indicative of this disease are, violent burning, itching, beating pains, deep in the ear; and finally, swelling and redness both inside and outside; great sensibility to noise; and more or less fever.

As a general rule, this disease can be controlled with one of the following remedies: —

Pulsatilla. — This is almost a specific remedy for this complaint; especially where the pain is of a burning, throbbing nature, and so severe as to cause the patient to be almost delirious.

Belladonna. — When the pain penetrates deep into the brain; twitchings of the hands, and of the muscles about the corners of the mouth, with great agitation; symptoms of convulsions.

Aconite, — in alternation with *Belladonna,* if the fever is very high.

Of the remedies chosen, — the most efficient being *Pulsatilla,* — dissolve six globules, in twelve spoonfuls of water, and administer one spoonful of the solution, at a dose, every hour. Should the pain be very severe, a dose may be given every half-hour.

It is neither judicious nor safe to be constantly introducing oil, laudanum, and the like, into the ear. Relief is often afforded by covering the ear with cotton, to protect it from the air and noise; and warm water, applied with a soft linen rag, will often ease the pain, without doing injury.

EARACHE.

This is a very frequent affection of young children, and, although resembling inflammation of the ear, is quite a different disease, — the one being accompanied by fever and the other not. The pain of earache is of a neuralgic or rheumatic nature, and generally arises from taking cold. The attacks come on suddenly, and are of short duration.

TREATMENT. — *Chamomilla.* — When the pains are acute, and shooting as if from a knife-wound, especially after taking cold, or suppressed perspiration ; with tearing pains which extend to the lobes of the ear, the patient being cross and irritable.

Pulsatilla. — When the pains are severe, darting and tearing, as though something would press out through the ear, or when the external ear is hot, red, and swollen, or there are itching and tearing through the whole side of the face, almost depriving the patient of reason.   This remedy is particularly beneficial to females and persons of chilly habit.

Belladonna. —When there is a determination of blood to the head, with stitches in and behind the ear ; tearing, boring, or shooting pains extending to the throat ; extreme sensibility to the slightest noise.

Hepar-s. — For the same symptoms as *Belladonna,* especially when the latter is insufficient; also, when there is throbbing and buzzing in the ears.

Mercurius. — After *Pulsatilla* or *Chamomilla,* if these are insufficient ; also, when there is a tearing pain extending to the cheeks ; chilly sensation in the ear ; pains worse when warm in bed.   When *Mercurius* affords but partial relief, a dose or two of *Hepar-s.* will generally subdue the remaining symptoms.

Nux-vomica. — When the pains are violent, of a tearing, stinging nature, and extend upwards to the forehead and temples. May follow *Chamomilla.*

Rhus. — For earache, arising from cold.   Other remedies such as *Cepa, Arnica, China, Calcarea, Platina,* and *Sulphur,* may be required in some cases.

ADMINISTRATION OF REMEDIES. — Of the remedy selected, dissolve six globules in twelve spoonfuls of water. Of this solution, give one spoonful every fifteen minutes, until better ; as the pain subsides, lengthen the intervals between the doses. When the case is not a very severe one, it may not be necessary to administer the medicine so often, perhaps once in an hour, or two hours, will be sufficient. If not convenient to dissolve the globules in water, they can be taken dry upon the tongue, about three at a dose. It sometimes happens, that, after the use of the above remedies, although the severe pain has passed away, there still remains some soreness, or slight, grumbling pain in the ear. This can be removed with *Sulphur* or *Calcarea*, one dose night and morning. In regard to external applications, which are so frequently used, and sometimes with apparent great advantage, I have no very serious objections to offer. For my own part, I would rather trust to the remedies above recommended, because the parts may be so injured, that restoration will be almost impossible. The safest way is to use nothing unless it be a little olive oil, or tepid water. Sometimes a sponge, or soft muslin, dipped in water, and applied to the ear, will mitigate the pain.

RUNNING OF THE EARS.

This troublesome and sometimes exceedingly offensive disorder arises from various causes. It frequently remains after inflammation, gatherings in the head, etc. Perhaps the worst form, and that which is the most difficult to cure, is that resulting from scarlet fever.

The attempts, which are frequently made to arrest the discharge by some local application, are greatly to be reprehended, as the most insignificant discharge may, when suddenly suppressed, produce most dangerous consequences. It is best to bear patiently with the affliction until a cure can be effected with the proper remedies. Never tamper with the eye or ear.

The principal remedies for this disease are *Pulsatilla*, *Mercurius*, *Hepar-s.*, *Belladonna*, *Lachesis*, *Calcarea-c.*, *Silicea*.

When it succeeds the measles, give *Pulsatilla* or *Sulphur*.

When it follows scarlet fever, *Belladonna*, *Mercurius*, *Hepar-s.*

When it follows small-pox, *Mercurius*, *Lachesis*, *Sulphur*, *Calcarea-c.*

When the matter discharged is purulent, give *Mercurius*, *Hepar-s.*, *Pulsatilla*, *Calcarea-c.*

Very offensive matter, *Mercurius*, *Hepar-s.*, *Lycopodium*, *Pulsatilla*, *Sulphur*.

If, upon a sudden suppression of this discharge, the glands of the neck should become hard and swollen, give *Pulsatilla* followed by *Mercurius* or *Belladonna*.

If severe headache and fever are present, give *Belladonna* and *Bryonia*, in alternation.

If, after such a suppression, the testicles begin to swell, give *Nux-vomica* and *Pulsatilla*.

ADMINISTRATION OF REMEDIES. — Dissolve, of the selected remedy, six globules in twelve spoonfuls of water, and give one spoonful of the solution, every four hours, for six days. If not beneficial, select another remedy, and use in like manner. Should it be more convenient to give the globules dry, put about four on the child's tongue, and let them dissolve. If, with these remedies, you do not soon perceive some improvement, you had better consult a Homœopathic physician.

BLEEDING FROM THE NOSE. EPISTAXIS.

DEFINITION. — CAUSES. — Bleeding from the nose is quite a common occurrence among the young people, and seldom amounts to more than a temporary inconvenience, rarely needing any remedial assistance. A slight blow, a fit of sneezing, or the summer heat, is sufficient, with many boys, to make the nose bleed. This is owing, perhaps, to an undue fulness of the blood-vessels of the head.

In young girls it sometimes comes on periodically, with or at the time the menses should appear; and frequently in fevers and many other diseases. It often relieves or cures headache and vertigo. A moderate hemorrhage from the nose is generally succeeded by a sense of relief and refreshment. Usually the blood flows in a succession of drops; it may, however, run in quite a stream. Generally, but a small quantity of blood is lost in this way; sometimes but a few drops; nevertheless, it does occasionally amount to quite a serious affair, the patient losing several pints of blood, producing pallor, faintness, debility, and, indeed, even death. Bleeding from the nose in young children is almost always salutary, and may be left to work its own cure.

TREATMENT. — The nursery remedy is to slip a cold key down the back, or to sprinkle the face with cold water. This sudden contact of some cold substance will often restrain the hemorrhage, perhaps by producing a contraction of the blood-vessels.

Very frequently severe hemorrhage can be stopped by causing the patient to hold his hands high above his head.

When it is caused by a fall, give *Arnica.*

When it results from a determination of blood to the head, give *Aconite, Belladonna,* or *Bryonia.*

When the hemorrhage arises from over-exertion, either at work or play, give *Rhus.*

If it occurs principally at night, give *Bryonia* and *Belladonna*.

If in the morning, *Nux-vomica* and *Bryonia*.

When it arises from over-heating, give *Aconite* and *Bryonia*.

Give *Cina* and *Mercurius*, when it occurs with children afflicted with worms. If the bleeding appears in females periodically, or with those who have a weak flow at the menstrual period, give *Pulsatilla*.

Sulphur, Carbo-veg., Lycopodium, will be found of service for those who are subject to frequent attacks of nose-bleed. They may be taken at intervals of eight or ten days between each remedy; six globules for a dose; continue four weeks.

ADMINISTRATION OF REMEDIES. — Of the selected remedies, dissolve six globules in twelve spoonfuls of water, and of this solution give one spoonful every ten or fifteen minutes, in cases of active hemorrhage. Where the attack is not so severe, a dose once an hour, or two hours, will be sufficient.

CHAPTER XI.

DISEASES OF THE URINARY ORGANS. — WETTING THE BED.

GENERAL REMARKS. — The generally received opinion in regard to this disorder is, that it is simply a "bad habit;" and all that is necessary for its prompt removal is a judicious administration of the rod. Every observant physician, however, knows full well that this is not simply a habit, but that in reality it is, in almost all instances, caused by some disease. This disease may be at the neck of the bladder, or in the bladder itself, or it may be in consequence of the secretion of acrid urine. Either of these conditions would produce the difficulty, and it would be just about as sensible to attempt to whip the measles or any other disease out of a child, as to attempt to eradicate *this* difficulty by such summary dealings. It is not a habit, of which a child can be broken, and those only who were unable to cure this disorder would ever bring into existence and disseminate such an utterly absurd notion.

I have a patient under my charge now, a young lady over sixteen, who, ever since she was a child, has been in the "habit" of nightly "wetting the bed," and many a severe whipping has she received for it, and that, too, by the advice of an "eminent and much-experienced physician." Her mother tells me, that the bitterest tears she ever shed are those which now flow as she recalls the scene, as, during the chastisement, the little one would writhe and cry, and say she would "never do so any more."

In the majority of cases, there is simply a weakness of the parts, and, as soon as the bladder fills up, the urine escapes; and, if the child is asleep or in a half-conscious state, without its knowledge or consent. Now, is the child to blame here?

Sometimes the weakness is so great as to amount almost to paralysis; in such cases, the bladder is unable to retain for a moment the urine secreted, which constantly dribbles away, very much to the inconvenience of the patient and the mother.

TREATMENT. — The following dietetic rules should always be adhered to. In the first place, all articles which have a tendency to increase the secretion of urine, must be avoided. Tea, coffee, and all salt and sour articles of food are objectionable.

The child should take a moderate supper, accompanied with as little drink as possible, — cold water or milk is preferable, — and should not be sent immediately to bed. Plenty of out-door exercise is always advisable. The child should be taken out every day, and permitted to run, hop, skip, and jump.

With a little care and attention, the time of these mishaps might be noted, and the child afforded an opportunity to relieve itself at stated intervals, if necessary, or shortly after retiring, and before the time of awaking in the morning.

It has been asserted that lying upon the back when asleep is the *cause* of this complaint, and the remedy, of course, is to turn the child over on its side. This is a mistake. No *healthy* child sleeps upon its back; and, when the child is found habitually sleeping in this position, no other symptom is necessary to assure the child's being in ill health, — the position itself being the result rather than the cause of disease.

The remedies for this complaint are, *Pulsatilla, Belladonna, Calcarea, China, Nux-vomica, Sulphur, Sepia, Causticum, Arsenicum, Silicea, Cina.*

Pulsatilla. — For children having slender frames, blond hair, and a mild disposition, inclined to weep; offensive urine. This medicine is better adapted to girls than boys.

Nux-vomica. — Symptoms similar to the above, and when the child is obstinate and cross, and inclined to lie with the arms above the head.

Belladonna. — When the urine passes in large quantities, and is pale and watery.

Sulphur. — This is the most important remedy, and, in the majority of cases, the one called for. It is especially applicable to thin, pale children, with protruding stomachs, — those who are always complaining.

Calcarea. — If Sulphur is not sufficient, give this remedy, particularly if the child passes water frequently and in small quantities. Best adapted to stout, or bloated-looking children.

Arsenicum. — When the urine is hot, and has a putrid smell, and if the child lies upon its back with the arms over the head.

Silicea. — Especially for children, in whom the slightest scratch amounts to quite a sore, not disposed to heal.

Causticum. — For children with black hair and eyes; also, when the urine is acrid, with frequent desire to urinate, both day and night.

Cina. — For "wormy children."

ADMINISTRATION OF REMEDIES. — The remedy chosen may be given about once in four hours; dose from three to six pills, according to the age of the patient.

If the difficulty occurs in young boys, ten or twelve years of age, it will be well to ascertain if their solitary habits are good; for too often, even at this early period of life, the habit of Onanism is taught these young lads by older offenders. When this practice gives rise to the difficulty under consideration, *China* is the best remedy that can be employed.

RETENTION OF URINE IN INFANTS.

As a general thing, new-born infants discharge the contents of the rectum and bladder shortly after birth; occasionally, however, it happens that the urine is retained for a longer period. Retention of the urine not unfrequently produces symptoms which demand our immediate attention. The infant is uneasy, nervous, and cries, especially when pressure is made on the region of the bladder. There is usually more or less fever; the little sufferer twists its body and draws up its legs. If relief is not soon afforded, convulsions and other dangerous symptoms follow.

TREATMENT. — The principal remedy for this difficulty is *Aconitum;* one dose may be given as frequently as every hour.

Pulsatilla — Is the next best remedy; and, should *Aconite* fail to afford relief, this should be administered.

When spasmodic symptoms develop themselves, give *Ipecacuanha.* A warm bath, or rubbing over the region of the bladder with the warm hand, is also of service.

ADMINISTRATION OF REMEDIES. — Put one globule of the selected remedy upon the child's tongue, and let it dissolve. This may be repeated every hour until relief is obtained.

INFLAMMATION OF THE PRIVATES.

A great source of annoyance sometimes befalls young girls in the shape of an inflammatory swelling of the private parts. The labiæ, or lips, of the vagina become swollen, hard, red, and very sensitive to the touch ; there is, also, more or less fever, accompanied by burning or shooting-pains. It may arise from cold, from excoriations or chafings, or from mechanical injuries. In women, it is at times caused by the rupture of the hymen, or it may result from difficult labor.

TREATMENT. — When the inflammation results from mechanical causes, give *Arnica*, internally, and apply a lotion of one part tincture in ten parts pure water, with which bathe the parts freely, night and morning. If the inflammation be the result of cold, give *Belladonna*, *Mercurius*, or *Rhus*. Should the swelling have the appearance of Erysipelas, give *Rhus* and *Belladonna* in alternation.

Boys are not unfrequently attacked with a similar inflammation, which affects the prepuce or foreskin. The same treatment may be used as for inflammation of the labiæ.

When, as sometimes happens, the foreskin becomes swollen or puffed up, like a bladder of water, give *Rhus*.

If the inflammation was caused by the touch of poisonous plants, first give a few doses of *Aconite*, after which *Belladonna* or *Rhus*.

ADMINISTRATION OF REMEDIES. — If the disorder be in an infant, give two globules of the selected remedy, on the tongue, and let them dissolve. Repeat the dose about once in two hours. If an older child or adult, dissolve six globules in twelve spoonfuls of water, and give one teaspoonful of the solution every two hours.

When the trouble is occasioned by some mechanical injury, such as friction of the clothing and the like, use a solution of the *Tincture* of *Arnica* as a lotion.

CHAPTER XII.

CASUALTIES.—EXTERNAL INJURIES. BURNS AND SCALDS.

Burns and Scalds, unless superficial, and of small extent, are always troublesome to manage. When extensive, covering a large surface, or when deep, they are both dangerous and troublesome. In the treatment of burns, two important points should always be kept in view. These are, first, careful attention to the constitutional symptoms, in severe cases, and the prevention of adhesions and contractions during the process of healing. When the shock is severe, and there is great constitutional depression, it will be necessary to give stimulants; these, however, should be given sparingly and with discrimination, for, when reaction takes place, it may proceed even to inflammation.

Burned surfaces, when they begin to heal, have a remarkable tendency to contract and adhere together. Therefore, strict attention should be paid during this time to keep the healing surfaces apart, to prevent deformity. When a burn is situated upon the hand, involving the fingers, the fingers should be widely separated and secured in such position by splints and bandages. If a burn takes place near or involving a joint, a splint and bandages should always be applied.

The treatment for superficial burns, of slight extent, is simple. It consists in the application of raw cotton, spirits of wine, strong brandy, lime-water, mixed with sweet-oil or soap-plaster. The most convenient way, and perhaps as good as any, to treat simple burns and scalds is, first to evacuate the serum from the blisters, cover the part with raw cotton, and apply a bandage firmly over it.

The following application comes highly recommended, from a source entitling it to great confidence, for burns of every degree of intensity : —

> Kreosote, one ounce;
> Alcohol, three ounces;
> Water, two quarts. — Mix.

After nipping the blisters with a pair of sharp scissors, cover the part with a soft cloth, kept saturated with the mixture. The cloth should not be removed to wet it, but the liquid should be dripped upon it.

Dr. P. P. Wells, of this city, says (*Homœopathic Examiner*, vol. ii. p. 19), — "For the cure of scalds and superficial burns, there is no remedy which can compare with local application of a saturated solution of *Nitrate of Silver*. It has been in use in my practice for the last six years, and its success has been such as to leave nothing more to be desired. In slight cases the cure is instantaneous. In more grave ones it is affected by a single application, and in a space of time incredibly short. A boy, five years old, received the boiling contents of a kettle on the top of his head. The scalp, face, arms, and upper part of the trunk were frightfully scalded. The mother applied lamp-oil, with no effect, to relieve his pain. Immediately, on applying strips of muslin dipped in this solution, he became perfectly tranquil. He had neither pain, anguish, trembling, shuddering, nor cold extremities; all which previously had been extreme. The accident occurred at four, P. M. The next morning at eight, the strips were removed, and with the exception of a spot between the right ear, and two or three others on the right arm, of the size of a quarter of a dollar, where the oil prevented the contact of the solution and skin, the child was perfectly well. A servant girl thrust her hand and two-thirds the length of the forearm into a kettle of boiling mush. The fingers were swollen so as to be rigid; and small blisters covered the skin of the immersed part. The solution was applied with a hair-pencil. The pain ceased instantly, and in three hours she was about her daily avocations."

Soap makes a good dressing for burns, and by many is much preferred to any other. It also has the advantage of being easily procured and applied. Take white soap (Castile is preferable), scrape it fine, and add sufficient warm water to make it into a thick salve; make a plaster of this, by spreading it upon pieces of muslin or soft linen, cover the burnt surface, and secure it with a bandage. Care should be taken that the whole of the injured surface is covered with the soap, and the bandage so applied as to exclude the air, by keeping the plaster in actual contact with the burn.

Where the cuticle is raised, forming a blister, the serum should be carefully evacuated, and the cuticle smoothed out, so as to protect the surface, and the above-named applications carefully applied.

Soap-dressings should be changed as often as once in twenty-four hours; carefully remove the old plaster, and *immediately* replace it (without washing the sore) by a fresh one.

Another good application for burns, and one usually easily obtained, is common flour; when dusted thickly over the sore, it absorbs all the discharge, and forms a good protection. The crust, when it becomes necessary, can be easily removed by poultices.

Linseed-oil and *lime-water* is a very good remedy; but it is not generally within easy access.

Cantharides. — A solution of Cantharides, ten drops to a tumbler half full of water, is also a good remedy. It may be applied by dipping linen cloths into it, and applying them to the burnt surface.

Arnica Tincture and *Tincture of Urtica Urens* are most efficacious remedies. They may be used the same as *Cantharides.*

In the treatment of all burns, it is of great importance to exclude the air, as much as possible, from the ulcer; therefore frequent changing of the dressing is objectionable. When saturated cloths are applied, — and it is linen ones that should always be used, if as convenient, — they should not be allowed to get dry; neither should they be removed to be re-moistened; but the liquid with which they are wet should be dripped upon them.

Always open the blisters, and evacuate the serum; but do not remove the loose cuticle.

Sores resulting from burns are often difficult to heal, owing to diminished vital force. And therefore stimulating applications are sometimes called for to promote healthy granulation. For this purpose we use *Kreosote-water.*

For the fever which sometimes attends burns give an occasional dose of *Aconite.*

When inflammation of an erysipelatous nature surrounds the burn, give *Belladonna* or *Rhus.*

Carbo-veg. and *Coffea* are said to be of great service, when there is great pain and restlessness.

If the burn ulcerates, wash it with a solution of *Causticum,* — twelve globules to a tea-cup half full of water, three or four times a day; and give internally *Sulphur* or *Silicea,* one dose every six hours.

For burns of the mouth or throat, caused by inhaling hot steam, or by partaking of hot liquids, give a dose of *Arsenicum* every half-hour, and gargle the throat with a solution of the same remedy. Should this not afford relief, try *Causticum* or *Rhus*. Or, dissolve *Castile soap* in whiskey, and give five or six drops of the solution, in a little water, every few minutes.

CONCUSSION OF THE BRAIN.

Concussion of the brain may arise from a fall or blow upon the head, or from some violent shock to the body. The symptoms will, in a measure, depend upon the severity of the shock. In cases where the violence has been comparatively slight, the disturbance of the intellectual functions will be transient. There will be, perhaps, some vertigo, dimness of sight, trembling of the limbs, and sickness of the stomach. In severe cases, loss of sensation may exist for many hours, and finally be followed by re-action, which, if not controlled by treatment, increases to inflammation.

Arnica is the appropriate remedy for accidents of this kind, and, when promptly administered, will often ward off the evil consequences of such misfortunes. Dissolve twelve globules in twelve spoonfuls of water, and give one spoonful for a dose, every hour. In case there is an external wound, a lotion made by a few drops of *Arnic. Tincture* to a little cold water, applied to the head, will also be of great service. After an injury of the head, the patient should be kept perfectly quiet, free from all kinds of excitement, until a complete recovery has taken place.

Should the case proceed to inflammation of the brain, notwithstanding the treatment recommended above, resort must be had to *Aconite* and *Belladonna*.

Consult a physician, when possible.

SPRAINS.

These are caused by falls, lifting heavy weights, jerks, false steps, etc. Sprains of joints, when severe, often arise from momentary displacement of the bones, which strains or perhaps partially tears the ligaments surrounding the joint. They are often troublesome, and require rest and bandaging a long time after the occurrence of the accident.

The treatment consists in bandaging the part with cloths wrung in cold water, to which a little *Tincture of Arnica* has been added.

Rhus or *Bryonia* should be given internally, a dose once in two hours.

If there is sickness of the stomach, give *Pulsatilla*.

WOUNDS.

Wounds are classified into incised, contused, lacerated, punctured, poisoned, and gunshot wounds.

INCISED WOUNDS — or clean cuts made in the soft parts with a sharp instrument. The troublesome feature of such wounds is hæmorrhage or bleeding. If an artery has been cut, the blood spouts in jets and has a bright-red color; if a vein, the flow of blood is gradual and purple.

Large arteries and veins are seldom wounded, and therefore the hæmorrhage from common cuts, although it is at times quite profuse, need cause no alarm.

The treatment consists in cleansing the wound by removing anything that may be left in it, arresting the hæmorrhage, bringing the cut surfaces and edges in close contact, and retaining them in such position.

Small, superficial wounds seldom require anything more than a bandage, snugly applied after the edges have been nicely adjusted. Wounds of greater depth sometimes require a stitch or two; but, as a general thing, small strips of adhesive or Arnica plaster, properly applied, will answer every purpose.

No more dressing should be applied than is actually necessary to keep the parts in co-aptation. After a wound is dressed the injured part should be kept in such a position that the wound will not gap. To accomplish this, it is sometimes necessary to apply a splint. To the cut surface, apply a solution of *Staphysagria*.

Surgeons, nowadays, apply nothing to wounds for the purpose of healing them, except cold water. Ointments, salves, and a host of other things once used, have long ago been abandoned.

After a proper dressing has been applied and placed in an easy position, if bleeding still continues, apply cold water or pounded ice: nothing more. Should this not arrest it, and the blood be of a bright-red color, spouting out in jets at intervals, as the pulse beats, endeavor to compress the artery between the heart and the wound. Feel for the artery on the interior part of the limb; you will know it by its beating when found. Place over it a large-sized cork, or a compress made by folding up a piece of cloth about two inches square and as thick as your finger, and bind it down firmly with a roller. This will arrest the bleeding until you can procure professional assistance.

CONTUSED WOUNDS. — BRUISES. — These are occasioned by blunt surfaces, by falls, or by forcibly coming in contact with some object. There is generally no break or division of the external surface; consequently, the hemorrhage is all internal and not great, forming what is usually termed a "black-and-blue" spot. The pain is in an inverse ratio to the extent and severity of the wound; for, the contusion being severe, the life of the part is destroyed and little or no pain results; but in slight cases the pain is intense. There is usually swelling, which is proportionate to the extent of the injury, and more or less discoloration of the skin, owing to the internal hemorrhage.

TREATMENT consists in the prompt and continued application of cold water. The water can be best applied by saturating linen cloths with it and keeping them applied to the parts.

Arnica should be administered internally.

Arnica lotion, made by putting a table-spoonful of *Arnica Tincture* to a pint of water, may also be used.

Hypericum-per., prepared like the Arnica, makes a good application; and, when early applied, is said to remove, in slight cases, all traces of the injury in a few hours.

In severe cases, where suppuration is about to take place, it should be hurried forward by poultices. When all the dead flesh separates and comes away, this application should be changed to simple dressing of water; perhaps it will also be necessary to support the parts by adhesive strips.

Hepar-sulph. must be given during the suppurative process.

Bruises about the eye should be kept *constantly* wet with a solution of *Arnica*.

Bruises upon the shin-bone sometimes cause severe and troublesome sores. The best application for such sores is *Collodium*, — a solution of gun-cotton in ether. It may be obtained at any drug-store, and is best applied with a soft brush. As it becomes dry, it forms a thin, transparent skin, which is not removed by water.

LACERATED WOUNDS are those where the soft parts are torn or rent asunder by violence, leaving a ragged, irregular wound. Hæmorrhage is usually slight. Treatment consists in cleaning out the wound, and adjusting the parts as near in apposition as possible, and securing them by as little dressing as possible. As these wounds generally suppurate, spaces should be left between the adhesive strips, to allow the matter to escape, and thus prevent abscesses from forming.

Cold water, to which a little *Calendula* has been added, should be constantly applied.

Lacerated wounds are prone to inflammation. Should there be fever, dryness of the skin, thirst, and restlessness, give an occasional dose of *Aconite* (three pills). If the inflammation has an erysipelatous appearance, give *Belladonna* or *Rhus*.

PUNCTURED WOUNDS. — Punctured wounds are made by some sharp, narrow instrument, as a needle, pin, thorn, splinter, piece of glass, etc. Slight wounds of this kind are seldom troublesome, provided the offending substance can be removed, unless, indeed, it extend deep down among the tendons and nerves, where matter may form, causing great pain, and even deformity.

When a person runs a nail or splinters of glass into the foot, they should always be removed, if possible. When this cannot be done, all that need be applied is a little Canada-balsam. This should be renewed every day. If there is much inflammation, apply cold water.

POISONED WOUNDS. — *Bites* and *Stings* of *Insects, Bees, Spiders, Bugs, Mosquitoes*.

The bites and stings of insects, though seldom dangerous, are often exceedingly troublesome.

The treatment for wounds of this kind consists in removing the sting of the insect, when it remains in the part, applying a plaster of damp earth, and keep it wet afterward with a mixture of *Arnica* and water.

Should inflammation and fever ensue, give internally *Aconite* or *Apis*.

For mosquito-bites, apply *Spirits* of *Camphor* or *Lemon-juice*.

DISLOCATION OF JOINTS.

To be skilful and successful in the reduction of dislocations, a perfect knowledge of the anatomy of the joints is indispensable.

A dislocation may be recognized by the following symptoms: in addition to the pain, there will be loss of motion; swelling; alteration in the shape, length, and direction of the limb.

The treatment consists in reducing the luxation as speedily as possible. This, however, a layman can seldom do. Still, you can make a trial. If you do not succeed, apply a solution of Arnica — a spoonful to half a pint of water — to the injured part, give *Arnica* internally, and await the arrival of a competent surgeon.

FRACTURES.

Fractures of bones may be recognized by the deformity, which, by comparing the sound limb with the injured one, will be readily recognized. The most certain sign, perhaps, is that of *crepitation*, which is a peculiar grating sound, distinctly heard, when the two broken surfaces of the bone are rubbed together.

In all cases of suspected fracture, place the limb in the most comfortable position, and keep it constantly bathed with a *solution of Arnica*, as recommended for dislocations; after which, send for a competent surgeon. Do not get so excited, and in such haste, as to accept the first doctor you can get, without regard to, or without any knowledge of, his abilities. The case will take no harm if a whole day should elapse before you get assistance.

If the patient should be faint and weak, give an occasional dose of *Aconite*.

FOREIGN SUBSTANCES IN THE EYE, EAR, NOSE, AND THROAT.

In the Eye. — No matter what has gotten into the eye, washing with cold water will always be beneficial.

Rubbing the eye only increases the irritation, and, therefore, should always be avoided. When lime or ashes get into the eye, a little cream or sour milk is the best remedy. If a hard subject or an insect has gotten into the eye, draw the eyelids apart, and turn the upper one over the lower one a couple of times until it is felt that the substance is removed.

If particles of iron have entered the eye, and are fastened, bathe with *Arnica-lotion*, — ten drops in a teacupful of water, — until you can have it extracted. Should there be much inflammation, take a few doses of *Aconite*.

In the Ear. — Insects sometimes find their way into children's ears; in such cases, lean the head to one side, and fill the ear, in which the insect is, with sweet-oil. This floats it to the surface, when it can easily be removed.

If a bean or any other substance, which will swell by heat and moisture, gets into the ear, the best way to remove it is to make a hook by bending a hair-pin into the right shape. This should be

cautiously introduced behind the substance, and an effort made
gradually to extract it.

After the operation, wash the ear out with a lotion of *Arnica*,
and give internally *Arnica* and *Pulsatilla*.   Should there be fever
and headache, give *Aconite* or *Belladonna*.

IN THE NOSE. — Foreign substances may be removed from the
nose with a small pair of forceps, or the same instrument recommended for the ear.

First, endeavor to eject it by sneezing, which may be excited
either with snuff, or by tickling the nose with a feather.   Sometimes the obstruction may be pushed back so as to fall into the
mouth.   If these means fail, apply to a surgeon.

IN THE THROAT OR WINDPIPE. — If a foreign substance lodges in
the throat, first examine closely, and, if within sight, endeavor to
extract it with the fingers.   If it is not visible, excite vomiting
immediately, by tickling the throat with a feather, or by putting
mustard or snuff far back upon the tongue.

Foreign substances have been removed from the windpipe by
gently turning the patient upside down.

INDEX.

448   INDEX.

I.

M.

W.

Z.